Robots Unlimited
Life in a Virtual Age

David Levy

A K Peters, Ltd.
Wellesley, Massachusetts

Editorial, Sales, and Customer Service Office

A K Peters, Ltd.
888 Worcester Street, Suite 230
Wellesley, MA 02482
www.akpeters.com

Library of Congress Cataloging-in-Publication Data

Levy, David N. L.
    Robots unlimited : life in a virtual age / David Levy.
      p. cm.
    Includes bibliographical references and index.
    ISBN 1-56881-239-6
     1. Robotics–History. I. Title.

TJ211.L456   2005
629.8'92–dc22

              2005045869

Printed in India
09 08 07 06   10 9 8 7 6 5 4 3 2 1

For Donald, who has made so many things possible.

Contents

Contents

Contents

Contents

Preface

This book is for everyone interested in the future. Although its subject matter, Artificial Intelligence, is one of the most avant-garde and exciting of sciences, you do not need any prior technical or scientific knowledge in order to understand what is written here. Nor do you need to know how computers work or how they are programmed. Everything is explained in easy-to-follow language. So any of you who consider yourselves technophobes should have absolutely no fear of delving further into these pages.

The first problem facing the author of any comprehensive book on AI is the surprisingly enormous breadth of the field, including, as it does, programs that play games, compose music and perform other creative tasks, programs that reason and learn. There are even programs and robots that exhibit emotions and can reproduce. A huge mass of source material exists on almost every topic within the field of AI, material which ranges in the depth of its coverage from the "relatively elementary" (which, in AI, tends to mean something written at a level suitable for first–year computer science students) to the detailed and technically advanced papers published in conference proceedings and learned journals and intended for higher level academics and research scientists. Had I been able to include all the material that I wished, this book would have extended to a series of several volumes, each of a similar size to this one. Instead I had to find a way, within this single tome, to convey not only the breadth of ideas and inventions within AI, but also the progress that has been made since 1955 when the field was conceived as a research subject in its own right,[1] and to explain the inevitable advances of the coming decades.

---

[1] Many sources give 1956 as the year in which Artificial Intelligence was conceived, that being the year of the Dartmouth workshop where John McCarthy's name for the subject was affixed. In fact the name "Artificial Intelligence" was coined by McCarthy in the proposal for funding the workshop, a document prepared and submitted in 1955. I therefore regard 1955 as the year of the conception of AI and 1956 as the year of its birth.

Even after I had decided on my choice of source material, a choice that often had to be fairly ruthless in its selectivity, my source papers for this book still comprised some 5,000 printed and photocopied pages. From all these it was necessary to distil the essence of each topic, while at the same time presenting the whole subject of AI in a way that could be understood by all, without the need for any previous Computer Science, Artificial Intelligence or other technical education or knowledge. The necessity of restricting the length of this work to "only" 450 pages or thereabouts, with all this source material sitting on my shelves, forced me to think about what would be really important in getting across to my readers the essence, excitement and potential of Artificial Intelligence. The approach I decided upon was this. In order to describe the foundations of AI as they existed pre-1955, Chapters 1 and 2 present an account of various artefacts dating back as far, believe it or not, as 1650,[2] all of which represent artificial forms of some kind of intelligence. These include methods for the composition of music, machines that played games, machines that could solve simple problems in logic, and devices for translating from one language to another. Many of these historical artefacts are remarkable for their anticipation of techniques that reached some level of maturity as computer programs in the late twentieth century.

After setting the stage in this way, the book continues with five chapters (Chapters 3–7) that divide up between them most of the exciting and important topics within AI, demonstrating the progress made in each of these domains during the 50 years spanning 1955 to 2004. In general I have described these progressions by first recounting the earliest computerized attempts within each topic,[3] and then by presenting one or more state-of-the-art exemplars, demonstrating how much progress has been made to date within each domain. Between these two stages of research—the seminal attempts and the current state-of-the-art

[2]It was in 1650 that Athanasius Kircher published details of his method for the automatic composition of music (see Chapter 5). I do not regard Ramon Llull's ideas on logic in the thirteenth century (see Chapter 1) as being of comparable import, as they did not lead directly to any useful process or artifact. Instead Llull's writings acted as a catalyst for others. There is also another claim made as to the earliest artificially intelligent artefact. Some researchers subscribe to the view that Eudoxus and Archytas, in ancient Greece, invented some sort of proof machine. That attribution stems from a passage in Plutarch's *Life of Marcellus* which, in fact, does not refer to any such machine, and authoritative sources such as the *Stanford Encyclopaedia of Philosophy* are therefore dismissive of the claim.

[3]Many of these early efforts were based on ideas and techniques that not only were seminal for their respective domains but which are still in use today.

achievements—there have been, in most domains, dozens or even hundreds of interesting research projects that are described in the literature. Having to exclude from mention a high proportion of those projects has been a regrettable but essential task during the writing of this book.

My state-of-the-art survey, split as it is amongst these five chapters, has its different strands drawn together somewhat in Chapter 8, which presents summary descriptions of robots that perform tasks as diverse as playing Soccer, helping to rescue disaster victims, attending conferences and driving an automobile. Chapter 8 serves as a prelude for what follows it, inasmuch as it heralds the implementation in robots of several different AI technologies. It is this transitional process, from desktop and laboratory computers into fully-fledged robots, that will transform ideas that many people currently regard as science fiction, into the science fact of the mid-twenty-first century. By presenting a sprinkling of the most advanced robotic achievements to date, Chapter 8 illustrates how the technologies of AI are being dramatically converted into intelligent artefacts that walk, talk, and perform all sorts of other tasks normally associated with human intelligence.

Thus, Chapters 1–8 present AI as it has evolved to the present day.

In order to help us assess the potential of AI for its second half-century, i.e., the period from 2005 to 2054, Chapter 9 reports on technologies that will bring about enormous increases in computing power in the foreseeable future. Already some of the most remarkable achievements of AI, including the discovery of previously unknown proofs of theorems in mathematics and logic and the defeat of World Chess Champion Garry Kasparov by a computer, owe much of their success to the great increases in the power and memory sizes of computers during the past 50 years. With an exponential rate of progress in scientific knowledge, and commensurate advances in the field of computing, it is to be expected that much faster and bigger computers will help to usher in many of the most dramatic achievements in AI that will come during the next few decades. Chapter 9 presents the foundations for this expectation.

It is by extrapolating from the state-of-the-art in AI as it was in early 2005, and by doing so in the light of the anticipated progress in computing technology, that this book justifies the principal assertions of Chapters 10 and 11, namely that the artificially intelligent robots of the mid-twenty-first century will exhibit awesome capabilities, both intellectually and physically, combined with the whole gamut of human emotional

attributes and responses plus some more of which we have not yet even conceived. The idea of falling in love with a robot and all that that entails, may be well outside the normal frame of reference of many readers, but such ideas will become common currency within 50 years. And please don't reject these predictions out of hand, using the argument that robots cannot be programmed to have emotions, to make love with humans and to reproduce. They can and will!

The two final chapters (Chapters 12 and 13) discuss the concept of robot consciousness and some other topics related to human-robot relationships—the religious beliefs of robots, their ethics, their civil and legal rights, and our own ethical attitudes as they pertain to our interaction with robots. There is much here to stimulate the thinking of everyone who is interested in the future of mankind.

I hope that the reader will find this book not only interesting but also extremely thought-provoking. It is, in some ways, a kind of personal snapshot into the future, an argued case for what I believe our grandchildren will experience as a result of the exponential growth in AI and related sciences. It is my firm belief that the robots of our grandchildren's generation will be so intelligent, so all-powerful and so sophisticated, that they will exhibit an unbounded range of emotions, that people will fall in love with them (and vice versa), that people will marry them and that robots will "have" (i.e., build) offspring which share some of the characteristics of each "parent".

The speculation within these pages, on where AI is going in the next few decades, is all well founded, being an extrapolation of past work in AI and on the current state of the art, making due allowance both for the ever-increasing speed of computer systems and the expansion of their memory capacities. Not everyone would be willing to believe, before reading this book, that robots will one day be able to write poetry and prose so beautiful that it will make us weep, to compose hundreds of symphonies or piano sonatas in the exact same style as Beethoven or Mozart, to carry on a conversation as though from the persona of a Nobel-winning scientist or a punk rocker or an Alaskan lumberjack, to judge a court case with absolute impartiality and fairness, or to have humans fall in love with and marry them. Thought-provoking and controversial? Certainly. But far-fetched? Not at all.

David Levy
London, April 2005

Acknowledgments

I would like to express my sincere thanks to the many people who have helped in the creation of this book. Firstly, I am indebted to Klaus Peters for accepting my concept for publication and for being supportive throughout the writing and publication process. Of the many at A K Peters who have been involved with this project, I must single out my editor, Charlotte Henderson, who has untiringly advised and assisted in all manner of ways. Special thanks are also due to Cynthia Guidici who copyedited my typescript.

During the writing stage, I had the benefit of the advice of friends and relatives who read some of the first draft: Jaap van den Herik, Alastair Levy, Anna Richey, Jonathan Schaeffer and David Tebbutt. And Christine Fox not only read and commented on the entire text, but also made invaluable suggestions that led to the addition of much of the material on ethics and consciousness.

Hans-Joachim Vollrath kindly provided me with a clear description of the functioning of Athanasius Kircher's *Arca Musurgia*—a seventeenth-century device for automatic music composition.

Thanks also to Murray Campbell, John McCarthy and Jonathan Schaeffer for assisting me in the location of some of the photographs.

The illustration of Artsrouni's translating machine (Figure 15) first appeared on the cover of *Automatisme*, vol. 5, number 3, 1960. The illustration of a cross section of the human eye (Figure 27) and that of the eye of a fish together with its numerical representation first appeared on a page entitled "Computer Vision" on the University of Sunderland Web site, credited to a James West. The illustration of the music of "Push Button Bertha", together with a photograph of a Datatron computer, first appeared in *Radio Electronics*, June 1957.

Part I

An Early History of Artificial Intelligence

I n which we encounter the earliest attempts to automate some of the processes of human thought. These two chapters span a period of almost seven centuries, witnessing the creation of mechanical devices that could talk, compose music and solve problems in logic. The advent of electrical technology made possible the creation of the first machines for playing games and translating languages, as well as the first autonomous robots, and by the end of this era, in the mid-twentieth century, the birth of the computer had spawned ideas for "thinking machines" that could outperform the best humans at a host of intellectually demanding tasks.

– 1 –

Early History—Logic, Games and Speech

Artificial Intelligence is the science of making machines do things that would require intelligence if done by men.

—Marvin Minsky [1]

The idea of "thinking machines" can be traced back to the ancient civilisations. Around 2,500 BC the Egyptians built statues containing hidden priests who espoused advice to citizens. A bust of one of these gods, Re-Harmakis, can be seen in the Cairo Museum. The secret of its "genius" is an opening at the nape of the neck, just large enough to hide the priest. The existence of these statues confirms the belief held by ancient civilisations, in the idea of man-made objects that can think.

But this book is not about hoaxes. The early history of genuine Artificial Intelligence begins almost seven centuries before the term was coined, placing the science of logic in context as a fundamental tool of AI and encompassing various early attempts to mechanize the processes of simple logical thought.

Early Logic Machines

Once the ancient Greeks had invented the science of logic, the idea of reducing all reasoning to some kind of calculation became a popular fascination amongst scientists and philosophers. Mathematical logic has its own language and that language has its own rules, just as there are rules (grammar) in English, French and Spanish. And just as anything we want to say can be said in the many different natural languages in use around the world, so any statement can be expressed in the language of logic.

Aristotle observed that the property that best distinguishes man from the rest of the universe is the faculty of reason. In the technical sense, reasoning is the science of logical argument. In a reasoned process one starts with two or more statements and uses the information in those

3

statements to construct a logical argument, proving whatever one wishes to prove or inferring whatever one wishes to infer. So if statements in English or some other spoken language can be translated into the language of logic, then the rules of logic can operate on those translations in order to reason.

Reasoning is arguably one of the most important, if not *the* most important, fundamental intellectual skill possessed by man. It allows us to deduce and infer certain conclusions from evidence we already have. It is the basis of being able to say that, because I am in London and there are trains that go from London to Edinburgh, if I get on one of those trains I can reach Edinburgh. It is the basis of being able to generalise, to say that because the sun rose today and yesterday and for every day in the past 100 years or more, that it will also rise tomorrow. These are simple exercises in reasoning for humans, but how about for computers? The creation of an artificial intellect on a par with or superior to our own requires, amongst other things, a method by which that artificial intellect can reason. The ability of computer programs to reason is absolutely vital to the foundations of Artificial Intelligence.

The nineteenth-century English mathematician George Boole developed a mathematical analysis of the laws of human logic. He also developed an algebraic language (or notation) in which it is possible to describe the interaction and relationship of *variables* that have only two states—"true" and "false". Boolean algebra, as it is now known, is based on the three logical *operators* **and**, **or** and **not**. The combination of Boolean algebra and the rules of logic enables us to prove things "automatically". That is to say, there is a process or method by which we can use Boolean algebra, on statements that are derived from the rules of logic, to create proofs—in other words, to reason.

Boolean Algebra provides part of such a method. It works on statements such as

A is true **and** B is true

A is true **or** B is false

Socrates is **not** a woman

"A", "B", "Socrates" and "woman" are the variables in these logical statements.

When we try to prove something using the rules of logic we start with two or more statements (called *premises*) and with a desired conclusion.

For example, we might start with the statements

> All men are human
>
> Socrates is a man

and the desired conclusion might be

> Socrates is human

Boole showed that if valid statements expressed in the language of logic are translated into Boolean algebra, it is straightforward to use them to generate other valid statements.

Ramon Llull

Probably the first person to attempt to mechanize human thought processes was the thirteenth-century Spanish hermit and scholar Ramon Llull, who was born in Palma de Mallorca some time between 1232 and 1236. At that time Mallorca was strongly multicultural, with perhaps a third of its population representing the Muslim faith and a smaller minority of Jews influencing the economic and cultural life of the island.

Llull became a Franciscan philosopher, one who has claimed a place in the history of logic even though his contributions were more notable for stimulating the ideas of later logicians than for achieving any useful proofs. He was something of a controversial figure in his day, claiming a logical basis for religious belief while being regarded by some as an alchemist and mystic, sometimes derided and sometimes pitied because his ideas were not properly understood.

Llull's life was one of contrasts. In his early teens he became a page boy to King James the First of Aragon and as a result achieved a position of some influence in the king's court. He took advantage of his position to lead a notoriously dissolute and lustful youth, only to see the light, confess his sins and convert to the life of a dedicated Christian. Legend has it that in 1274, aged about 40, Llull climbed Mount Randa, near the city of Palma de Mallorca, in search of spiritual refreshment. There, after a few days of contemplation, he decided to attempt to convert the entire Muslim world, and as many Jews as possible, to Christianity. The crusades had failed to achieve this, as had all attempts to convince the "infidels" that their own religious beliefs were false. And so Llull set about developing a system based on logic, aimed at proving that the beliefs of Christianity were true.

Llull's idea was to try something that was not based on the specific beliefs of any one religion, but instead based only on whatever beliefs or areas of knowledge all three religions had in common. For example, all three religions supported the idea of there being only one God, and no-one could deny that his own God possessed several positive attributes: goodness, greatness, eternity, power, wisdom, will, virtue, truth and glory. All three religions also shared a common heritage of Greek science, from which they knew that the Earth is at the centre of our universe, with seven planets rotating around it, and that our planet is composed of four elements: fire, earth, air and water. The philosophies of all three religions were based on that of Aristotle. And all three were in general agreement about what constituted virtues and what were vices.

Llull then set out to show how one could combine these theological, scientific and moral components to produce arguments that at least could not be rejected outright by his opponents. The components of these arguments would be finite in number and clearly understood in very much the same way that the components of arguments in formal logic are finite in number and are clearly understood by logicians. Each domain of knowledge in Llull's system therefore involved a finite number of basic principles, so that by creating the permutations of these basic principles in pairs, triples, and larger combinations, a list of the basic building blocks for theological discourse could be assembled.

Llull's writings in his *Ars Magna (Great Art)* used geometrical diagrams and primitive logical devices to try to demonstrate the truths of Christianity. He mechanized the process of forming these permutations by constructing devices with two or more concentric discs, each listing the basic principles around the circumference. The permutations could then be formed by spinning the discs so as to line up different permutations.

One such device was used for studying the divine attributes. Each of its two discs contained the fourteen accepted attributes (goodness, greatness, eternity, power, etc.), and the device allowed the user to create the 196 (i.e., 14 × 14) permutations, for example

> God is good and God is eternal
>
> God is eternal and God is great
>
> God is eternal and God is wise

Similar devices were constructed for the study of the soul and the seven deadly sins. Although these devices did not really offer any additional logical powers, the *Ars Magna* was admired by many Renaissance clerics and commented on by such noted scholars as Wilhelm Leibniz.

Llull's writings advanced the idea that logical reasoning could be performed, or at least assisted, by a mechanical process, but as Martin Gardner points out, none of Llull's scientific writings added to the scientific knowledge of his time. Nevertheless, Llull firmly believed that, by using a mechanical device to combine the terms of a logical expression, it would be possible to discover the building blocks that could be employed to construct logical arguments, and his beliefs inspired other pioneers in this field.

Gottfried Wilhelm von Leibniz

Gottfried Wilhelm von Leibniz (1646–1716) was a great mathematician and philosopher, one of the first to try to build a mechanical calculator. He was also the first to appreciate that Llull's method of logical argument could be applied to formal logic. Although Leibniz had little regard for Llull's work in general, he believed there was a chance it could be extended to apply to formal logic. In a rare flight of fancy, Leibniz conjectured that it might be possible to create a universal algebra that could represent just about everything under the sun, including moral and metaphysical truths. If this were possible then the rules of the universal algebra would allow the creation of logical arguments to prove anything that could be proved, to disprove anything that could be disproved, and to deduce and infer. In summary, it would make it possible to reason about anything, using the language of the universal algebra as the language of reasoning:

> All our reasoning is nothing but the joining and substituting of characters, whether these characters be words or symbols or pictures . . .

> . . . If we could find characters or signs appropriate for expressing all our thoughts as definitely and as exactly as arithmetic expresses numbers or geometric analysis expresses lines, we could in all subjects, insofar as they are amenable to reasoning, accomplish what is done in Arithmetic and Geometry.

For all inquiries which depend on reasoning would be performed by the transposition of characters and by a kind of calculus,[1] which would immediately facilitate the discovery of beautiful results...

Moreover, we should be able to convince the world what we should have found or concluded, since it would be easy to verify the calculation either by doing it over and over or by trying tests similar to that of casting out nines in arithmetic. And if someone would doubt my results, I should say to him: "let us calculate, Sir", and thus by taking to pen and ink, we should soon settle the question. [2]

In some of his later writings Leibniz described his method as combining logic with algebra ("algebraico-logical synthesis"). Thus, in 1666, at the age of 19, Leibniz wrote his *Dissertio de Arte Combinatoria*, in which he attempted to formulate a *Mathesis universalis*, a sort of scientific language which would permit any two disputants to settle their differences with pen and ink, as he describes above. One problem with Leibniz' and Llull's thinking is that mankind's ideas of what is right and wrong are constantly changing and, therefore, cannot always be treated as scientific facts that are subject to logical proof. And even if methods of logical proof *were* applicable to their ideas, the twentieth-century Austro-Hungarian mathematician Kurt Gödel has since shown that it is not possible to settle all differences of opinion by Leibniz-like proof. [2]

How Logic Machines Work

It was not until almost sixty years after Leibniz' death that the first machine of any kind was constructed that was able to solve problems in logic. Most of the early (mechanical) logic machines were based upon the same logical principles, employed in a process called the *method of elimination*. The first stage of this process consists of enumerating all the possible statements that can apply to the matter in question. For exam-

---

[1] A calculus is a mathematical system of calculation using symbols.

[2] In 1931 Gödel demonstrated that, within any given branch of mathematics, there would always be some propositions that could not be proven either to be true or to be false using only the rules and axioms of that branch of mathematics itself. For example, it might be possible to prove every conceivable statement about numbers *within* a system by going *outside* that system to find new rules and axioms, but by doing so one only succeeds in creating a larger system with its own unprovable statements. The implication of Gödel's work is that *all* logical systems of any complexity are, by their very definition, "incomplete". In other words, each of them contains, at any given time, more true statements than it can possibly prove according to its own defining set of rules.

ple, if we are discussing an object which might (or might not) be a book and which might (or might not) be blue, then there are four possibilities:

1. The object is a book and it is blue

2. The object is a book and it is not blue

3. The object is not a book but it is blue

4. The object is not a book and it is not blue

The process then examines each of these possibilities and eliminates all those that are not consistent with any information already known about the object.

Logic machines work by having some sort of physical mechanism for eliminating the statements that cannot be valid. The mechanisms for accomplishing this task vary, but they all employ some physical system for generating all of the theoretically possible different combinations of the various terms and their negatives (such as our example of the four listed above). Once all of these combinations have been generated, there is some physical process, such as pressing a key or punching a lever, which mechanically eliminates the ones that are inconsistent with any known facts (the premises).

The Stanhope Demonstrator

The inventor of the world's first real logic machine (one that could solve problems in formal logic), as opposed to the devices described by Llull (which could not), was the Englishman Charles Stanhope, the third Earl of Stanhope (1753–1816). The first version of his logic machine, which became known as the Stanhope Demonstrator, was built in 1775, and he consructed a more advanced version two years later.

The Demonstrator consisted of a brass plate, fixed to a block of mahogany, in the center of which there was a square opening or depression. Two coloured slides made of transparent glass, one red and one gray, were pushed by the user into slots in the sides of the box. By manipulating the demonstrator it was possible to show the validity of simple deductive arguments involving two assumptions and a single conclusion. This might not sound very impressive, but in Stanhope's day logic existed only in the

DEMONSTRATOR,

INVENTED BY

CHARLES EARL STANHOPE.

The right-hand edge of the gray points out, on this upper scale, the extent of the gray, in the logic of certainty.

The lower edge of the gray points out, on this side scale, the extent of the gray, in the logic of probability.

The area of the square opening, within the black frame, represents the holon, in all cases.

The right-hand side of the square opening points out, on this lower scale, the extent of the red, in all cases.

The right-hand edge of the gray points out, on the same lower scale, the extent of the consequence, (or dark red,) if any, in the logic of certainty.

Rule for the Logic of Certainty.

To the gray, add the red, and deduct the holon: the remainder, (or dark red,) if any, will be the extent of the consequence.

Rule for the Logic of Probability.

The proportion, between the area of the dark red and the area of the holon, is the probability which results from the gray and the red.

PRINTED BY EARL STANHOPE, CHEVENING, KENT.

Figure 1. The Stanhope Demonstrator as it appeared in *Mind*, a philosophical journal, in April 1879

limited form of syllogistic logic[3] as it had been handed down by Aristotle 2,000 years earlier.

Stanhope's Demonstrator could solve numerical syllogisms in the following way. Let us consider the syllogism

Eight of my ten cats are male

Four of my ten cats are black

Therefore at least two of my male cats are black

---

[3] A syllogism is the basic form of a logical argument. It consists of two statements (called premises in logic parlance), for example, "All cats are furry" and "Tom is a cat", and one conclusion, for example, "Tom is furry". In each of these statements there are two *terms*. In "All cats are furry", the terms are "cats" and "furry", and in "Tom is a cat" the terms are "Tom" and "cat".

If the red slide is used to represent my male cats and the gray slide to represent my black cats, Stanhope's method was to push the red slide, from one side of the Demonstrator, eight units across the window, and the gray slide, from the other side of the Demonstrator, four units across the window. In the window there will be two units that overlap—these represent the number of my male cats that are certain to be black.

Although Stanhope's brainchild was limited to logical expressions with only two terms (for example, "cats" and "furry" in footnote 3), the Demonstrator was a first practical step on the road towards the mechanisation of thought processes.

Stanhope started work on a book that he intended to be called *The Science of Reasoning Clearly Explained upon New Principles*. In 1800 he printed some of the early chapters of the book on his own hand-operated printing press, and sent copies to only two friends, with the request that they not discuss his invention with anyone in case "some bastard imitation" were to appear before he could publish his own full description of the Demonstrator. Around 60 years after his death the Earl's notes and one of his devices came into the possession of the Reverend Robert Harley, who published an article on the Stanhope Demonstrator in 1879.

Charles Babbage

Charles Babbage (1791–1871) was a prolific inventor. His inventions included a dynamometer, the standard railway gauge, the heliograph ophthalmoscope, the Greenwich time signals and a lighting system for lighthouses. But it was his designs for the Difference Engine and Analytical Engine for which he is best known today, the latter providing his credentials for being the inventor of the first digital computer.

Babbage was the first to conceive of a general-purpose programmable computer but was never able to secure sufficient funding to be able to complete the project. He received a grant of £1,500 from the British Chancellor of the Exchequer to enable him to develop the first of his calculating machines, the Difference Engine, and when that money ran out Babbage financed the continuing development using his own money, until he was granted a further £3,000 of public funds. By then his plans for the Difference Engine had become more ambitious. Instead of calculating to six decimal places, as had been his original intention, he now wanted to calculate to 20 places. This extra complexity meant that Babbage's estimate of the cost of completing the Difference Engine would

be more than his total estimate for the cost of developing his new idea, the Analytical Engine, which he believed would be able to perform various tasks that normally require human thought, such as playing games of skill, including Checkers, Tic-Tac-Toe and Chess. But to his dismay Babbage was unable to obtain further funding from the state, partly because his failure to complete the Difference Engine had badly dented his credibility. As a result the Analytical Engine remained on his drawing board until his dying day.

Babbage planned that the Analytical Engine would be able to reason with abstract concepts and not just with numbers. Together with his collaborator, Lady Ada Lovelace,[4] Babbage realized that a mechanical device (or "engine", as he called it) could manipulate the symbols of a mathematical formula. Its mechanism could embody the rules for logic or calculus and output the results.

Babbage considered the possibility of programming Chess and recognized that a Chess automaton would require a basic framework of rules and objectives to define its actions or moves. Babbage theorized a general set of guidelines based on several conditional statements of the form

if this condition is true... then

Such statements are commonplace in the programming languages of the electronic computer age.

The conditional statements Babbage had in mind for a Chess machine included

1. Does the current state of the game agree with the rules?

2. If yes, then has the automaton lost?

3. If not, then has the automaton won?

4. If not, can the automaton win with the next move? If yes, make that move.

5. If not, can the opponent win with the next move?

6. If yes, stop the opponent from winning.

---

[4]Lady Lovelace was the daughter of Lord Byron and the person after whom the Ada programming language is named.

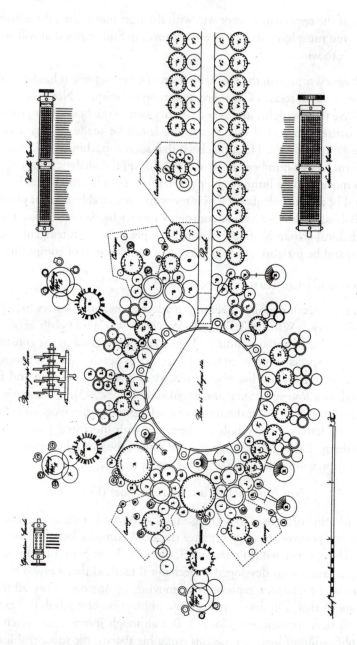

Figure 2. Plan of Babbage's Analytical Engine (1840) redrawn in 1858 as it appears in *Babbage's Calculating Engines* (London, E. and F. N. Spon, 1889) by H. P. Babbage, his son

7. If the opponent cannot win with the next move, then the automaton must look ahead successive moves to find moves that will allow it to win.

Babbage's writings on the mechanisation of Chess represent hardly a drop in the ocean of today's Chess programming techniques. Nevertheless, he did have the foresight, just over a century before the first Chess program was written, to realise that a machine could be made to play a complete game of Chess. He appreciated that such a machine would need to perform a large number of calculations during the analytical process for each move, and he humbly accepted that not even his Analytical Engine would be able to calculate a Chess move in a reasonable amount of time.

Babbage's designs had an enormous impact by demonstrating that a mechanical system could perform what appeared to be intelligent operations, and he is rightly regarded as one of the founders of computing.

Nineteenth-Century Logic Machines

Two nineteenth-century inventors made significant advances on Stanhope's work. William Jevons (1835–1882) was an English academic whose first attempt at building a logic machine resulted in the construction of a special type of abacus as an aid to teach logic. It "consists of a common school blackboard placed in a sloping position and furnished with four horizontal and equidistant ledges". [3] Jevons devised a method of representing the terms of a syllogism by thin wooden rectangles with letters and symbols on them and with short steel pins inserted into them. Each rectangle represented one of the possible combinations of an expression with four terms, for example:

(Not A is true) (B is true) (C is true) (D is true)

The location of a pin in its rectangle indicated whether the corresponding term was positive (true) or negative (false). A complex logical expression could be constructed from these rectangles on the uppermost ledge of his abacus, and Jevons developed a mechanical method that allowed him to simplify the complex expression by moving, in one operation, all those rectangles that contained a particular combination of symbols (for example, all those containing "Not A"). But although Jevons could teach the simplification of logic expressions using his abacus, the successful use of the device depended on the user understanding the process, which was quite complicated.

In 1869 Jevons produced a more advanced logic machine, with a keyboard added, a machine that was able to properly analyse syllogisms. This machine used an ingenious system of rods, wires, levers, springs and latches to enable the wooden rectangles to be moved automatically, eliminating the cumbersome and complicated method used for the abacus, and the machine was limited to expressions with four or fewer terms and their negatives: A, not A, B, not B, etc.

Jevons' construction was the first machine that could solve logical problems at an acceptable speed. The machine was about three feet tall and, because of its appearance, it was described as a "logic piano".

The other major figure in the development of nineteenth-century machines designed specifically to solve problems in logic, was Allen Marquand (1853–1924), the son of a New York banker. He studied at Johns Hopkins University with Charles Peirce, who has been described by Max Fisch as "the most original and the most versatile intellect that the Americas have so far produced." [4] Peirce well understood Babbage's plan for the Analytical Engine and was also interested in logic machines, so it was probably from Peirce that Marquand acquired his own interest in logic machines, which was further developed after Marquand was appointed tutor of logic at the College of New Jersey, later renamed Princeton University.

Marquand first constructed a crude but technically improved version of Jevons' machine in 1881 and, a few months later, with a Princeton colleague, Charles Rockwood, he created a second more advanced design, using a mechanical action with rods and levers connected by pins and catgut strings, this design requiring the use of only ten keys rather than the 20 employed in Jevons' machine.

Marquand's machines were limited in the complexity of argument they could handle and both could only produce logical combinations *consistent with* the concluding proposition, rather than producing the proposition itself. He proposed a third version of his machine that would have changed the action from mechanical to electro-mechanical (i.e., using components that were partly electrical and partly mechanical), but difficulties with the new electrical technology prevented him from advancing beyond building a prototype out of a hotel annunciator, a device that indicated which of the hotel guests has rung a bell from their room. Although his invention worked well, Marquand abandoned his mechanical logic machine project when he was offered a position as a professor of art and archaeology at Princeton.

Peirce later wrote to Marquand suggesting that he resume work on the project, but this time relying on the use of electricity. Peirce could see the parallel between the on or off status of an electrical switch and the truth or falsity of a logic statement, and his suggestion was akin to using Boolean algebra in the design of circuits based on relays.[5] In wanting Marquand to build a relay version of his machine, Peirce may well have had in mind a version of Babbage's Analytical Engine, realising that a system of relays, each of which switches according to whether a statement or a particular term in a statement is true or false, can be employed in much the same way as the mechanics of the older logic engines.[6]

Following Peirce's suggestion, Marquand drew up a wiring diagram, in or around 1885, for an electro-mechanical version of his logic machine, showing 16 relays, but the relay machine was never built. Nevertheless, Marquand's ideas were well in advance of his times and, had he constructed the relay design, his work might well have sped the development of electronic computing by a few decades.[7]

Electrical Logic Machines

Benjamin Burack

In 1936, at the first annual open house meeting at Lewis Institute in Chicago, the psychologist Benjamin Burack demonstrated the first electrical logic machine, one that could detect and demonstrate fallacies in deductive reasoning. Burack's work predated by two years Claude Shannon's 1938 paper on the use of relays to represent truth and falsehood,[8] and was contemporaneous with the more advanced work of Konrad Zuse.[9] Burack deserves considerable credit for being the first to both invent and build an electrical logic machine, but his work received no publicity apart from a short article he wrote about the machine in 1949.

The earlier logic machines described in this chapter, as well as those designed and constructed by Khrushchev (around 1900), Pastore (in

---

[5] A relay is an electrically operated switch. When it is turned on it switches one way and when it is turned off it switches the other way.

[6] With this realisation, Peirce anticipated Claude Shannon's work of 50 years later.

[7] The intended method of operation of Marquand's relay machine has been described by George Buck and Stephen Hunka.

[8] See the next section, "Claude Shannon and Electrical Logic".

[9] See the section "Konrad Zuse and the First Computer" later in this chapter.

1903), Macaulay (in 1910) and Shchukarev (in 1914), were designed to prove conclusions from a given set of premises. But none of these machines could discover fallacies in a deductive process. Burack's machine was therefore unique in two distinct ways—it was the first electro-mechanical logic machine to be built and the first logic machine of any sort that could demonstrate fallacies in logical arguments.

Burack's machine was mounted inside the top half of a sturdy suitcase and weighed about 25 pounds. Burack used a flat wooden block to represent each logical statement (the premises and the conclusion). In order to test the validity of reasoning of a syllogism, Burack would select the blocks representing each of the premises and the conclusion, and place these blocks onto three spaces provided on the panel of the machine. If the proof contained one or more fallacies, i.e., if it was not a valid proof, one or more light bulbs would be illuminated. Each of the light bulbs indicated a different type of logical error in the proof, making it easier to trace the fallacy.

The machine operated by establishing metal contact areas on the backs of the blocks, to convey to the machine information concerning the statement printed on the front of the block. For example, the block representing "Some M is not S" (which could be employed in "some cats are not black" or any similar statement), had separate contact areas on the back of the block to represent information about the statement. One of these items of information on this particular block was that the statement is negative (because of the logical operator **not**); another was that the statement refers to "some" rather than "all" or "none".

When the blocks were set into the front panel of Burack's machine, the metal contact areas on the backs of the blocks were connected to plugs in the panel. Certain combinations of blocks would close one or more of the electrical circuits that were wired from the panel to the bulbs. When a fallacy was detected, the appropriate bulb would be illuminated.

Claude Shannon and Electrical Logic

Claude Shannon (1916–2001) came from a fairly well-educated family and his early environment was one of intellectual stimulation. In fact, he spent much of his youth working with radio kits and Morse code. In 1938, while a graduate student at MIT, Shannon published his master's thesis, which became a seminal paper on the application of symbolic logic to relay circuits. He showed that the TRUE and FALSE aspects

of Boolean logic could be seen as the ON and OFF states of an electric switch. For this pioneering work, important for the functioning of both telephones and computers, Shannon received his MSc[10] in 1940, at the same time as he was presented with his PhD for a thesis on the application of Boolean algebra to genetics.

Shannon was also the pioneer of Information Theory, which underlies all modern electronic communication, and it is his work in that field for which he is best known. *Scientific American* went so far as to refer to his paper, "A Mathematical Theory of Communication", as the "Magna Carta of the information age". Shannon also made enormously significant contributions to AI by his research on computer learning and game-playing. His seminal paper on computer Chess and its influence on the field are discussed in Chapter 3.

Early Games Machines

Von Kempelen's Chess Automaton

The earliest "thinking machines" appeared to play Chess, and play it rather well, but they were elaborate hoaxes. The best known of these was the Chess Automaton, designed and built in 1790 by Baron Wolfgang von Kempelen, who held the title of Counsellor on Mechanics to the Royal Chamber in Vienna. When he first had his new invention wheeled into court, the amused spectators saw a life-sized figure, dressed as a Turk, seated behind a chest that was about four feet long, two feet wide and three feet high. On top of the chest was screwed a chessboard. Baron von Kempelen proudly announced that his Automaton, without any help whatsoever from himself, would play and probably defeat any member of the audience. Like a conjuror, he would open and close doors in the machine, to "prove" that no person could be possibly be concealed inside it.

Von Kempelen toured Europe with his machine, playing in Paris, London, Berlin and other cities. Critics suggested, quite correctly of course, that someone was hidden inside the machine, though no-one was able to guess all the details.

---

[10] Shannon's MSc thesis is widely regarded as one of the most important master's theses ever written, and was awarded the annual Alfred Noble Prize of the combined engineering societies of the United States.

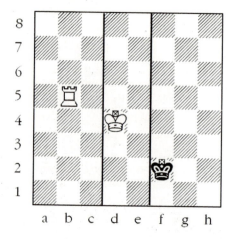

Figure 3. Chess position with King and Rook against King

Although the Turk was a hoax, von Kempelen made an important in-direct contribution to Artificial Intelligence by stimulating people's imag-ination about the idea of a Chess-playing machine. Goethe had already described Chess as "the touchstone of the intellect", a view that has not dimmed with time. And because it is generally accepted that playing Chess well requires intelligence, the creation of a machine that could play Chess well would in fact be the creation of an artificial intellect.

Torres y Quevedo's Chess Endgame Machine

Von Kempelen was followed by others who devised machines that worked along similar lines to his own, but it was not until the Spanish inventor Leonardo Torres y Quevedo (1852–1936) designed and built an electro-mechanical Chess player in 1890 that a genuine artificially intelligent game-player was born. Torres y Quevedo was a prolific and highly cre-ative inventor. He invented a type of cable-car that was installed in Bil-bao, Rio de Janeiro, Chamonix and at Niagara Falls where it is still in use. He pioneered the radio control of machines; he designed and built a series of airships; he built a machine for solving algebraic equations; he published a paper in 1914, demonstrating the possibility of designing an electro-mechanical digital computer 20 years before Konrad Zuse; and in 1920 he designed a precedessor to the digital computer in the form of a calculating machine connected to a typewriter.

Figure 4. Torres y Quevedo's Chess endgame machine (from "Les Automates" by H. Vigneron, *La Natura* 1914, page 57)

Torres' Chess machine was not intended to play a complete game of Chess but to win the ending where one player (the machine, playing white) has a king and a rook against the other player's lone king, as in Figure 3.

Torres divided the chessboard into two zones, one consisting of the a-, b- and c-files, and the other of the h-, g- and f-files. The machine would decide on its move by choosing from one of six classifications, depending upon the position of the black king.

Torres' Chess ending machine (see Figure 4) is still in good working order and can be seen in the museum at the Polytechnic University in Madrid. Its design was remarkable for two reasons. It was the first genuine machine ever to play even part of a game of Chess or any other thinking game. And it was the first machine to operate as a rule-based *expert system*,[11] making its decisions entirely on the basis of a set of rules that incorporate human expertise, the rules given in Figure 5.

---

[11] See the section "Expert Systems" in Chapter 7.

The black King					
is in the same zone as the rook	is not in the same zone as the rook and the vertical distance between the black king and the rook is				
	more than one square	one square, with the vertical distance between the two kings being			
		more than two squares	Two squares, with the number of squares representing their horizontal distance apart being		
			odd	even	zero
The rook moves away horizontally	The rook moves down one square	The king moves down one square	The rook moves one square horizontally	The white king moves one square towards the black king	The rook moves down one square
1	2	3	4	5	6

Figure 5. The six possible operations of Torres y Quevedo's Chess endgame machine (originally printed in the author's book *Chess and Computers* (Computer Science Pr, Rockville, MD, 1976))

The Nimotron

In 1938 the first machine was built that could play a complete game of skill. The machine was designed and patented by Edward Condon, Gerald Towney and Willard Derr and it played the game of Nim. Their machine was called the Nimotron and was exhibited by their employer, Westinghouse, at the 1939 World's Fair in New York.

Nim is a two-player game that starts with a number of piles of objects, such as matches, and with a number of objects in each pile. Typically there might be four piles, as in the Nimotron, but any number of piles can be chosen and any number of matches in each pile. (Four piles with 1, 3, 5 and 7 matches makes for an entertaining game.) The players take turns to move. In making a move a player must choose one of the piles and remove at least one match from that pile. (S)he can remove more than one, up to the whole pile—any number that (s)he wishes. The player who removes the last match of all loses the game. (There is another version of Nim in which the player who removes the last match of all wins the game.)

There is a method of selecting the best move in Nim from any position. Anyone who follows this method will always win the game provided that at some point (s)he has a winnable position. Such a method is called an *algorithm*, which means a method that will find a correct solution to a problem if such a solution exists. In Nim the winning algorithm is based on the use of binary numbers (made up of 1s and 0s). The first step in the algorithm is to convert the number of matches remaining in each pile after your next move into binary. So if you make a move that creates the

position

Pile A   6

Pile B   4

Pile C   3

Pile D   2

the binary representations[12] of the numbers in each pile are

Pile A   110

Pile B   100

Pile C   011

Pile D   010

The next step of the algorithm is to add each column of binary numbers, but to write the totals as decimal numbers, which gives us

$$\text{Total of first column} = 1 + 1 + 0 + 0 = 2$$
$$\text{Total of second column} = 1 + 0 + 1 + 1 = 3$$
$$\text{Total of third column} = 0 + 0 + 1 + 0 = 1$$

The algorithm specifies that if all the totals are even numbers, then the player who moves next will lose, assuming correct play by both sides.[13] But if all the totals are not even numbers, as is the case here, then the player who moves next can win if (s)he finds the correct move. (The winning play is to remove all 3 matches from pile C, leaving the configuration 6, 4, 2, which is a loss for the player who must make the next move.) If no move exists that leaves all the column totals even, then it is not possible to play a move that guarantees a win. In this case the best play is probably to remove a single match from the largest pile, thereby giving your opponent the most complex choice possible and hence maximizing his chances of making a mistake.

Because there is an algorithm for finding the correct move, perfect play in Nim can be "programmed" quite easily. But the Nimotron was not a programmable machine so the "programming" had to be accomplished using relays. There was a lamp corresponding to each "match"

---

[12] For those readers unfamiliar with binary numbers, the binary 110 of pile A represents: $1 \times 2^2 + 1 \times 2^1 + 0 = 4 + 2 + 0$. The binary number 100 in pile B represents: $1 \times 2^2 + 0 + 0 = 4 + 0 + 0$, etc.

[13] There is an exceptional case. If all the remaining piles have one match each, the next player to move will lose if the decimal total is odd and will win if the total is even.

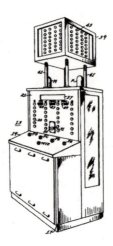

Figure 6. The Nimotron as shown in its patent document (U.S. patent # 2,215,544)

(see Figure 6) and the lamps were illuminated to show which matches were still in the game. When the human opponent made a move by taking away some of the matches from a pile, the corresponding lamps were switched off by the relay circuit. And when the Nimotron made its reply move, more lamps would go out, corresponding to the latest matches removed by the machine.

Konrad Zuse and the First Computer

Konrad Zuse (1910–1995) was a German who invented the world's first digital computer that functioned on electricity. He started working on the design of computers in 1934, at a time when the computing industry was limited to mechanical calculators using decimal arithmetic. At first Zuse worked completely independently and without knowledge of any of the other developments in the field going on in the world around him. "In fact, I hadn't even heard of Charles Babbage when I embarked on my work." [5]

Zuse's research was initially aimed at pure number calculation, but in 1935 and 1936 it led to new ideas about computing in general, causing him to reminisce in later life, "Personally, I believe that was the birth of modern computer science." Zuse realised that all computing operations could be carried out by electro-mechanical relays, and he was thus able to forge a link between engineering and mathematical logic. His

former mathematics teacher showed Zuse that calculations in Boolean algebra are identical with the so-called propositional calculus of mathematical logic. This revelation encouraged Zuse to develop a system, corresponding to Boolean algebra, of *conditional propositions* for relays. (A conditional proposition is one of the form: "If this is true then that is true".)

It appears that Zuse's work with Boolean algebra predates that of Claude Shannon by two or three years. Shannon's master's thesis, "A Symbolic Analysis of Relay and Switching Circuits", was written in 1936 and published in 1938, and it presented a method for representing any circuit consisting of combinations of switches and relays by a set of mathematical expressions. Shannon also developed a calculus for manipulating these expressions, his calculus being exactly analogous to the calculus of propositions used in the field of symbolic logic. Shannon did not know about Zuse's research in Nazi Germany—they were simply researchers in different locations, independently developing the same ideas at around the same time.

Along with Burack's much simpler machine, Zuse's work with relays was the first creation of a working device that converted logic problems into a form in which they could be solved using electro-mechanical means. Although Marquand had already designed a relay-based machine in 1885, his drawings from that year were not published in his lifetime; they lay in a collection of his papers in Princeton University library until they were discovered by Alonzo Church in 1953. It is indicative of how far ahead of their time the ideas of Peirce and Marquand were, that they were suggested in 1877, drawn in 1885, but nothing similar was built until Zuse's admittedly more sophisticated concept 50 years later.

Having solved the problem of how to handle logical deductions using electro-mechanical relays, Zuse was faced with a different problem—a form of high capacity computer memory was needed.

> One device that could deal with this type of operation was the electro-magnetic relay, which can adopt two positions, "open" or "closed". However, at the time I felt that the problem could be better solved mechanically. I played around with all sorts of levers, pins, steel plates, and so on, until I finally reached what was a very useful solution, for those days. My device consisted mainly of pins and steel plates, and in principle could be extended to

1,000 words.[14] A proper machine using telephone relays would have needed 40,000 relays and filled a whole room.

The basic principle was that a small pin could be positioned right or left of a steel lug, thus memorizing the value 0 or 1. This was something new on the Babbage designs. It was clear that programs could be stored provided they were composed of bit combinations—one reason why programmable memory had already been patented by 1936. [5]

One of Zuse's friends helping with the project was Helmut Schreyer who, at that time, was working on the development of electronic relays. Schreyer had the bright idea of using vacuum tubes (or valves as they are called in the U.K.). At first Zuse thought the idea to be a joke, as Schreyer was full of fun and given to fooling around. But after thinking about the suggestion Zuse decided that the idea was definitely worth a try. Thanks to a discipline called *switching algebra* (i.e., Boolean algebra translated into electrical switching), Zuse had already married together two different types of technology—mechanics and electromagnetics. But vacuum tubes could switch a million times faster than mechanical and electro-magnetic components, so Zuse realised that with vacuum tubes the possibilities were staggering.

Zuse was working on a completely private basis, with no government or corporate support but just the help of some friends, and it took him two years to develop a semi-functioning electro-mechanical computer. The Z1 computer (see Figure 7), which weighed about 500 pounds, employed thin metal plates for its memory that had 64 cells each of 22 bits. The machine itself proved somewhat unreliable but, by employing switching algebra, it proved easy for Zuse to develop a relay-based version.

Zuse's Z2 computer employed the same type of mechanical memory as the Z1, but Zuse used old relays that he obtained from telephone companies to construct the arithmetic and control units. Subsequently he also built his Z3 computer completely out of relays, 600 of which were used for the arithmetic unit and 1,800 for the memory and control units. A page from Konrad Zuse's diary recalls that on 12 May 1941 he presented the working Z3 to scientists in Berlin. It is now undisputed that the Z3 was the first reliable, freely programmable, working com-

---

[14]Zuse's use of "word" refers to the unit of storage in a computer.

Figure 7. The Z1 computer in the apartment of Konrad Zuse's parents in 1936 (Courtesy of Horst Zuse)

puter in the world based on *floating-point* binary numbers.[15] The Z3 was destroyed in a World War II bombing raid but a replica is on show in the Deutsches Museum in Munich.

Around 1942 Zuse decided to build an even more powerful computer designated the Z4. Based on his experiences with the Z1, Z2 and Z3 machines, Zuse decided that the Z4 needed much more memory than he had previously employed. He concluded that constructing the memory from metal sheets was far less expensive than building a memory using relays and it was already clear to him that a memory of 1,024 words,[16] each of 32-bits, would, if it consisted of relays, be much too big—he would need more than 32,000 relays (1,024 × 32). But Zuse's mechanical memory system, which he had patented in 1936, worked very reliably, and for 1,024 words the system did not need more than 35 cubic feet of space. Zuse also estimated the costs of one 32-bit word of his mechanical memory as being only five Reichmarks, approximately $2.50 in 1942.

---

[15]A floating point binary number looks like this: 11.1011×2^3. The 11 before the decimal point is the binary for 3 (because it represents $1 \times 2^1 + 1 \times 2^0$). The 1011 after the decimal point is the binary for $11/16^{ths}$ (because it represents $1/2 + 0/4 + 1/8 + 1/16$). It is called a floating point number because the decimal point moves as the power of 2 increases or decreases (in this case it is 2^3)—as this power increases the decimal point moves to the left, as this power decreases the decimal point moves to the right.

[16]The memory in a computer system is often provided in blocks that are measurable as a power of 2. Hence 1,024 words (i.e., 2^{10}) rather than (say) 1,000.

Zuse originally believed that he could complete the development of the Z4 in 12 to 18 months, but in fact it took more than four years to build. The Z4 ended up being much smaller than was originally planned. Zuse intended to have, in addition to a mechanical memory with a capacity of 1,024 words, several card readers and punches and various features to facilitate the programming of the computer. Construction of the machine started well but it was not long before the war imposed its delays. The procurement of staff and materials became increasingly difficult and, around 1943, the Berlin blitz began, with heavy bombing raids on an almost daily basis. Several times Zuse and his friends had to move the machine to a new location, eventually finding refuge in Göttingen, giving its first demonstration on 28 April 1945, just ten days before the German surrender.

Zuse's computers could solve different types of problems, not only those from the field of logic. During the years 1943–1945 Zuse developed a programming language called *Plankalkül* (plan calculus), designed to help in the programming of solutions to combinatorial problems. [17]

For Zuse, Chess was the archetype of a combinatorial problem: "I even went as far as learning to play Chess in order to try to formulate the rules of the game in terms of logical calculus." [5] His monograph on the *Plankalkül* includes a 44-page chapter devoted to programming Chess. Zuse optimistically believed that his machines would eventually be able to play the game better than humans, but his design for a Chess program was never implemented and the Plankalkül was not published until 1972. "I remember mentioning to friends back in 1938 that the World Chess Champion would be beaten by a computer in 50 years time." In this prediction Zuse was not so very far out, but "The importance of my Chess program, as an example of applied logic, was simply ignored." Thus, Zuse's contribution to the early history of AI came on two distinct fronts. By being the first to build a fully functional electrical digital computer he also became one of the first two people to solve logic problems using an electrical machine (Benjamin Burack being the other, earlier inventor). And Zuse was the very first to design a program to play Chess, preceding Turing's hand simulation by seven years. [18]

---

[17]A combinatorial problem is one that grows dramatically in difficulty, just as the number of Chess positions to be analyzed grows dramatically as a program looks further and further ahead in the game.

[18]See the section "From Patzer to World Champion" in Chapter 3.

Mechanical Speech Synthesis

The ability to communicate our thoughts, feelings and desires effectively to other human beings through language is one of the characteristics of being human. Any self-respecting super-robot of the future, one that can communicate with us in our own way, must therefore possess the facility of speech (as well as the ability to understand what is said to it). The history of "talking machines" dates back to the ancient Greeks and Egyptians and their talking oracles. In order to impress citizens with devices that appeared to be able to communicate through some divine means, huge idols such as the "speaking head" of Orpheus at Lesbos had long talking tubes connected to them, giving the impression that the head was talking.

The science of speech synthesis, as it is nowadays called, has its foundations in the eighteenth century, with the first attempt to simulate the human speech mechanism using a bellows and vibrating reeds. A few such devices were developed around that time, the earliest of which appears to have been by none other than Baron Wolfgang von Kempelen, he of the Chess Automaton. In 1771 Erasmus Darwin, the grandfather of Charles Darwin, announced that he had "contrived a wooden mouth with lips of soft leather, and with a valve over the back part of it for nostrils." Darwin's contraption had a larynx made of "a silk ribbon...stretched between two bits of smooth wood a little hollowed." It could say "mama", "papa", "map" and "pam" in "a most plaintive tone". Other early devices included one fabricated by Christian Kratzenstein, who was a professor of mechanics in St. Petersburg and professor of physics and medicine in Copenhagen, and one by Abbé Mical in France.

Von Kempelen's Talking Machine

Von Kempelen purportedly began work on his speech device in 1769 but it was not completed until 1791. His work started two years ahead of Darwin's announcement and ahead of the completion and demonstration of Mical's machine in 1778 and Kratzenstein's in 1779, the latter winning the annual prize of the Imperial Academy of St. Petersburg. Although von Kempelen's machine received considerable publicity, it was not taken as seriously as it should have been, possibly because of doubts and rumours concerning whether or not his Chess player was genuine science or a fraud. His speaking machine, however, was a completely

legitimate device. The speech sounds it produced were the forerunners of the concatenated electronic speech systems in use today,[19] and the machine was not only able to produce speech sounds (19 of them), but was also the first that could speak whole words and short sentences. It could say "mama", "papa", "You are my friend—I love you with all my heart", "My wife is my friend" and "Come with me to Paris", but not very distinctly.

Von Kempelen's machine consisted of a bellows that simulated the human lungs and was designed to be operated with the right forearm. A counterweight allowed for "inhalation". There was a "wind box" that was provided with some levers to be actuated with the fingers of the right hand, a "mouth", made of rubber, and a "nose". The two "nostrils" had to be covered with two fingers unless it was intended to produce a nasal sound. The whole speech production mechanism was enclosed in a box, with holes for the hands and additional holes in its cover.

The final version of von Kempelen's talking machine is preserved in the Deutsches Museum in Munich, in the department of musical instruments. Its voice production mechanism, including the pitch control, is still functional, and reportedly sounds like that of a child or of an adult speaking quite loudly.

Joseph Faber's "Wonderful Talking Machine"

Joseph Faber was a German inventor who developed a talking machine called the Euphonia. In 1840 he demonstrated his machine in Vienna where it attracted very little public interest, causing him to leave Europe for America. In December 1845 Faber exhibited his "Wonderful Talking Machine" at the Musical Fund Hall in Philadelphia. The machine was similar to a small chamber organ but possessed only a single pipe. It consisted of a bizarre-looking talking head that spoke in a "weird, ghostly monotone," as Faber manipulated it with foot pedals and a keyboard. By pumping air with the bellows, and using different combinations of 16 keys to manipulate a series of plates, chambers, and other apparatus including its artificial tongue, the operator could make the machine speak any European language.

When the Euphonia was exhibited in London in 1846, the machine produced not only ordinary and whispered speech but also sang the national anthem "God Save the Queen". But after Faber spent 17 years

---

[19]See the section "Text-to-Speech Synthesis" in Chapter 7.

perfecting the machine, he discovered that few people showed any interest in it. In America Faber achieved little more success with his machine than he had experienced in Europe and so, with the state of his mind also affected by the fact that his eyesight was failing, he destroyed his machine and then committed suicide around 1850.

Electrical Speech Synthesis

The first electrical speech synthesis device was introduced by John Stewart in 1922, while he was employed in the Development and Research Department at the American Telephone and Telegraph Company (now known as AT&T). This synthesizer had a buzzer to simulate the human vocal cords and two resonant circuits to model the acoustic resonances of the throat and mouth. The machine was able to generate single vowel sounds, but not any consonants or connected utterances. The same kind of synthesizer was also built by Harvey Fletcher of the Bell Telephone Laboratories and demonstrated to the New York Electrical Society in February 1924—it could utter a limited number of sounds including the words "mama" and "papa".

The Voder

The first electrical synthesizer which attempted to produce connected speech was the Voder (Voice Operated Demonstrator), developed by Homer Dudley, a research physicist at Bell Laboratories in New Jersey. It was first demonstrated on 5 January 1939 at the Franklin Institute in Philadelphia (see Figure 8) and presented later that year on the Bell System exhibit at the San Francisco and New York World's Fairs. The device synthesized the entire spectrum of human speech sounds and was operated by a young woman using a finger keyboard with additional wrist and foot controls.

The Voder was a compact machine with a pair of keyboard units, more than a dozen other controls and an electrical circuit featuring a vacuum tube and a gas-filled discharge tube. It built up its speech from the same fundamental sounds from which the human speech organs create spoken words. A trained operator would analyse each word the machine was required to speak, then duplicate the sounds (and therefore the words) by pressing a combination of the keys and the other controls. When the Voder was demonstrated at the World's Fair, members of the

Figure 8. The Voder being demonstrated at the Franklin Institute in 1939 (Courtesy of the *Journal of the Franklin Institute*, Philadelphia, PA)

audience were asked to suggest phrases for it to speak. Even such difficult foreign words and phrases as "hasenpfeffer" and "Comment allez-vous?" were convincingly spoken by the machine.

The Voder functioned by imitating the two fundamental types of human speech sounds: the relatively musical note of the vocal cords, which in the Voder came from a vacuum tube, and a sibilant hiss which can be recognised most easily in a whisper, and which, in the Voder, was produced in a gas-filled tube. The tubes each produced an electric wave whose pattern corresponded to the sounds in question, the wave being converted into sound in an amplifier.

In one way the Voder was actually superior to any human being. Ordinarily it spoke in a firmly masculine baritone voice, but it could also speak in tones ranging from the lowest bass to the highest soprano voice. But the Voder was quite difficult to play. It took skilled telephone operators six months to complete the course of 40 lessons devised to train them on the use of the Voder, and a year to become proficient. Nevertheless, to this day the device has been a major influence on the science of speech synthesis, and the staff at Bell Labs still refer to it with reverence.

Franklin Cooper's Pattern Playback

A speech synthesis system of quite a different kind was completed in 1950 by Franklin Cooper at Haskins Laboratories in Connecticut, a Yale-affiliated institute for speech research. The device created the speech from a sound spectrogram which, like a musical score, is a visual representation of the sounds.

The horizontal axis on the spectrogram (Figure 9) corresponds to time (the further to the right you look along the time axis, the later it is). The vertical axis corresponds to frequency (or pitch), with the higher pitched sounds appearing higher on the diagram. The dark patches on the spectrogram indicate relatively intense sounds. A spectrogram provides precise information about the sound because it is based on accurate measurements of the changing frequency content of a sound over the relevant period of time.

Pictures of sound are useful for describing and transforming sounds. Audio researchers often use visual representations of sound to gain a better understanding of the components of the sound and to transform the sound in some way, for example by adding an echo effect or changing the pitch. Cooper realized that visual representations of sound can be turned back into the original sounds themselves, a process of inversion that he called Pattern Playback.

In general this inverse process is not straightforward. The task is essentially one of finding a sound waveform, a representation of the sound wave's amplitude during a certain period of time, that comes closest to generating the original picture (the spectrogram or "voiceprint"). Cooper's device converted the spectrograms into sound, using either photographic copies of actual spectrograms or synthetic patterns that had been painted by hand on a cellulose acetate base (rather like the material in a film negative). The speech sounds were recreated by passing light

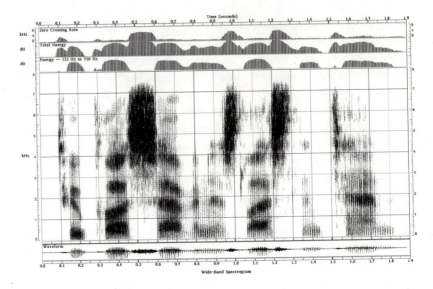

Figure 9. A spectrogram of a male human voice saying "Two plus seven is less than ten" (Courtesy of Victor Zue)

from a lamp, through a rotating disk, and then through a spectrogram pattern, into photo-electric cells. The amount of light at each frequency band corresponded to the amount of acoustic energy at that band, so by converting the light energy impinging onto the photo-electric cells into sound energy, the original speech sounds could be heard.

Speech Recognition

> How often, on waking of a cold morning, we have wished for a robot we could command to shut the bedroom window. To design a machine that could "hear" and "understand" speech is certainly an old dream of mankind. In our age of technology we can think of myriads of ways to put such a machine to work—from printing a dictated speech to answering the telephone or operating a factory on spoken commands. [6]

Thus began a 1955 article in *Scientific American* by Edward David describing the world's first speech recognition machine. It was designed by Stephen Balashek and Keith Davis at Bell Telephone Laboratories and was called Audrey.[20] This machine was intended to facilitate voice dial-

---

[20] For Automatic Digit Recognizer.

ing by being able to recognize only the words for the digits 0 to 9. After a brief training session with a new voice, Audrey would dial correctly, most of the numbers much of the time, but it was rarely able to cope with a complete seven-digit telephone number correctly.

Audrey was trained on Balashek's voice and had stored in the machine the speech patterns (spectrograms) of those ten words, as he spoke them. When Audrey heard a spoken word, the incoming sound was first analyzed by its electronics, and then a spectrogram pattern for the word was produced. Then this pattern was automatically compared with the stored spectrogram patterns for Balashek's voice for each of the ten digits, in order to find the nearest match. Within about 0.2 seconds the machine illuminated a bulb to indicate which of the ten digits it thought had been spoken, choosing the most probable word (i.e., the closest match), unless the incoming sound bore almost no resemblance to any of the ten stored patterns (in which case Audrey simply did nothing).

Audrey performed quite well when Balashek spoke to the machine but with other male speakers Audrey made mistakes ten to thirty percent of the time, depending on the characteristics of the speaker's voice. And with female and children's voices the results were rather poor because of the different pitches and other voice characteristics.

– 2 –

Early History—Robots, Thought, Creativity, Learning and Translation

Robot Tortoises

In his *History of the Second World War*, Sir Winston Churchill wrote a chapter on the "wizards" who had helped Britain to win the war in the air by the development and use of radar. William Grey Walter (1910–1977) was one of those young wizards.

Walter was born in Kansas City but was educated in England, where he chose to remain. During the late 1940s Walter, who was also a renowned neuro-physiologist and had been a leading figure in the development of electro-cephalogram technology, carried out pioneering research on mobile autonomous robots. He started building three-wheeled, tortoise-like robots at the Burden Neurological Institute in Bristol, as part of his attempts to model the function of the brain. He wanted to study the basis of simple reflex actions and to test a theory of complex behavior arising from the interconnection of neurons. He was convinced that even organisms with extremely simple nervous systems could show complex and unexpected behavior.

Walter conceived and created the first autonomous robotic "animals", who he named Elmer (Electro-mechanical Robot) and Elsie. They were built in the guise of tortoises, inspired by the turtle who taught the other turtles in *Alice's Adventures in Wonderland* and who was therefore called a tortoise.[1] Walter's robots were unique because, unlike the few robotic creations that had preceded them, they did not have a fixed behavior. His robots had reflexes which, when combined with their environment, resulted in them avoiding the repetition of the same actions. This life-like behavior was an early form of what we now call Artificial Life.

Walter's robot design was rather crude (see Figure 10). The tortoise's "muscles", which caused it to move, consisted of three wheels, of which two were for propelling the tortoise forwards and backwards while the

---

[1] Get it? "Taught" becomes "tort".

Figure 10. William Grey Walter soldering one of his tortoise robots (circa 1948) from *The Robot Book* by Richard Pawson (Windward, U.K., 1985, page 14)

third was for steering. The tortoise's "sense organs" were extremely simple, consisting only of a photo-electric cell[2] to give it sensitivity to light and some electrical contacts mounted on the tortoise's surface and serving as touch sensors, which enabled the creature to respond to anything with which one of its sensors came into contact. A telephone battery provided the tortoise's power and a plastic shell covered the whole assembly and provided a measure of protection against physical damage to the tortoise's innards.

The tortoise's "nervous system" was also rather primitive. An electronic circuit incorporating two vacuum tubes[3] controlled the wheel motors and the tortoise's sense of direction, using information about its surroundings that had been acquired by the sensors. The tortoise could perform only two actions: it could avoid big obstacles, which it detected by a contact sensor, retreating when it hit one; and it was attracted to sources of light that were detected by its photo-electric cell, though if the light was too intense the robot backed off rather than moving towards the light. Until it saw a light, the photo-electric cell was constantly rotating, scanning its horizon for light signals.

---

[2] A photo-electric cell is a device for converting light into electrical energy.

[3] Vacuum tubes (called valves in the U.K.) were glass tubes containing complicated-looking arrangements of electronic components in the form of metallic plates. They were employed in amplifiers, radios and other electronic products before the introduction of transistors.

The scanning process was linked with the tortoise's steering mechanism in such a way that the photo-electric cell was always looking in the direction of movement, so that when a light was detected in any direction the tortoise could respond without too much manoeuvring. The photoelectric-cell and motors were connected in such a way that when an adequate source of light was detected, the tortoise would turn towards it and approach it. When the tortoise was moving towards an attractive light source and met an obstacle on the way, the attraction of the light would be reduced so the tortoise would lose interest in the light until after it had dealt with the obstacle. In addition to avoiding bright lights, the tortoise also disliked steep gradients and all but lightweight physical obstacles. And while it would go around heavy objects, it could also push lightweight obstacles out of the way by a process of repeated butting and withdrawing.

Walter's tortoises possessed some measure of self-recognition. They were each fitted with a flash-lamp bulb on the head, which was switched on whenever the photo-electric cell received an adequate light signal. When a tortoise encountered a mirror or some other reflective surface, its photo-cell could "see" a reflection of this light signal, making the tortoise head towards its own reflection. But as it did so this light source appeared brighter, causing the tortoise to move away from it. As the tortoise got further away the intensity of the reflected light was reduced below the tortoise's "I don't like it" threshold and, once again, the tortoise started to head towards the reflected light. If such behaviour was observed in an animal it would be accepted as being caused by some degree of self-awareness. In this respect Walter's tortoises appeared to be exhibiting an intelligence superior to that of those animals who treat their own reflection as though it is another animal.

Thus Walter had produced a kind of "social behaviour" in which the tortoises "danced" around each other, in movements of attraction and repulsion, which Walter likened to mating. This prompted him to write

> I noticed uncertain, randomic, free-will or independent characteristics [which constituted] aspects of animal behavioral and human psychology... Despite being crude they conveyed the impression of having goals, independence and spontaneity. [1]

Walter considered his tortoise to be a new "species" of animal and gave the species the Latin name *Machina Speculatrix* because of their exploratory, speculative behavior. In a second series of experiments, Walter cre-

37

ated a version of his tortoise that was capable of learning, which he called *Machina Docilis* (machine capable of being tamed). His idea for the *Machina Docilis* was to train it in much the same way as the Russian physiologist Ivan Pavlov had trained his dogs.[4] Pavlov employed two reflexes with his dogs, the ringing of a bell and the arrival of food. The dogs became accustomed to getting food whenever the bell was rung and so they associated the sound of the bell with being fed. As a result, whenever they heard the bell, Pavlov's dogs would salivate even when there was no food in sight or within smelling distance.

Walter's *Machina Docilis* was created by grafting on to the back of his *Machina Speculatrix* what he called a Conditioned Reflex Analogue (CORA)—a learning box. This provided the robot with the ability to form conditioned reflexes, associating the physical action of being kicked with the hearing of a whistle. This conditioning could be reinforced by giving the robot five or six kicks, each one straight after a whistle, but if the conditioning was not continually reinforced it wore off, so that subsequent whistles would not create the same reflex action in the robot.

The learning box created a connection in the robot's "brain" between the robot's light reflex and a whistle, and another connection between the robot's contact reflex and the whistle. The robot could be trained so that, for example, whenever the whistle was blown the *Machina Docilis* turned and backed away from an obstacle which it imagined to be in its path, just as the conditioned reflexes of Pavlov's dogs caused them to respond to the sound of a bell by expecting the arrival of food.

Walter's tortoises had a tendency to explore the environment actively rather than to wait passively for something to happen. They were never still, except when "feeding", i.e., recharging their batteries which they did in a specially built hutch. At the top of the hutch was a lamp. Walter noticed that the tortoises wandered around a lot, but eventually, as their battery gradually ran down, they would turn towards the hutch in order to recharge themselves. Once a tortoise had "fed" itself in this way it would leave the hutch to search its surroundings for new sources of light:

---

[4]Pavlov is renowned for his landmark study on conditioning. When he launched the study Pavlov was actually observing the physiological effects of eating in dogs. He began by studying digestion, but began to observe that the salivation of the dogs was very curious. He would place meat powder or some other food morsel on the dog's tongue, waiting for the salivation to occur, and soon observed that the dogs were salivating as soon as he entered the room, which was before any food was even in sight. Since salivation in any animal is a reflex, Pavlov decided to probe deeper into the conditioning of the dogs. This later became known as classical conditioning.

> It bustles around in a series of swooping curves so that in an hour
> it will investigate several hundred square feet of ground. In its
> exploration of an ordinary room it inevitably encounters many ob-
> stacles, but apart from stairs and fur rugs, there are few situations
> from which it cannot extricate itself. [1]

The behaviour of the tortoises established Walter as the pioneer of electronic robots. His were among the first man-made creatures that possessed some of the properties typical of living beings, such as behaviour
and self-organization. His work attracted considerable media and public attention, resulting in several requests to "adopt" Elsie and Elmer as
pets. Walter himself developed a feeling of affection for them, referring
to them as "little beasts that seem to have a personality of their own". His
were also the first free-ranging, autonomous robots capable of exploring
their own limited worlds, and they showed what can be achieved with
relatively simple electronic brains. They were so life-like in their behaviour that an old lady who felt pursued by one of them ran upstairs and
locked her door!

Although the results of Walter's research into robotics were quite spectacular for his day, the significance of his work was not widely recognized
at the time. Sadly his career was cut short by a tragic motorcycle accident
when he was 60, in which he sustained massive brain damage, leading to
his death seven years later. A specimen of one of his second generation
tortoises is in the collection of the Smithsonian Institution.

Alan Turing

Alan Turing (1912–1954) was a brilliant British mathematician and one
of the founding fathers of computer science (see Figure 11). Very early
in life, Turing showed signs of the genius he was to display more promi-
nently later. He is said to have taught himself to read in three weeks and
to have shown an early affinity for numbers and puzzles, and when he
was only 16 Turing discovered and understood Albert Einstein's work,
even extrapolating Einstein's Law of Motion from a text in which it was
never properly explained.

During World War II Turing was a major participant at Bletchley
Park, in the British efforts aimed at breaking the German Enigma code.
He also contributed several mathematical insights on the Fish teletype
ciphers, encoding machines used by the Germans. Turing's work on Fish

Figure 11. Alan Turing (Courtesy of the Archive Centre, King's College Cambridge)

proved useful in the development of the special-purpose digital computer Colossus, which was built in London in 1943 and used to crack the Fish cyphers.

From 1945 to 1948 Turing worked at the U.K. National Physical Laboratory on the design of a computer called ACE.[5] In 1949 he became Deputy Director of the computing laboratory at the University of Manchester, and worked on software for one of the earliest computers— the Manchester Mark I. During this period he continued research into more abstract projects and, in the journal *Mind*, he published his seminal article: "Computing Machinery and Intelligence". In that article Turing discussed artificial intelligence (though not under that name—the term was not coined for another five years), and proposed what is now known as the Turing Test, an attempt to define a standard that a machine had to achieve if it were to be described as having intelligence and being aware of its own existence.

"Computing Machinery and Intelligence"

The repercussions of Turing's famous paper, and in particular its first section on the Imitation Game, have resonated throughout the world of Artificial Intelligence ever since. This, arguably, is the most significant publication in the entire history of AI.

---

[5] Automatic Computing Engine.

1 The Imitation Game

I PROPOSE to consider the question, "Can machines think?" This should begin with definitions of the meaning of the terms "machine" and "think".

The definitions might be framed so as to reflect so far as possible the normal use of the words, but this attitude is dangerous. If the meaning of the words "machine" and "think" are to be found by examining how they are commonly used it is difficult to escape the conclusion that the meaning and the answer to the question, "Can machines think?" is to be sought in a statistical survey such as a Gallup poll. But this is absurd. Instead of attempting such a definition I shall replace the question by another, which is closely related to it and is expressed in relatively unambiguous words.

The new form of the problem can be described in terms of a game which we call the "imitation game". It is played with three people, a man (A), a woman (B), and an interrogator (C) who may be of either sex. The interrogator stays in a room apart from the other two. The object of the game for the interrogator is to determine which of the other two is the man and which is the woman. He knows them by labels X and Y, and at the end of the game he says either "X is A and Y is B" or "X is B and Y is A".

The interrogator is allowed to put questions to A and B thus:

C: Will X please tell me the length of his or her hair?

Now suppose X is actually A, then A must answer. It is A's object in the game to try and cause C to make the wrong identification. His answer might therefore be "My hair is shingled, and the longest strands, are about nine inches long."

In order that tones of voice may not help the interrogator the answers should be written, or better still, typewritten. The ideal arrangement is to have a teleprinter communicating between the two rooms. Alternatively the question and answers can be repeated by an intermediary.

The object of the game for the third player (B) is to help the interrogator. The best strategy for her is probably to give truthful answers. She can add such things as "I am the woman, don't listen to him!" to her answers, but it will avail nothing as the man can make similar remarks.

We now ask the question, "What will happen when a machine takes the part of A in this game?" Will the interrogator decide wrongly as often when the game is played like this as he does when the game is played between a man and a woman? These questions replace our original, "Can machines think?" [2]

Later in the paper Turing discusses his own belief, namely that the answer to his question is "yes".

It will simplify matters for the reader if I explain first my own be-liefs in the matter. Consider first the more accurate form of the question. I believe that in about fifty years time[6] it will be possible to program computers with a storage capacity of about 10^9 to make them play the imitation game so well that an average interrogator will not have more than a 70 per cent chance of making the right identification after five minutes of questioning. [2]

Turing's figure of 10^9 is 1 gigabyte in today's terminology, roughly the memory capacity of a typical PC's hard disk in the closing years of the twentieth century. By the year 2000, although computers had not been programmed to enable them to play the "imitation game" so well that an average interrogator had no more than a seventy percent chance of making the correct identification after five minutes of questioning, some of the programs entered into the annual Loebner Prize competition[7] are not so very far away from this goal. My own program Converse, which won the 1997 Loebner competition in New York, convinced one of the five judges after his first ten-minute session with the program that it was in fact human, only giving itself away in a subsequent five-minute session later that day.

There are other predictions made in Turing's 1950 paper that already appear to have been vindicated. In particular, educated opinion has al-ready been altered to the extent that one can now speak of the possibility of machines thinking and learning, and the term "machine intelligence" is not the oxymoron that it might have been taken for when Turing first started thinking about the subject.

Turing's paper continued:

The original question, "Can machines think?" I believe to be too meaningless to deserve discussion. Nevertheless I believe that at

---

[6]I.e., around the year 2000.
[7]See the section "Passing the Turing Test" in Chapter 7.

the end of the century the use of words and general educated opinion will have altered so much that one will be able to speak of machines thinking without expecting to be contradicted. I believe further that no useful purpose is served by concealing these beliefs. The popular view that scientists proceed inexorably from well-established fact to well-established fact, never being influenced by any unproved conjecture, is quite mistaken. Provided it is made clear which are proved facts and which are conjectures, no harm can result. Conjectures are of great importance since they suggest useful lines of research. [2]

Having explained his own position on the question, Turing then considered "opinions opposed to my own", the first of which he dismisses out of hand:

(1) **The Theological Objection.** Thinking is a function of man's immortal soul. God has given an immortal soul to every man and woman, but not to any other animal or to machines. Hence no animal or machine can think.

I am unable to accept any part of this. [2]

The next objection considered by Turing is a natural but purely emotional one, again dismissed in short order.

(2) **The "Heads in the Sand" Objection.** "The consequences of machines thinking would be too dreadful. Let us hope and believe that they cannot do so."

This argument is seldom expressed quite so openly as in the form above. But it affects most of us who think about it at all. We like to believe that Man is in some subtle way superior to the rest of creation. It is best if he can be shown to be necessarily superior, for then there is no danger of him losing his commanding position. The popularity of the theological argument is clearly connected with this feeling. It is likely to be quite strong in intellectual people, since they value the power of thinking more highly than others, and are more inclined to base their belief in the superiority of Man on this power.

I do not think that this argument is sufficiently substantial to require refutation. Consolation would be more appropriate: perhaps this should be sought in the transmigration of souls. [2]

Turing's opinion on this particular contra-argument was later endorsed by another of the founding fathers of Artificial Intelligence, Nobel laureate Herbert Simon of Carnegie Mellon University.

AI has been thought controversial because it challenged the uniqueness of human thought, as Darwin challenged the uniqueness of human origins. The boundaries of AI continue to expand rapidly, settling the controversy for those who know the evidence. [3]

The Turing Test

The Imitation Game described in Turing's paper has since become the basis for what is known as the *Turing Test*. This phrase has been used more generally to refer to some kinds of behavioural tests for the presence of mind, or thought, or intelligence in entities that are intended to appear intelligent. The modern form of the Turing Test, as employed for example in the Loebner Prize competitions, calls for human judges to be given the opportunity to interact via a computer terminal with whatever entity is at the other end of the communications link, be it human or computer program. Typically each judge is given a ten-minute session on each terminal and, once all of the terminals have been interrogated by all the judges, each judge is given a further five minutes with each terminal. The judges may interact in any way they wish, asking whatever questions and making whatever statements they like. At the end of the day every judge ranks the terminals from "most humanlike" to "least humanlike" and the program with the highest average ranking is the winner of the competition. Each judge is also asked to draw a line in his or her ranking list—above this line are those terminals which the judge believes to have a human at the other end.

This form of the test is very much as Turing himself suggested, using conversation in natural language as the domain in which to determine whether a computer program can justifiably be said to be intelligent. If a program can convince the judges that it is "above the line" (i.e., human), then that program will pass the Turing Test. But it is not only in the realm of human-computer conversation that the Turing Test applies. Indeed, it is relevant to just about every realm in which human intelligence is a prerequisite for performance at a high level. An example from a different realm is Chess. As the twentieth century drew to a close the world's leading Chess programs performed at the upper echelons of human playing strength. It seems perfectly reasonable to argue that, by achieving, in this highly intellectual pursuit, the level of the world's top human players, the best programs had passed the Turing Test for Chess because their standard of play was so high. Yet Kasparov and a few other

top human players are sometimes able to tell, from an occasional "typical computer move", that their opponent's brain is made of silicon. Does this negate the programs' performances with respect to the Turing Test? I think not. After all, if a conversational program were occasionally to use poor grammar in communicating with a human competition judge, would that necessarily give the game away? Humans too are perfectly capable of making grammatical mistakes.

Since Turing's 1950 paper was published, the Turing Test has had its fair share of criticism as a yardstick of intelligence in computer programs. The test's critics have put forward various reasons why the test is inadequate for demonstrating AI, and they have often suggested "better" alternative tests, but it is far from clear that any of these alternatives actually proposes a better goal for research in AI than is set by the Turing Test itself.

It is sometimes suggested that the Turing Test was anticipated some 300 years earlier by René Descartes, one of the most important Western philosophers of the past few centuries and often referred to as the "father" of modern philosophy. In an essay of 1637, "Discourse on the Right Method for Conducting one's Reason and Discovering the Truth in the Sciences", Descartes wrote

> If there were machines which bore a resemblance to our bodies and imitated our actions as closely as possible for all practical purposes, we should still have two very certain means of recognizing that they were not real men. The first is that they could never use words, or put together signs, as we do in order to declare our thoughts to others. For we can certainly conceive of a machine so constructed that it utters words, and even utters words that correspond to bodily actions causing a change in its organs (for example, if you touch it in one spot it asks what you want of it, if you touch it in another it cries out that you are hurting it, and so on). But it is not conceivable that such a machine should produce different arrangements of words so as to give an appropriately meaningful answer to whatever is said in its presence, as the dullest of men can do. Secondly, even though some machines might do some things as well as we do them, or perhaps even better, they would inevitably fail in others, which would reveal that they are acting not from understanding, but only from the disposition of their organs. For whereas reason is a universal instrument, which can be used in all kinds of situations, these organs need some particular action; hence it is for all practical purposes impossible for a machine to have enough different organs

to make it act in all the contingencies of life in the way in which our reason makes us act. [4]

From this it appears that Descartes believed machines would never be able to think, hardly a surprising opinion in the mid-seventeenth century. But what is most interesting about Descartes' comments is his reason for denying the possibility of thinking machines—he believed that machines could not pass what we now know as the Turing Test.

As can be imagined, Turing's proposal of his test created considerable controversy amongst scientists and philosophers, controversy that has lasted even to today. One of the more intriguing critiques was an article published the following year, also in *Mind*, entitled "Do Machines Think About Machines Thinking?" Its author, Leonard Pinsky, wrote somewhat in tongue-in-cheek style, proposing an experiment:

> Let us take one of Mr. Turing's highly complex electronic or digital computers and, for a Christmas gift, send it a subscription to *Mind*, retroactive to October 1950. This means that the first article which will become part of its "store", and so part of its experience, will be Mr. Turing's article, on the problem "Can Machines Think?" The machine finds the article stimulating, probably, and a thought (the term is used loosely with no intention to prejudge the issue) runs through its wiring—it is thinking about the possibility of machines thinking! Since this is the very sort of thing which led philosophy astray for so many centuries, it will not surprise us when we discover that the machine suffers a nervous breakdown. (According to Norbert Wiener[8], machines have breakdowns under pressure which cannot be distinguished from the nervous breakdowns of human beings.) Its efficiency is greatly decreased, the answers the machine gives are paradoxical, and the engineer is worried. Presumably, the engineer can fix the machine by ordinary means. In this instance, however, the machine fails to respond to the customary electronic therapy; the engineer is forced to call in assistance. Since the machine can be regarded as a nervous organism, the nerve specialist comes to the engineer's aid. After a thorough inspection and consultation with other specialists, the conclusion is reached that he machine's difficulty is psychosomatic. The clinical psychologist is called. After a reasonable length of time, it is evident that the Rogerian[9] non-directive technique is

---

[8] Wiener would later become known as the founding father of Cybernetics, the science concerned with control systems in electronic and mechanical devices.

[9] Rogerian psychiatry is explained very briefly in the section "The First 50 Years of NLP" in Chapter 7.

of no avail, and the psychoanalyst insists that he be allowed to try his hand. The psychoanalyst finds no syndrome for which he can account in his system. The machine has no Oedipus complex, for it had no parents. Since it had no childhood, plunging into its past does no good. The machine has failed to respond to any of the traditional forms of psychotherapy.

Where do we turn now? To none other than the Therapeutic Positivist, or T.P. The T.P. sits down near the machine, asks it a few questions and discovers that it is perplexed about the problem "Can Machines Think?" The job of the T.P. is to show the machine that it has been making "metamechanical" statements. In other words, in discussing the problem with itself, it has really only been recommending changes in its calculus, in the binary code.[10] This revelation should make it clear to the machine that it was only tussling with a pseudo-problem, and it will thereupon desist from making "metamechanical" utterances. If the T.P. can perform this therapy upon the machine, we have, *a fortiori*, shown that the machine *does* think, since it has been able to misuse its thinking powers! This, I suggest, is the experimental crucial. We may leave it to the engineer to work out the technical elaborations. [5]

Turing's research into thinking machines quickly advanced from the theoretical and the philosophical into the practical. In 1952 he "wrote" a Chess program. Lacking a computer powerful enough to execute it, he himself simulated the computer, taking about half an hour to calculate each move. Only one game was recorded, in which the "program" lost to one of Turing's colleagues. During the Turing era games became a subject of attention amongst those interested in computing, and since then playing games of skill has consistently attracted Artificial Intelligence researchers. From the mid-1950s onwards Chess in particular became a popular measure of how much progress had been made in AI, the reason being that Chess and other strategy games are generally regarded as pursuits that require intelligence in order to play them well.

Turing never saw his ideas in Chess reach the embodiment of a computer program. Within two years of his hand simulation he was dead, at the age of 41. He had been prosecuted as a result of his homosexual activities and had become viewed as a security risk by G.C.H.Q., the centre of Britain's post-war work on code breaking. The depression he went through following that conviction is assumed to be the main

---

[10]I.e., its programming.

cause of his suicide in 1954, when he ate an apple soaked in potassium cyanide.

Creativity

The earliest examples of artificial creativity came in the field of music composition.

Kircher's Device

In *Musurgia Universalis*, an encyclopaedic scientific work published in Rome in 1650, the famous German Jesuit scholar Athanasius Kircher[11] provided two illustrations of his *Arca Musurgia*, a device by which a non-musician could compose a chorale, for four voices, by selecting at random prearranged musical fragments that were inscribed as tables on wooden strips.

The *Arca Musurgica* comprised a wooden box with several compartments housing wooden strips, each strip being inscribed on its top and bottom face as seen from the example strip in Figure 12.

Examining one of these strips provides an illustration of Kircher's method for the automatic composition of music. The wording at the top of the strip is *"Iambica Euripedaea"*, which indicates that this particular strip is for composing a chorale in which each line contains six syllables. An example of this type of chorale is

> *Ave maris stella*
> *DEI Mater alma,*
> *Atque semper Virgo*
> *Felix coeli porta.*

The roman numerals on the strip indicate possible differences in the character of the chorale (reverent, joyful, sorrowful, anguished), with the key of the composition being determined according to the character of the song.

Next there are six tables, each containing four rows with six numbers per row. Each number stands for a musical note. The number 1 denotes the keynote; the number 2 denotes the following note, and so on.

---

[11] Kircher's (1602–1680) oeuvre extended to some 44 books and more than 2,000 letters that are now in the Vatican Museum in Rome.

Figure 12. Strip from Kircher's Arca Musurgica (Courtesy of the Herzog Anton Ulrich-Museum, Braunschweig, Germany)

G	A	Bb	C	D	Eb	F	G
1; 8	2	3	4	5	6	7	8; 1

Figure 13. Notes from the reference list on the lid of the box of Kircher's Arca Musurgica

In each case the first row of a table sets out the notes for the first voice (soprano), in other words the melody. The second row gives the notes for the second voice (alto), the third row is for the tenor voice, and the fourth row is for the bass voice. Thus the strip offers the choice between six different melodies, each with its three other associated voices. The choice as to which of the six tables should be used for a particular composition can be made by the user at random, for example by throwing a die.

Next on the strip we come to a section with eight rows of notes in which the lengths of the individual sounds vary. Here the user may again choose at random how long the individual notes in the melody will be.

Looking a little more closely at the example *Ave maris stella*, we find that in keeping with the character of the song, Kircher selects the key of G minor, which determines the numbers corresponding to the individual pitches. With the help of the reference list on the lid of the wooden box, we arrive at Figure 13. If the user happens to select the third table on the strip, that table indicates that the melody, which is sung by the soprano, is represented by the row of numbers

3 2 3 5 4 5

for which the corresponding notes (taken from Figure 13) are

Bb A Bb D C D

Similarly, the numbers and the corresponding notes for the other three voices (also taken from the third table on the strip) are

Second voice (alto):	5 7 1 7 7 7	viz.:	D F G F F F
Third voice (tenor):	3 4 1 3 2 3	viz.:	Bb C G Bb A Bb
Fourth voice (bass):	8 7 6 3 7 3	viz.:	G F Eb Bb F Bb

Lower down the strip we find the rows corresponding to the various arrangements for the durations of the notes. If the user happens to choose the second row (a choice that could also be made using a die or some other method of random selection)

this corresponds to note durations of

$$\frac{3}{4} \quad \frac{1}{4} \quad \frac{1}{2} \quad \frac{1}{2} \quad 1 \quad 1$$

Thus, by making a random choice of which table to use for the musical notes and which row to use for their duration, the first line of the chorale can be composed. The process continues, line by line, until the composition is complete.

Mozart's Dice Game

Several other similar systems were developed by authors and composers interested in *chance music*, including Philip Kirnbirger in 1757 and Josef Haydn's *Philharmonic Joke* in 1790. In the introduction to his book *The Ever Ready Composer of Polonaises and Minuets*, Kirnbirger explained that the readers "will not have to resort to professional composition" but could now compose their own music. His book met with such success that in 1783 he published a more complex system, enabling the reader to compose sonatas, overtures and even symphonies.

The best known work in this genre was Mozart's *Musikalishes Würfelspiel (Musical Dice Game)*,[12] which appears to have first been published anonymously in 1787, and then by Mozart's publisher, Nikolaus Simrock, in Berlin in 1792, with instructions in German, French, and English. Mozart's idea was to enable his readers to compose "... without the least knowledge of music, German waltzes, by throwing a certain number with two dice."

Mozart provided 176 bars of music, arranged in 16 columns with 11 bars in each column. To select the first bar of a piece of music the reader would throw two dice and, according to which of the 11 possible totals resulted, the corresponding bar would be selected from the first column. Then the dice would be thrown again and the new total used to select a bar from the second column, and so on. The number of possible pieces composed in this way is $11 \times 11 \times 11...$, a staggering $46 \times 1,000,000,000,000,000$ combinations. The method works sufficiently

---

[12] There have been some doubts cast upon its authenticity but most Mozart scholars now accept it as authentic—his handwritten manuscript for at least part of the publication is listed as K516F in the index compiled by Köchel, the standard authority on Mozart's work.

well that every piece it creates sounds Mozartian in style, due to Mozart's deep understanding of the musical style of his period. Implementations of the dice game that have been made available on the Internet demonstrate that the music sounds quite natural.[13]

Several musical dice games were printed and reprinted in many parts of Europe during the latter part of the eighteenth century and represent the first efforts at algorithmic composition—the creation of music using a clearly defined algorithm. This idea was taken further in 1821 by the Dutch inventor Diederich Winkel, who had previously built some weaving machines and discovered a way of producing an almost infinite variety of fabric patterns. Winkel designed and built a mechanical device for algorithmic composition, called the Componium (see Figure 14), endowed not only with a sound of excellent quality, but also with the capability of repeatedly improvising new variations on a given musical theme of 80 measures.

The improvisation process was based on two complementary synchronised musical cylinders, each having its musical measures alternating so that two measures of silence were followed by two of musical notation, arranged so that one cylinder played the two measures of music while the other was silent. Both cylinders continued to move throughout the composition process, with the "non-playing" cylinder at any moment using its silent time to move, under the control of a randomly moving "programmer", to a different position, while the other cylinder was playing its two measures.

The cylinders were studded with pins in such a way that, to the right of the pins that represented a given theme, there were pins representing seven variations on that theme. The alternation of the cylinders was arranged in such a way that two measures of a variation of one of the eight motifs[14] on one of the cylinders could be followed by two measures of a variation of one of the eight motifs from the other cylinder. Gears and spiral movements of the mechanism, together with variations in timbre as well as of the measures of music, meant that so many different musical combinations could be created by the Componium that, if five minutes was allowed for the performance of each piece, it would have taken more than 138 trillion years for all the musical combinations to be played.

---

[13] For example at http://sunsite.uniview.ac.at/Mozart/dice/.
[14] The eight motifs were the original and the seven sets of variations.

Figure 14. Componium, D. N. Winkel, Amsterdam, 1821 (inv M456), Musical Instrument Museum, Brussels, copyright MIM-Luc Schrobiltgen

Machine Learning

An essential aspect of building machines that think and reason is to enable them to learn from experience and to improve their performance. Possibly the earliest learning program was one written in 1954 by Russell Kirsch at the National Bureau of Standards in Washington. Kirsch's

program was designed to learn a successful strategy for playing a coin-tossing game that had been suggested by Shannon. In this game one computer would choose heads or tails and another computer would try to guess which. The guessing computer attempts to detect a pattern in the choosing computer's choices, while the choosing computer attempts to detect a pattern in the guessing computer's guesses. The game is made more complicated by the fact that each of the computers will attempt to change its pattern sufficiently often to flummox its opponent.

Kirsch devised a learning mechanism that seemed to him like a model of the way in which an animal learns. His animal model was then used to play the coin-matching game against a choosing program. Evidently a learning mechanism must respond to some sort of stimulus, which, in the case of the coin tossing game, was defined as follows. At each matching of the coin two numbers were generated, one corresponding to the chooser's move and one corresponding to the animal's guess at the opponent's move. Since there are only two possibilities, heads (H) and tails (T), binary numbers were used (1s and 0s).

Kirsch's "animal" learning program was designed to react to the past four pairs of moves. A pair of moves provides four possible combinations (HH, HT, TT, TH), and four pairs of moves provides $4 \times 4 \times 4 \times 4 = 256$ possible combinations (called stimuli).

The learning mechanism consisted of the animal becoming conditioned to a certain move by the choosing opponent after each of the 256 possible stimuli. For each stimulus, the animal noted what move the opponent made next. Then, the next time that same stimulus occurred, i.e., the same sequence of four pairs of moves, the animal duplicated the move of the opponent that followed that same stimulus on the previous occasion it occurred. The more the choosing opponent repeated the same move after any given stimulus, the more the animal program became "conditioned" to that move.

For each of the 256 possible stimuli the animal program stored a number, called the conditioning number, which varied between $+1$ and -1. Whenever the animal scored a success in predicting the opponent's choice, the number in the animal program's storage location for the appropriate stimulus was increased.[15] Thus, if the opponent always fol-

---

[15] The size of this increase was a fixed fraction of the difference between its previous value and $+1$. Thus, if this fraction were $1/2$ and if the value associated with a particular stimulus was -1, then the next increase to the conditioning number would be $1/2 \times 2$, since the difference between -1 (its previous value) and $+1$ is 2. So the increase would be 1 and the new value would therefore be

lowed a particular stimulus with the same choice, the conditioning number would rise towards 1, quickly at first and then more slowly. If the opponent should suddenly start making the opposite choice, the animal's conditioning number would decrease in the same way, moving rapidly at first towards −1 and more slowly later. Thus, when a particular pattern of moves by each player was suddenly followed by a different response from that which the choosing opponent had been making in the past, the learning program's conditioning number, like the conditioning of a real animal, went into reverse. This approach gave more weight to recent opponent's moves than to more distant ones, causing the animal to "forget" older information.

This work is still relevant today, and has been extended to algorithms that cope with choosing from three options rather than two.[16]

Machine Translation

The earliest plan for an automatic translation process was put forward in 1629 by René Descartes. He advocated a universal language of symbols, so that an idea in one language would share a common symbol with the same idea in a different language. In this way the language of symbols would provide an intermediate step in the translation process, allowing for the translation from one spoken language to the symbol language and then from the symbol language to a different spoken language.

Descartes' concept was taken further in 1661 when a German monk, Johannes Becher, living in the town of Speyer, wrote a booklet about his invention of a mathematical intermediate-language that was designed to describe the meaning of sentences written in any natural language. Becher's intermediate-language consisted of strings of numbers representing the meanings of words, with other numbers expressing the meanings of the various inflections (word endings). The rules of Becher's intermediate language included lists of equations that assigned words in different spoken languages to mathematical expressions representing the meanings of these words. Becher's theory was that sentences in one of these natural languages could be translated into another natural language in a

---

0 (i.e., −1 + 1). The next increase after that would be $1/2$ (i.e., $1/2 \times 1$), making the next value $0 + 1/2 = 1/2$. The increase after that would be by $1/4$ (i.e., $1/2 \times 1/2$), which takes the value to $3/4$, and so on.

[16]World Championship tournaments have been held between programs playing the game of Rock-Paper-Scissors, for which variations on this algorithm are often employed.

"mechanical" way, via the equations, but in reality his booklet offered little more than a Latin vocabulary with a numerical notation.

The most detailed of the various intermediate languages proposed around that time was the *Real Character* published by John Wilkins in 1668. Wilkins (1614–1672) was the Bishop of Chester, a founding member of the Royal Society and one of the most influential British thinkers of the seventeenth century. This essay was Wilkins' attempt at creating a universal language. In it Wilkins maintained that because all people's minds functioned in the same way and had a similar "apprehension of things", it should be possible to cultivate a rational universal language. Wilkins work was the most detailed attempt up till then to construct a rational "universal" notation for common concepts, a proposal for an intermediate-language. But it was not a method of automatic translation.

Proposals for machines to perform dictionary consultation or translation did not come until the technological developments in the early twentieth century. The earliest report of a translating machine is of a model of a proposed typewriter-translator that was reputedly demonstrated by an Estonian, A. Vakher, in February 1924. An article in the Estonian newspaper *Vaba Maa* reported the "demonstration of a model of a translating typewriter" by its inventor, who planned to develop a prototype. The article was reproduced in the proceedings of a Machine Translation conference held in the Estonian capital, Tallinn, in 1962, but the editors of the conference proceedings added that no more was heard of Vakher's machine, so presumably the prototype was never constructed.

The first genuine forerunners of Machine Translation were two devices patented almost simultaneously in 1933, one in France by Georges Artsrouni and the other in Russia by Petr Trojanskii. In both cases the patents were for electro-mechanical devices capable of being used as translation dictionaries.

Artsrouni's Machine

Georges Artsrouni was a French engineer of Armenian extraction who had been a student in St. Petersburg. His patent was for what he called a mechanical brain, a general-purpose machine which could also function, with some equipment added, as a mechanical multilingual dictionary. He suggested various applications for his mechanical brain, such as the automatic production of railway timetables, telephone directories, commer-

Figure 15. Georges Artsrouni's translating machine (this photograph first appeared on the cover of *Automatisme*, vol. 5, no. 3, 1960) (Courtesy of the University of Birmingham library)

cial telegraph codes and bank statements. He also claimed that it would be particularly suitable for cryptography—deciphering and encrypting messages—and that it was also a device for translating languages. Artsrouni's design was actually a storage device on paper tape which could be used to find the equivalent of any word in another language.

At the Universal Exhibition in Paris in 1937 Artsrouni's device attracted much attention (see Figure 15). Many demonstrations were given and the machine received a prize in the section for data processing. From the beginning, Artsrouni foresaw one of its main applications as a mechanical dictionary for producing crude word-to-word translations across four languages. As a mechanical dictionary, the "brain" had four basic components: a "memory" of words in the four languages, an input device consisting of a keyboard activating a reading head, a search mechanism,

and an output mechanism that was also activated by the reading head. The four components were driven by a motor, and the whole apparatus was contained in a rectangular box measuring 10 inches × 16 inches × 8 1/2 inches.

The memory was the core of the device. It consisted of a roll of paper in the form of an "endless" 131 foot band, 16 inches wide, that moved over two rolling drums and was held in position by perforations on the edges, in rather the same way as a 35mm film is held in position in a camera. The dictionary entries were recorded line by line in five columns. The first column was for the word in the source language (the language from which the translation was being made), the other columns for equivalents in three other languages and for additional useful information. By employing a Varityper[17] to type the words, the paper band could contain up to 40,000 lines, which could be doubled if both sides of the band were used.

In order to achieve an even greater capacity, Artsrouni proposed that words could be printed in two different colours (red and blue) superimposed on each other on the same lines, and read by switching from one to the other by changing coloured filters. Since the machine could use several bands, and since the width of the bands could also be increased, the amount of dictionary information stored on the bands could be enormous.

A selector mechanism enabled the perforation band to locate the corresponding word in the memory band. The whole line of five columns was then displayed in the row of five slits at the top of the operator board. These slits represented the output mechanism: the first slit showed the source word and the four other slits showed the translations and other information. The slits were provided with windows of red and blue glass, allowing users to select either blue or red entries. As well as this visual display of the machine's results, the "brain" could be provided with a printer to create typed output. Even more ambitiously, Artsrouni envisaged spoken output by means of a special mechanism (presumably using recordings.)

Artsrouni claimed that the selector and the memory could search for words at a speed of 40,000 lines per minute. If the search began midway on the band, the average speed would be doubled. In fact, a special brak-

---

[17] The Varityper was a highly ingenious word processor whose heyday was in the 1920s and 1930s. This machine could set text in more than 300 different typefaces, it could adjust the spacing between characters and could even produce right-justified copy.

ing and acceleration device was suggested that could reduce the search of a full band to 10 or 15 seconds. In a later model of the machine the friction between the reading head and the selector was eliminated and the search speed claimed was further reduced to three seconds.

When he announced his invention, Artsrouni was not thinking of fully automatic translation and certainly not of high quality translation. But he did believe that his device could be used for producing quick, rough translations, as indeed hand-held electronic translators are used today by millions of travellers. He did not expect his machine to replace human translators but rather that it could act as an aid to communication.

Trojanskii's Machine

Petr Trojanskii (1894–1950) also applied for a patent in 1933 for a mechanical dictionary for use in multilingual translation, but he went much further than Artsrouni with his proposals for coding and interpreting grammatical functions. Trojanskii's plan was to use "universal" symbols and to develop a complete translating machine. Partly because of the political tension between Russia and the West at that time, Trojanskii's work did not become known in the U.S.A. until the late 1950s, but when his work was "discovered", the logician Yehoshua Bar-Hillel described Trojanskii as the "Charles Babbage of Machine Translation". Just as Babbage had constructed an early form of calculating machine and had made suggestions about programming, in both cases using the limited technology of his day (the mid-nineteenth century), so Trojanskii had described how a translating machine might be constructed using the electro-mechanical technology of his own times (the 1930s and 1940s).

Trojanskii's patent described a machine consisting of "a smooth sloping desk, over which moving easily and freely in different directions is a belt provided with perforations which position the belt in front of an aperture". This belt was provided with the words of a large dictionary, with entries in six languages in parallel columns, much as Artsrouni had done for four. The operator would locate a word of the source language (the language *from* which the translation was to be made) and then move the belt so that it displayed, in the aperture, the corresponding word of the target language (the language *to* which the translation was required). The operator would then type in a code indicating the grammatical category or role of the word in question—codes that Trojanskii referred

to as "signs for logical parsing"—and the combination of target word and the code were printed onto a tape. Then the next source word would be located and "translated" in the same way. From a tape of the target language words in sequence, a typist would produce a "coherent text" so that a "reviser" could substitute the correct word endings for each word based on the assigned codes. As a final stage a "literary editor" would improve the style of the translation to produce the final target text.

What sets Trojanskii's proposal apart from that of Artsrouni was that he went beyond the mechanization of the dictionary by his clear enunciation of some basic processes of translation and by his proposals for "logical parsing symbols". These symbols were intended to represent "universal" grammatical relationships, and would therefore be applicable to any language and when translating between any two languages.

Trojanskii's invention was received by Russian linguists with profound scepticism. It was considered impractical and quite unnecessary, even though in 1944 he was able to give a demonstration at the Soviet Academy of Sciences in which his machine performed a translation of a Russian sentence into French. And when Trojanskii died in 1950 he had not been able to demonstrate his ideas on a real computer because the electronic computer was still virtually unknown in the Soviet Union.

The Start of the Modern Age of Machine Translation

Near the end of the 1940s, when the first large electronic calculating machines began to be used for mathematical tasks, scientists also began to think about using these machines for non-numerical purposes. One example of a possible use was the decoding of encrypted messages and translating them into natural language. On 4 March 1947 Warren Weaver, the Director of the Natural Sciences Division of the Rockefeller Foundation, wrote to his mathematician friend Norbert Wiener suggesting the possibility of translation by computer. Weaver's letter, and his memo that followed it two years later, were to provide the impetus needed to kick-start Machine Translation into a serious topic for scientific research. Weaver discussed his ideas with Wiener and then, in July 1949, wrote a memorandum, entitled simply "Translation", which he sent to 200 leading scientists. In this memo Weaver suggested that computers could be programmed to translate language without being able to "understand" the meanings of words.

The impact of Weaver's memorandum is attributable not only to his widely recognized expertise in mathematics and computing, but also, and perhaps even more so, to the influence he enjoyed with major policy-makers in U.S. government agencies. Weaver put forward four proposals in his memorandum. The first addressed the problem of multiple meanings of words (polysemy), and how it might be tackled by an examination of the immediate context where the word occurs:

> If one examines the words in a book, one at a time through an opaque mask with a hole in it one word wide, then it is obviously impossible to determine, one at a time, the meaning of words. "Fast" may mean "rapid"; or it may mean "motionless"; and there is no way of telling which.

> But, if one lengthens the slit in the opaque mask, until one can see not only the central word in question but also say N words on either side, then, if N is large enough one can unambiguously decide the meaning... [6]

The problem was, of course, to determine how much context would be required in order to determine the meaning of a word, and Weaver expected this to vary from one subject to another. However, Weaver thought that "relatively few nouns, verbs and adjectives" were actually ambiguous, and therefore that the problem was not a major one. How wrong he was!

Weaver's second proposal started from the assumption that there are logical elements in language. The mathematical possibility of computing logical proofs suggested to Weaver that "insofar as written language is an expression of logical character", the problem of translation can be solved using formal logic.

The third proposal concerned the possible applicability of cryptographic methods. Weaver had been impressed at the success of cryptography based on, as he put it, "frequencies of letters, letter combinations, intervals between letters and letter combinations, letter patterns, etc. which are to some significant degree independent of the language used."

For his fourth proposal, Weaver became more utopian. It was based on the belief that, just as there may be logical features common to all languages, there may also be linguistic features common to all languages. Earlier in his memorandum he commented on a paper by a sinologist, Erwin Reifler, who had remarked that the Chinese words for "to shoot"

and "to dismiss" show a remarkable phonological agreement (i.e., they had similar sounds) and a remarkable graphical similarity (i.e., the Chinese pictograms for the words appeared similar). Weaver's comment was: "This all seems very strange until one thinks of the two meanings of 'to fire' in English. Is this only happenstance? How widespread are such correlations?" Clearly Weaver thought that such universals might be very common.

At the end of his 1949 memorandum, Weaver asserted his belief in the existence and applicability of language universals with what is one of the best-known metaphors in the literature of Machine Translation:

> Think, by analogy, of individuals living in a series of tall closed towers, all erected over a common foundation. When they try to communicate with one another, they shout back and forth, each from his own closed tower. It is difficult to make the sound penetrate even the nearest towers, and communication proceeds very poorly indeed. But, when an individual goes down his tower, he finds himself in a great open basement, common to all the towers. Here he establishes easy and useful communication with the persons who have also descended from their towers.
>
> Thus it may be true that the way to translate from Chinese to Arabic, or from Russian to Portuguese, is not to attempt the direct route, shouting from tower to tower. Perhaps the way is to descend, from each language, down to the common base of human communication—the real but as yet undiscovered universal language—and then re-emerge by whatever particular route is convenient. [6]

The response to Weaver's memorandum was mixed. Some rejected the very idea of mechanizing the complexity of translation, in much the same terms as many professional translators reject Machine Translation today. Others, however, were considerably less negative, and ultimately his seminal 12-page memo had the effect of launching the field of Machine Translation as a subject for serious scientific research. Perhaps the most significant outcome of the memorandum was the decision in 1951, at the Massachusetts Institute of Technology to appoint the logician Yehoshua Bar-Hillel to the first-ever research position in Machine Translation. Bar-Hillel wrote the first report on the state of the art and convened the first conference on Machine Translation in June 1952, a conference that was to have a highly catalytic effect on the field.

One of the participants at Bar-Hillel's conference was Leon Dostert, who went away convinced that "rather than attempt to resolve theoretically a rather vast segment of the problem, it would be more fruitful to make an actual experiment, limited in scope but significant in terms of broader implications". [7] Dostert's aim was a system requiring no pre-editing of the input, and producing "clear, complete statements in intelligible language at the output", although "certain stylistic revisions may...be required..., just as when the translation is done by human beings."

In 1953, a team at Georgetown University, led by Dostert, collaborated with IBM to create the world's first working MT program, one that could translate from Russian into English. They chose Russian for their first experiments because it was a difficult language and they believed that a system that could translate Russian could handle anything. For the demonstration program a corpus of 49 Russian sentences with a dictionary of only 250 words had been carefully selected. The Russian words were coded in a system employing three types of symbols (called diacritics). One type of symbol indicated which of six rules of operation was to be applied in the translation. Another type indicated what information about the context of a word should be sought to determine the choice of output. And a third type indicated the location of the storage area of the English equivalents.

The program had six rules that governed its operation:

1. In the simplest case the source word is replaced by the target word, and the word order of the source sentence is followed.

2. This case involves a change in the order in which the words appear in the target sentence—the order of the words is swapped around.

3. There is a choice between different equivalent target words. The program determines which word is chosen by referring to a symbol attached to the *next* word.

4. There is a choice between different equivalent target words. The program determines which word is chosen by referring to a symbol attached to the *previous* word.

5. A word in the source sentence is omitted because it is superfluous, and therefore might be confusing if it appeared in the target

sentence. In this case no corresponding word appears in the target sentence.

6. A word is inserted in the target sentence which has no corresponding word in the source sentence. This might, for example, be to clarify the meaning of the target sentence.

On 7 January 1954, the Georgetown/IBM team unveiled the Russian-to-English program publicly, running on an IBM 901 computer, at the company's Technical Computing Bureau in New York. Although the project had little scientific value, the program's achievement in that demonstration caused the idea of Machine Translation to catch fire in the press. The following day the program made headline news when the *New York Times* carried a front-page report of the demonstration, describing how several short messages, within the 250-word range of the device, were translated. Included were brief statements in Russian about politics, law, mathematics, chemistry, metallurgy, military affairs and communications. The machine would ring a bell to indicate its rejection of any incoherent statements or when it encountered a misprint. The Russian sentences were all turned into good English almost instantaneously, and without any human intervention.

Dostert described the 1954 demonstration as "a Kitty Hawk of electronic translation." [8] The success of the project promised enormous implications for both linguistics and electronics, and the general public was impressed with the program's performance; Machine Translation was now seen as a feasible objective and the translation quality was certainly acceptable. The demonstration undoubtedly encouraged U.S. government agencies to support research on a large scale for the next decade, and it stimulated the establishment of MT groups in other countries, notably in the U.S.S.R.

The 1956 Dartmouth Workshop

If Turing's article "Computing Machinery and Intelligence" was the catalyst that helped conceive Artificial Intelligence as a science, the birthplace of this embryonic science was Dartmouth College in New Hampshire. In 1955 John McCarthy, then an assistant professor of mathematics at Dartmouth, had the idea for a summer workshop to study the state of the art in a discipline that he had just christened Artificial Intelligence,

Figure 16. John McCarthy circa 1957 (Courtesy of John McCarthy)

and to make proposals for its future development. McCarthy solicited Claude Shannon as co-organiser of the workshop and together they approached the Rockefeller Foundation, which initially expressed a certain amount of scepticism. The foundation made some suggestions, including that McCarthy and Shannon widen the range of participants. As a result Marvin Minsky and Nathan Rochester also became co-organisers and in August 1955 the four of them submitted a proposal for $13,000 of funding for the project.[18] McCarthy recalls that the Rockefeller Foundation gave $7,500 in support.[19] The Dartmouth proposal document represents an important milestone in the history of AI and the following extract provides an excellent snapshot of the state of AI thinking at that time.[20]

---

[18]The items listed in the published version of the proposal add to $13,000 but the total given in the proposal was $13,500.

[19]John McCarthy, private communication.

[20]To put into perspective the magnitude of the problem of creating an artificial intelligence that matches our own, it is helpful to consider what computing resources might be necessary to achieve that goal. The prolific inventor and author Ray Kurzweil has estimated the memory capacity of the human brain to be approximately 10^{15} bytes. This assumes a brain size of 100 billion neurons (i.e., 100,000,000,000), with each neuron connected to 1,000 other neurons and 10 bytes of information being stored per neuron. Multiplying these three numbers together gives an estimate for the total memory capacity of the brain of 10^{15} bytes (100,000,000,000 × 1,000 × 10). Kurzweil further estimates the total computational capacity of the human brain to be in the region of 20 billion million calculations per second (i.e., 2×10^{16}). This assumes a brain size of 100 billion neurons, 1,000 connections per neuron, and a brain speed of 200 calculations per second per neuron (so 100,000,000,000 × 1,000 × 200). When compared to the typical memory capacities and speeds of the computers of the early 1950s such numbers are astronomic. The IBM 704, for example, ran at 2,000 operations per second and had a core memory capacity of 12,000 bytes in its earliest version, rising later to 96,000 bytes.

Chapter Two

A PROPOSAL FOR THE DARTMOUTH SUMMER
RESEARCH PROJECT ON ARTIFICIAL INTELLIGENCE

J. McCarthy, Dartmouth College
M. L. Minsky, Harvard University
N. Rochester, I.B.M. Corporation
C.E. Shannon, Bell Telephone Laboratories

August 31, 1955

We propose that a 2 month, 10 man study of artificial intelligence
be carried out during the summer of 1956 at Dartmouth College
in Hanover, New Hampshire. The study is to proceed on the basis
of the conjecture that every aspect of learning or any other feature of intelligence can in principle be so precisely described that
a machine can be made to simulate it. An attempt will be made
to find how to make machines use language, form abstractions and
concepts, solve kinds of problems now reserved for humans, and
improve themselves. We think that a significant advance can be
made in one or more of these problems if a carefully selected group
of scientists work on it together for a summer.

The following are some aspects of the artificial intelligence
problem:

1. **Automatic Computers**

   If a machine can do a job, then an automatic calculator can
   be programmed to simulate the machine. The speeds and
   memory capacities of present computers may be insufficient
   to simulate many of the higher functions of the human brain,
   but the major obstacle is not lack of machine capacity, but our
   inability to write programs taking full advantage of what we
   have.

2. **How Can a Computer be Programmed to Use a Language?**

   It may be speculated that a large part of human thought consists of manipulating words according to rules of reasoning
   and rules of conjecture. From this point of view, forming
   a generalization consists of admitting a new word and some
   rules whereby sentences containing it imply and are implied
   by others. This idea has never been very precisely formulated
   nor have examples been worked out.

3. Neuron Nets

How can a set of (hypothetical) neurons be arranged so as to form concepts? Considerable theoretical and experimental work has been done on this problem by Uttley, Rashevsky and his group, Farley and Clark, Pitts and McCulloch, Minsky, Rochester and Holland, and others. Partial results have been obtained but the problem needs more theoretical work.

4. Theory of the Size of a Calculation

If we are given a well-defined problem (one for which it is possible to test mechanically whether or not a proposed answer is a valid answer) one way of solving it is to try all possible answers in order. This method is inefficient, and to exclude it one must have some criterion for efficiency of calculation. Some consideration will show that to get a measure of the efficiency of a calculation it is necessary to have on hand a method of measuring the complexity of calculating devices which in turn can be done if one has a theory of the complexity of functions. Some partial results on this problem have been obtained by Shannon, and also by McCarthy.

5. Self-Improvement

Probably a truly intelligent machine will carry out activities which may best be described as self-improvement. Some schemes for doing this have been proposed and are worth further study. It seems likely that this question can be studied abstractly as well.

6. Abstractions

A number of types of "abstraction"' can be distinctly defined and several others less distinctly. A direct attempt to classify these and to describe machine methods of forming abstractions from sensory and other data would seem worthwhile.

7. Randomness and Creativity

A fairly attractive and yet clearly incomplete conjecture is that the difference between creative thinking and unimaginative competent thinking lies in the injection of some randomness. The randomness must be guided by intuition to be efficient. In other words, the educated guess or the hunch include controlled randomness in otherwise orderly thinking. [9]

In addition to the above collectively formulated problems for study, we have asked the individuals taking part to describe what they will work on.

One of the products of the Dartmouth workshop was a list of goals for AI. Sadly there is no record of the complete list but it is known to have included the creation of a program that could defeat the World Chess Champion in a match, which was achieved in 1997.

During the workshop McCarthy became attracted to the idea of developing a programming language specifically suited to Artificial Intelligence work. After the workshop he joined IBM as a consultant where he helped to develop the language LISP. The design of LISP was very much geared to solving the type of problems that were creating interest within the AI community and the language has since held a strong niche position in AI.

While McCarthy was working with IBM at MIT in 1958–1959, Nathan Rochester was a visiting professor there and he helped McCarthy with the development of LISP. Rochester also apparently lent his support to the creation in 1958 of the MIT Artificial Intelligence Project. In spite of Rochester's early activity and that of other IBM researchers, the corporation's interest in AI cooled. Although work at IBM on Checkers and Chess programming continued, with Arthur Samuel and Alex Bernstein respectively, an internal IBM report prepared around 1960 took a strong position against the idea of the company giving broad support for AI. It was to be the leading academic research establishments, MIT, Stanford University and Carnegie Institute of Technology,[21] that were to run with the AI baton after IBM's loss of interest in the field.

---

[21] In 1967 the institute was renamed Carnegie Mellon University.

Part II
Fifty Years of Progress

In which the nascent science of Artificial Intelligence develops many branches, each corresponding to a different human thought process, during which development the state of the art in some of these branches grows from novice to expert. Over a time span of only half a century, technologies have been developed that can compose music sounding like that of Chopin or Mozart, play Chess and other challenging games better than a world champion and demonstrate inventiveness and expertise in many domains at the highest levels of human intellectual endeavour. The combination of all these technologies places mankind, today, at the start of an era of almost unimaginable achievement.

- 3 -

How Computers Play Games

I n Chapter 1 we saw how the game of Nim can be programmed by
means of fixed rules—a method called an algorithm—that guarantees to find the solution to a problem if a solution exists. Some other,
relatively simple games can also be played perfectly using algorithms designed for the purpose. But for most interesting games, the problem of
finding the best move or even a good one is much too complex, and
the number of possibilities that need to be examined is too vast, for an
algorithmic method to provide a practical solution within a reasonable
amount of time. Instead it is necessary to use rules of thumb, called
heuristics, to enable a program to choose its move. Heuristics, unlike
algorithms, do not guarantee to find the solution to a problem. What
heuristics contribute is the means to make a hopefully intelligent guess
at which moves are good and which are less good, and to enable a program to disregard certain moves from serious consideration. Chess is one
example of a game whose programming derives great benefit from the
use of heuristics.

The section that follows not only provides an account of how computers play Chess, it also serves as an induction into some of the most important techniques employed to program many other two-person games.

Chess

The problem of how to program thinking games has attracted the interest
of computer scientists ever since the birth of computers, and even earlier.
In the world of Artificial Intelligence it has long been argued that, since
Chess is a game that requires intelligence to play it well, building a strong
Chess program is synonymous with the creation of an artificial intellect.
Charles Babbage thought about "programming" Chess on his ill-fated
Analytical Engine, and Konrad Zuse described the basis for program-
ming Chess in his *Plankalkül*. But it was not until Claude Shannon's
seminal paper in 1950, "Programming a Computer for Playing Chess",
that anyone provided a lucid explanation of how the problem might be

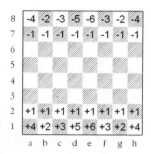

Figure 17. The starting position in a game of Chess

Figure 18. Numerical representation of the starting position

attacked.[1] The basic approach described in Shannon's paper has been employed in almost all of the Chess programs written since then, as well as in programs for many other classic two-person games such as Checkers (Draughts) and Reversi (Othello ^{TM}).

How Computers Recognize a Chess Position

Computer programs know nothing of the different shapes and sizes of the Chess pieces so they must be given some way to represent a Chess position—a way that a program can understand. A simple representation assigns different numerical values to the various Chess pieces, designating White's pieces by positive values and Black's by negative. Thus the starting position in a game of Chess (see Figure 17) could be represented numerically as in Figure 18.

In the right-hand diagram, the pawns are represented by 1 (White's pawns are +1 and Black's are –1), the knights by 2 (and –2), the bishops 3 (and –3), rooks 4 (and –4), queens 5 (and –5) and kings 6 (and –6). Empty squares are represented by 0.

The Moves of the Game

The first thing that a human beginner learns at Chess is how the pieces move. Similarly, a program cannot begin to play the game until it can work out which moves are possible from any Chess position. The detail of how this is accomplished varies from one program to another but the essence is the same. As a simple example, a program could be given a

---

[1] Shannon's paper was originally presented at a conference in New York, on 9 March 1949— Bobby Fischer's sixth birthday.

list of which squares a particular piece type may be able to move to from a certain square: a white pawn on the square a2 can move to a3 if that square is empty, it can also move to a4 if both a3 and a4 are empty, while if the square b3 is occupied by a black piece then the white pawn on a2 may capture it. All this is taught to human beginners and is easy enough to program.

How Computers Evaluate a Chess Position

Once a program has the ability to generate a list of all the legal moves from any position, it needs a way to decide which of these legal moves is the best one. What it does, in simple terms, is to examine each of the positions that can arise after its next move and compare the merits and demerits of these positions, employing an algorithm called an evaluation function.[2] What this function does is to assign a numerical score to a Chess position in such a way that a high score indicates a good position for the program and a low or negative score indicates a bad position. The program then chooses the move leading to the position with the highest score.

An evaluation function determines the score for a position by taking into account various features of the game. The most important feature in Chess is what is known as *material*. It is clear that the different pieces each have their own values—Chess players call the total value of their pieces the material value. One of the simpler heuristics used by Chess players is that it is normally a good idea to capture one's opponent's pieces for nothing whenever possible, or for pieces of lesser value. Experience has shown that the following numerical values are a very reasonable indication of the relative worth of the different pieces, and these were the values proposed by Shannon:[3]

Pawn = 1

Knight = 3

Bishop = 3

Rook (castle) = 5

Queen = 9

---

[2]Sometimes called a scoring function.

[3]The kings are beyond value because a player who is faced with the certain loss of his king automatically loses the game.

A very simple evaluation function for Chess, a function with only one feature, would add up the values of the pieces on each side and assign a score corresponding to the total net value. But Chess is much more complicated than that, and besides, in many situations a player must choose between two or more moves that lead to positions with the same material balance, so material itself is normally not sufficient as a distinguishing factor.

Another important heuristic employed by human Chess players is "mobilize your pieces", and the second most important feature in Chess, after material, is the relative mobilities of the two armies. Other things being equal, the player with the more mobile army has the advantage. In Chess we measure mobility by the number of legal moves that can be made by all the pieces in a position. Simply add up how many moves can be made by each white pawn, knight, bishop, rook, queen and the king, and the total is a reasonable measure of White's mobility. Carry out the same calculation for Black and the difference in mobility scores will generally indicate which player has the advantage and by how much.

With two features in its evaluation function, material and mobility, a Chess program will need numerical weights to indicate the relative importance of each feature. Material is much more important than mobility so its weight must be large enough to reflect this. A ratio of between 5-to-1 and 10-to-1 would probably be appropriate here, meaning that an advantage is mobility of somewhere between five and ten moves is equivalent to an advantage in material of one pawn. The exact ratio chosen for the weighting of each feature will affect the "style" in which the program plays, by making it more likely or less likely to be willing to sacrifice a pawn here and there in return for a significant advantage in mobility.

After material and mobility there are several other features of a Chess position that often contribute to a player's advantage, for example control of the central squares. This can be measured by assigning different squares on the board different values, according to how near or far they are from the centre. It is possible to create an evaluation function for Chess with almost any desired level of sophistication. Bonuses can be given for "developing" one's pieces early in the game (i.e., bringing them out from their original squares so they are taking some part in the action), for attacking the area around the opponent's king, for protecting one's own king, and so on. The more sophisticated the evaluation function becomes, the more accurate will be its numerical estimate of the merit of a position. But there is also a down side—the more complex

an evaluation function is, the slower it will be in its computation of the score for a position. Determining the best balance when faced with such choices is one of the factors that makes games programming such a challenging and interesting task.

How Computers Look Ahead

Just as human Chess players look ahead when they are planning their strategy and trying to defend against their opponent's plans, so a Chess-playing computer program needs to "think ahead", to analyse the future consequences of its choice of moves. Otherwise the program may well make a move which appears superficially to be quite strong but which, in fact, is easily refuted by an even stronger counter-move that the opponent could play in reply. When a human Chess grandmaster looks ahead, his whole analytical search rarely encompasses much more than 100 positions. His expertise is such that he can immediately discard from his thought processes almost all of the possible moves in any position, knowing from experience that they cannot possibly be relevant. He can do this because his own evaluation function, i.e., his instinct, allows him to concentrate on only those moves that might be fruitful.

The device employed by many game playing programs to look ahead is called a search tree.[4] Computer trees are not peculiar to game-playing programs but are used to help solve a variety of decision-making computational problems. Like the arborial variety, computer trees have roots and branches, but traditionally they are drawn as growing downwards rather than up towards the sky.

The position in which the program is considering its move is the root of the tree. Each branch of the tree represents one legal move, and the position at the lower end of that branch arises on the Chessboard when the move corresponding to that branch is made. A simple tree is shown in Figure 19.

Here the position P_0, the root of the tree, is the one from which the program must select a move. If the program follows the branch that represents the move of its rook from the square a5 to the square a6, then it will reach position P_1. If it chooses the branch representing the move of its rook to the square a7, then it will reach position P_2, and similarly for the move by the rook to the square a8 (position P_3) and the capture by the rook of the black pawn on the square a4 (position P_4). These

---

[4]The word "tree" reflects its branching nature.

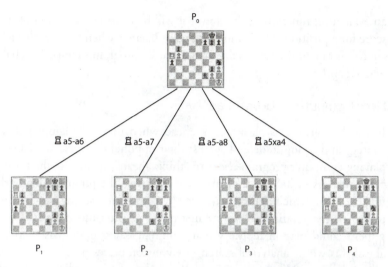

Figure 19. A Chess tree for a position in which White has four possible moves (originally printed in the author's book *Chess and Computers* (Computer Science Pr, Rockville, MD, 1976))

five positions and four branches make up a small tree, whose *depth* is one half-move,[5] also called one *ply*. By evaluating all the positions at the lower extremities of the tree (called the *leaf* positions) the program can decide which move to make. In this example the decision is simple—the score for position P_3 is beyond the values of the other three positions (P_1, P_2 and P_4) because the move by the white rook to a8 is checkmate; the capture of the black pawn on a4 gains one unit of material; whereas neither of the other moves gains any material. Clearly the correct choice of move is that of the white rook to a8.

Now let us grow a slightly larger tree (see Figure 20) in order to see how a program looks ahead more than one half-move. Each Chess position on a diagram of a look-ahead tree is represented by a large dot.[6]

This tree has been grown to a depth of two half-moves. The program is thinking about which move to choose from the root position P_0. It can move to position P_1, when its opponent will have the choice of moving to P_{11}, P_{12} or P_{13}, or the program could move from the root to position P_2, when its opponent will have the choice of moving to P_{21}, P_{22} or P_{23}. Associated with each of these six positions at a depth of two-ply there is a score, obtained from the evaluation function, that measures how good or

---

[5] A half-move in Chess program parlance means a move by one player, White or Black.

[6] In computer parlance this is called a node.

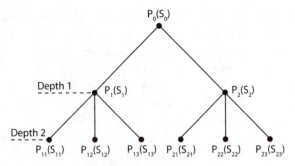

Figure 20. A Chess tree of depth two half-moves (originally printed in the author's book *Chess and Computers* (Computer Science Pr, Rockville, MD, 1976))

bad that position is from the perspective of the program. These scores are indicated in Figure 20 as S_{11}, S_{12}, S_{13}, S_{21}, S_{22} and S_{23}. If the program's opponent had to choose a move from position P_1, and if it chose its best option, it would move to whichever of P_{11}, P_{12} and P_{13} carried with it the best score from the opponent's point of view. This would be the minimum (lowest score) of S_{11}, S_{12} and S_{13}, and this is the score (S_1) associated with position P_1 because it is the best score the program can hope for, assuming correct play by the opponent, if it chooses the move from the root position to P_1.

Similarly, if the program's opponent had to choose a move from position P_2, and if it chose its best option, the opponent would move to whichever of P_{21}, P_{22} and P_{23} carried with it the best score from the opponent's point of view. This would be the minimum score of S_{21}, S_{22} and S_{23}, and this is the score (S_2) associated with position P_2 because it is the best score the program can hope for, assuming correct play by the opponent, if it chooses the move from the root position to P_2.

Clearly, when searching the tree to find its best move, and assuming correct play by both sides, a program should choose to move to whichever of the positions P_1 and P_2 has the highest score, so once it knows the scores S_1 and S_2 it chooses the maximum of these two and makes the corresponding move. The maximum of S_1 and S_2 becomes the score (S_0) associated with the root position P_0 and represents how well or badly the program's evaluation function thinks the program is doing in the root position, assuming best play by both sides. This process of choosing the maximum of the minimums (of the maximums of the minimums ...) is called, not surprisingly, the *minimax* method of tree searching. The score for the root position is called the backed-up or minimax score.

The trees shown above are very small, but the sizes of the trees examined by the Chess programs of today are huge. There are 20 legal moves in the starting position and this number normally rises to an average of around 37 during the middle phase of the game, so after just one half-move by White and a reply move by Black there are typically approximately 1,000 different positions to consider. Two half-moves by each side means approximately one million positions. This number becomes enormous when the depth of the tree reaches double figures, and some of today's leading Chess programs typically search many parts of the tree to a depth of 20 half-moves or more along the most important branches. To put the size of the problem into perspective, consider the fact that the number of possible different games of Chess, which has been estimated at 10^{120}, is greater than the number of atoms in our universe. This means that even if every atom in the universe were to be replaced by a computer, it would take the combined power of all these computers a huge amount of time to calculate the first move in a perfectly played game of Chess. Fortunately there are clever techniques for performing deep tree searches, up to 20 ply or more.[7]

In 1951 Shannon built a computer for playing Chess endgames (see Figure 21) but he published nothing about its design or how it worked.

Shannon's original paper described three different search strategies for programming Chess. The simplest of these was to consider all moves by both sides up to a certain depth, as in the example above for a tree of depth two. A much improved strategy, which Shannon called "type B", is more selective and is the one employed in virtually all Chess programs since Shannon's day:

1. Examine forceful variations out as far as possible and evaluate only at reasonable positions, where some quasi-stability has been established.

2. Select the variations to be explored by some process so that the machine does not waste its time in totally pointless variations.

---

[7] The most powerful of these is called the alpha-beta algorithm, the workings of which are beyond the scope of this book.

Figure 21. Claude Shannon (right) demonstrating his Chess endgame computer to International Master Edward Lasker,[8] circa 1951 (Courtesy of Claude Shannon)

Openings Databases

Tournament Chess players from club level up to the world champions place a great deal of emphasis on the study of books and magazine articles on the Chess *openings*—the many different ways of playing the first several moves of the game. At the latest count more than three million

---

[8] Edward Lasker maintained a keen interest in computer Chess from its earliest days. He celebrated his 93rd birthday, in 1978, at the ACM North American Computer Chess Championship in Washington, where he would animatedly discuss the games with the programmers.

games played by Chess masters and grandmasters are available on disk—this is the twenty-first century form of openings book most popular with serious human players and is *de rigeur* as part of the openings arsenal of the most competitive Chess players, man or machine.

Endgame Databases

In 1970 Thomas Ströhlein created the first Chess endgame database for his PhD project at Technische Hochschule München, and within a few years Ken Thompson[9] had pioneered the use of such databases in competitive Chess programs.

Thompson realised that the number of possible Chess positions in which are very few pieces on the board is a manageable number in terms of computer storage. For example, if we ignore the fact that some positions will not be feasible for reasons connected with the rules of the game, for the endgame in which one player has a king and a rook while his opponent has only a lone king, this configuration can be represented in fewer than 250,000 positions.[10] In fact the number of positions that needs to be stored to represent this configuration completely is considerably smaller, partly because of symmetry and partly because of considerations relating to the rules of Chess, such as a position being illegal if the side that just moved is in check.

Thompson's aim was to build a database of perfect knowledge about Chess endgames in which there are very few pieces on the board. His method was to back up from all positions in which one player has been checkmated, in order to create a database of all the positions that are won for the side with the extra rook, together with a note of how many half-moves are required from each position to checkmate. Then, when faced with any winnable position, the program simply looks into the database and chooses a move that leads to a win in the smallest number of half-moves.

In practical terms, using a database to win the ending of king and rook versus a lone king is sheer overkill, since a relatively simple evaluation function can be designed to win this ending, hardly surprising given

---

[9]Thompson was one of the co-developers of the BELLE Chess computer that won the World Computer Chess Championship in 1980. Thompson was also co-author of the Unix operating system for computers, an accomplishment that earned him the U.S. National Medal of Technology in 1999.

[10]The rook can be on any of 64 squares, one player's king on any of the remaining 63 and the other king on any of the remaining 62, hence $64 \times 63 \times 62$.

that there are only three pieces on the board. But where the database technique becomes of real use is in some four-piece endings, such as king and queen versus king and rook, which is not trivial to play, even for some humans of master strength. And for endings with five pieces and more, the endgame databases make a huge difference in computer versus human play, enabling programs to play these endings perfectly and even teaching human grandmasters lessons in Chess. There is also another reason why these databases are useful for a competitive program. If the program sees, during its search of the game tree, that it can reach a position stored in its endgame database, the program knows what the result of the game will be if that position is reached, and uses the result to decide whether or not it wishes to aim for that position. This factor has been of enormous help in programming Checkers.[11]

From Patzer to World Champion

It took almost 50 years from the time of Shannon's first publication on computer Chess in 1950 until a program was able to defeat the human World Champion in a match. Progress during most of that period was painfully slow. Much of the research in computer Chess during the 1950s, 1960s and 1970s was carried out by academics who lacked the resources to play many games between their programs and human opponents, and it is only by playing such games and studying printouts of the resulting program analysis that programmers are able to make significant strides.

Since Shannon's kick-start of research in this field, each decade has been characterised in a way that reflects a quantum leap forward in understanding, if not results. The end of the 1940s and the whole of the 1950s were notable for a small number of research efforts: hand-simulations by Alan Turing and Donald Michie; a program developed at the Los Alamos Scientific Laboratory for playing Chess on a 6 x 6 board (omitting the bishops); the efforts of Allen Newell, Herb Simon and John Shaw at Carnegie Institute of Technology; and the strongest, a program developed by Alex Bernstein at IBM (see Figure 22). None of these efforts resulted in a program that could play well enough to challenge any but the very weakest human players.

The biggest step forward of the 1960s, and the most publicized, was the program MacHack VI, written by an MIT student, Richard

---

[11] See the "Chinook" section later in this chapter.

Figure 22. Alex Bernstein with chessboard and the IBM 704 (Courtesy of IBM Corporate Archives)

Greenblatt. Work on the program began in November 1966 and within three months it was ready to play in a human Chess tournament, the first program ever to do so in any game. MacHack's first tournament result was hardly spectacular—it played five games, losing four and drawing the other one, for which it achieved a rating of 1243 on the U.S. Chess Federation scale and a lowly Class D designation from the federation. Work continued on the program and in April 1967 MacHack played in a four-

game tournament, scoring two wins and two losses for a performance rating of 1640, corresponding to Class B. In recognition of this achievement Greenblatt's program was made an honorary member of both the U.S.C.F. and the Massachusetts Chess Association, and for the first time computer Chess attracted some attention in the media.

The following year, 1968, saw the birth of serious human versus computer challenges at Chess when I started what became known as "the Levy bet". During one of the annual Machine Intelligence Workshops, organized in Edinburgh by Donald Michie, I was invited to a cocktail party along with several luminaries of the AI world, including John McCarthy. During the evening McCarthy invited me to play a game of Chess, which I won. He then said: "You may be able to beat me, David, but within ten years there'll be a program that can beat you." I was incredulous. As the reigning Scottish Champion I had no doubt at all that McCarthy was wrong, and offered to make a bet to that effect. McCarthy asked me how much and, brashly, I suggested £500 (around $1,200 at that time). I was in my first job, teaching programming at Glasgow University and earning the magnificent sum of £895 per year, so £500 represented a significant wager. McCarthy called over to Michie, who was sitting on the floor a few feet away from us, and asked Donald what he thought about my challenge. Michie's reaction was to ask McCarthy if he could take half of the bet, and so it started, with the two of them wagering £250 each that I would lose a match to a computer program before the end of August 1978. The following year Seymour Papert of MIT came in for another £250 and in 1971 the total stake rose to £1,000 when Ed Kozdrowicki, a Computer Science professor at the University of California at Davis and co-author of COKO, one of the early competitive Chess programs, joined the fray. As time went on my confidence in the outcome of the bet remained undimmed, so I was delighted when, in 1974, Donald Michie offered to up the stakes, making the total £1,250, and that is how the bet remained.

The start of the 1970s saw another important event in the history of computer Chess. The Association for Computing Machinery, at its annual convention in New York, organized the first ever Chess tournament in which all of the participants were computer programs. There were six entries and the event was won by a program called Chess 3.0, written at Northwestern University by David Slate, Larry Atkin and Keith Gorlen. Chess 3.0 evaluated approximately 100 positions per second and played at the 1400 level on the U.S. Chess Federation rating scale.

The ACM tournament became an annual event and ran for 24 years. These tournaments, and other computer versus computer events, including the triennial[12] World Computer Chess Championship, had a decisive influence in the rate of improvement in the playing strength of the best programs. The programmers would sit at the chessboard, typing the moves of their opponent's program into their own computer and making their own program's moves on the board. During the gaps between moves the programmers would chat to each other in a friendly atmosphere, very unlike human tournaments, and would often exchange ideas on Chess programming. In this way the programmers would learn from each other and by studying the play of other programs. At the end of one year's tournament the programmers would retreat to their ivory towers and return the next year with a stronger program. Thus was progress made, year by year.

In August 1978, at the Canadian National Exhibition in Toronto, I played a six-game match to determine the outcome of my bet. My opponent was the Northwestern University program, the reigning World Computer Chess Champion, now called Chess 4.7. During the match I sat, dressed in a tuxedo, in a soundproof glass box, facing a young lady who made the program's moves on the chessboard and pressed its button on the chess clock (see Figure 23).

I managed to beat the program fairly convincingly, by three wins to one with one game drawn (the sixth game did not need to be played), and with this match I won my bet.

When sending me his cheque John McCarthy wrote that, when he made the bet, he had expected I would lose to an intelligent program. But because the leading programs of the day used the "brute force" approach (searching very large trees with very little Chess intelligence), rather than the "selective search" approach (searching very small trees with a lot of Chess intelligence), McCarthy said that he did not mind losing the bet.

In 1980 Ken Thompson's program BELLE, co-developed with Joe Condon at Bell Laboratories in New Jersey, won the World Computer Chess Championship which was held in Linz, Austria. By then BELLE was very different from other Chess-playing computer systems—it relied for its strength on the speed it acquired from special-purpose computer hardware that was designed specifically to play Chess. Thompson had concluded that "...computer Chess belonged to the fastest computer"

---

[12]This has since become an annual event.

Figure 23. David Levy playing Chess 4.7 in Toronto, August 1978, to decide the result of his 10-year bet

so he designed a special purpose Chess chip to replace some of the time-critical aspects of his program. Instead of having a program module that generates a list of all the legal moves in a Chess position, this task would be undertaken by the silicon chips,[13] which could do it much faster. Likewise, the evaluation function was also made in silicon for much faster execution. By executing these tasks in hardware rather than software the whole system was speeded up significantly.

The first version of the BELLE Chess hardware, in 1977, had incorporated 25 chips but produced virtually no speedup over Thompson's earlier, purely software version of BELLE, which searched some 200 positions per second. The next version of BELLE, a year or so later, contained 325 chips and searched approximately 5,000 positions per second. The version that won the World Championship in 1980 employed 1,700 chips and searched 160,000 positions per second. BELLE's success in Linz vindicated Thompson's conviction that speed is almost everything in computer Chess. In 1983 BELLE's rating surpassed the 2200 mark, earning it the title of U.S. Master.

The 1980s saw the first truly impressive results by programs against grandmasters when playing at tournament speed.[14] Having won my bet in 1978 I was happy to repeat the challenge for another five years, this time making a $1,000 wager with Dan McCracken, a leading authority on Computer Science and the author of many books on the subject. In 1984 I defended this second bet in a four-game match in London against the then World Computer Chess Champion program, Cray Blitz, and crushed it 4-0. By then I could see that steady progress was being made, with many more enthusiasts joining the ranks of the competing programs at the various regular computer championships, and I therefore resisted the temptation to make a third bet. But rather than remove myself as a target altogether, I joined forces with *Omni* magazine to announce a $5,000 prize for the authors of the first program to win a match against me, whenever that might be. I put up $1,000 of the prize money and *Omni* contributed the other $4,000. Soon a program called Deep Thought, developed at Carnegie Mellon University and running on spe-

---

[13]Thompson's design required several silicon chips. Later researchers were able to take advantage of smaller silicon chip technology to cram all the necessary functionality onto a single chip.

[14]At fast games, sometimes called blitz Chess, in which each player might have only five minutes for all of their moves in the game, programs perform considerably better against strong human players than they do at tournament speed (in which the players have an average of three minutes per move or thereabouts). This is because, when playing at speed, humans are more likely than their computer opponents to make careless oversights.

cial purpose Chess hardware, was crushing many top human players and even tied for first place in a tournament at Long Beach, California, ahead of ex-World Champion Mikhail Tal. Ken Thompson's work on BELLE had led the way in special purpose Chess hardware—the Carnegie Mellon group, and in particular Feng-hsiung Hsu who designed their Chess chip, were Thompson's most successful disciples.

Deep Thought's success at Long Beach prompted Donald Michie to organise a match between the program and myself, in which Hsu and his colleagues would have the opportunity to pick up the *Omni* prize. The program "trained" by playing a two-game match against World Champion Garry Kasparov in New York in October 1989, losing both games convincingly. My own match was played in London in December 1989 and I was very horribly crushed, losing 4-0. I had survived the ten-year period of the original bet and another 11 years afterwards, but now had to accept that I had been well and truly eclipsed. Fortunately, Kasparov himself was ready and enthusiastic to take up the challenge.

Deep Thought had, not surprisingly, won the World Computer Chess Championship in Edmonton, Alberta, in May 1989. But the real news being discussed during the Edmonton event was that the development team had just been hired by IBM, whose financial and technical muscle were to support the programmers' efforts to defeat Garry Kasparov in a match. It was rumoured that $5 million had been set aside for the project and that a three-year time frame was the target. The technical basis for the project was not the sophisticated software written by very smart programmers, although that also played a crucial role, it was the special purpose hardware on which the program was run. The Deep Thought chip was redesigned by Hsu and the new version provided enormous computational power, supplanting up to 40,000 general-purpose program instructions with hardware that can execute the same tasks 2 to 2.5 million times per second.

It was this combination of ultra-fast hardware and extremely sophisticated software, plus the Chess knowledge given to the programmers by various human grandmasters, that enabled the IBM system to take on Kasparov and have a genuine chance of success. Deep Blue, as it was now known, played its first serious match against Kasparov in Phildelphia, in February 1996, and lost $3^1/2$–$2^1/2$, after surprisingly winning the first game of the match. The following year in New York it was a different story.

Through a combination of more advanced chip design and faster chip speeds, the IBM group created a machine for the 1997 match employing 480 Chess chips that could analyse a total of some 200 million moves per second. This raw computing power, combined with a big increase in Chess knowledge from its grandmaster "advisors", most notably a former U.S. Champion, Joel Benjamin, enabled Deep Blue to play remarkable strategic Chess. When the program outplayed Kasparov in the second game of the match, slowly but steadily manoeuvering its pieces onto better and better squares, the price of IBM stock immediately rose by $4. Kasparov's morale was shot to pieces and he was unable to play at his best for the remainder of the match. After five games the score was still level at $2^1/_2$–$2^1/_2$ but Kasparov was in no fit psychological state to play the sixth and final game (see Figure 24), with the result that he was forced to resign after no more than one hour's play. AI had made history. Feng Hsu and his colleagues, Murray Campbell and Jospeh Hoane, had achieved one of the original goals of Artificial Intelligence as defined during the Dartmouth workshop of 1956.

Despite Kasparov's loss to Deep Blue in 1997, it appears that the very best human Chess players, Kasparov included, are still (in 2005) a little stronger than the best programs, but not sufficiently so to be able to outperform a top program under match conditions. One reason why

Figure 24. Garry Kasparov in play against Deep Blue, operated by Feng-hsiung Hsu, the sixth and decisive game, New York, 1997 (Courtesy Murray Campbell)

the programs are not yet dominating is that, although the strength of the best programs has been rising slowly and continues to rise, human Chess strength is also rising. But it can only be a matter of a few years at most before Kasparov[15] and his super-grandmaster colleagues will be pleased to win a single game out of six or eight in a match against the world's strongest programs. Kasparov himself forsees the day when one such win will be treated by the human player as a triumph!

Checkers (Draughts)

The game of Checkers[16] is very much simpler than Chess, with fewer pieces, fewer piece types and fewer squares of the board on which to play. Because of these smaller numbers there are far fewer possible board positions in Checkers than there are in Chess and the goal of creating a World Championship level program was therefore, for many years, widely assumed to be much easier to achieve in Checkers than in Chess.

The first Checkers program to attract the attention of the public was developed by Arthur Samuel at IBM in 1952. His program was initially written for an IBM 701, the corporation's first computer that could store a program in its memory. When Samuel's program was about to be demonstrated for the first time, Thomas J. Watson, the founder and president of IBM, remarked that the demonstration would raise the price of IBM stock 15 points. It did.

Samuel upgraded his program in 1954 when he added two learning mechanisms. The program was able to learn from experience and thereby to improve its own evaluation of positions. The simpler of Samuel's two learning methods was rote learning. Whenever it carried out a search of the game tree, Samuel's program stored the current position, together with its score. If the same position arose as a leaf position of the tree during a later search, even in a later game, the program would already know the score for that position as derived from the earlier tree search. This meant that evaluating the same position again was both unnecessary, because its value was already known, and less accurate, because the known value was based on a tree search rather than merely the result of a "static" evaluation.

---

[15] In March 2005 Kasparov announced his retirement from professional tournament play, but said that he still planned to play in some events for fun. Hopefully he will, from time to time, accept the challenge to play the top programs.

[16] In the U.K. and the British Commonwealth this game is called Draughts.

Figure 25. Arthur Samuel with IBM computer (Courtesy of IBM Corporate Archives)

The second method of learning devised by Samuel was designed to enable the evaluation function to improve itself by improving the weightings assigned to each of its features. The method is based on the realisation that the backed-up score for the root position in the game tree should ideally be the same as the score found when the evaluation function is applied directly to that same position. During play, Samuel's program would keep track of how much each of the features in its evaluation function had contributed to the overall score for a position, and by how much the backed-up score differed from the static evaluation of the same position. These differences were used to correct the weightings for each of the features in the evaluation function, which tended to make future differences smaller. This particular approach is today called Temporal Difference learning[17] and was employed, almost 40 years later, by another IBM researcher, Gerald Tesauro, in his world class Backgammon program.[18]

---

[17] Technically Samuel's technique was slightly different from Temporal Difference learning, but very close.

[18] See the sections on NeuroGammon and TD-Gammon later in this chapter.

Samuel's program was extremely sophisticated for its time and played a mean game of Checkers. It won a demonstration game played in the summer of 1962 against Robert Nealey, a strong tournament player.

Chinook

It remains as a testament to Samuel that little new was achieved in computer Checkers until Jonathan Schaeffer and his colleagues at the University of Alberta[19] developed the Chinook program. Schaeffer's goal with Checkers was simply to "solve" the game. By this we mean to develop a program that can play perfectly from any position, always winning a winnable position in the shortest possible number of moves, always putting up the stiffest resistance when faced with a lost position, and never allowing a winning advantage to be dissipated or a draw to become a loss. It was Thompson and Stiller's work on Chess endgame databases that provided the inspiration for Schaeffer.[20]

Schaeffer's idea was to create a Checkers endgame database so big that it would allow his program to search, in effect, to the end of the game tree, along all variations, from the very start of the game. His program would conduct as deep a search of the game tree as was possible in the time available for it to choose its move and then, instead of evaluating the leaf positions on the tree using an evaluation function, Schaeffer's program would look up the leaf positions in the database and assign the corresponding score (win, draw or loss) to that position in the tree. This approach would guarantee perfect play, provided that every leaf position on the game tree could be found in the database. For this to be possible the database would need to be huge and the program would need to be able to search very deeply. The bigger the database, the shallower the program could search, while still playing perfect Checkers.

---

[19] Schaeffer heads a research group specializing in programming various strategy games. His team, and a similar group at the University of Maastricht, in the Netherlands, is home to many of the world's leading experts in this field. Schaeffer himself had long been one of the world's top Chess programmers and had written the Sun Phoenix program that tied for first place at the 1986 World Computer Chess Championship in Cologne, Germany.

[20] While the number of possible Chess positions has been estimated at approximately 10^{44}, the estimate from Schaeffer's group of the corresponding number for Checkers was "only" 5×10^{20}. To be precise, Schaeffer has calculated it to be 500,995,484,682,338,672,639, and explains that "To put a number this big into perspective, imagine the surface of all the earth's land mass as being equivalent to the number of possible Checkers positions. Then one position is roughly equivalent to one-thousandth of a square inch." [1]

[21] Because the match was played in the U.K. the anglicized name of the game was used in the name of the event, the World Draughts Championship.

Figure 26. Marion Tinsley (right) in play versus Chinook (operated by Jonathan Schaeffer), the World Checkers[21] (Draughts) Championship, London 1992 (Courtesy of Jonathan Schaeffer)

In 1990 Chinook came second to the legendary Marion Tinsley at the U.S. Open Checkers Championship, proving itself ready to challenge Tinsley to a match for the World Checkers Championship (see Figure 26). Winning a match against Tinsley was an intermediate goal on the road to solving the game completely—Tinsley had lost only three games in a span of 40 years. The match was played in London in 1992, with Chinook having the benefit of an openings book developed by Martin Bryant, whose research had found some flaws in the established Checkers openings literature.

Although Chinook took the lead in the match by winning the eleventh game of the 40-game series, eventually Tinsley drew level, then went ahead and remained ahead until the end. Schaeffer and his team had come very close to dethroning Tinsley but, in the end, Tinsley's faith was vindicated. In an interview with *The Independent* before the match he had said: "I can win. I have a better programmer than Chinook. His was Jonathan, mine was the Lord."[22]

There can be little doubt that, since Tinsley's death in 1995, Chinook has been the world's strongest Checkers player. All that now remains

---

[22]A rematch was started in Boston in 1994, but after four games, all of which were drawn, Tinsley was hospitalized and diagnosed with cancer, from which he died six months later.

for the program to achieve is Schaeffer's goal, solving the game. As of January 2005 the Edmonton group had completed the compilation of all databases up to nine pieces: roughly 4.5 trillion positions. Of the ten-piece database, five pieces against five pieces and six pieces versus four had been completed (22.8 trillion positions). The total size of the databases at that time was 30 trillion positions and counting, as a result of which Chinook had already proved that one of the most popular openings in Checkers, known as the White Doctor, is a draw with perfect play.[23]

Using Databases to Solve Other Games

Once the power of endgame databases was demonstrated in Chess and Checkers it was not long before programmers in other game domains jumped on the bandwagon. Some classic strategy games have already been solved using this approach. As computers become faster and acquire more memory, so it will become easier for programmers to develop databases for games that are currently beyond solution. Within the next two decades I would certainly expect Xiang-Qi (Chinese Chess, with 10^{45} possible positions) and Dames (Checkers on a 10 x 10 board, with 10^{35} positions) to be solved, amongst others. The games that have already succumbed to databases include Connect-4^{TM}, Nine Men's Morris and Awari.

Connect-4^{TM}, which is also known as four-in-a-row, is marketed by Milton Bradley. The game is played on a vertical array with seven columns and six rows. The players take turns to drop coloured discs into whichever of the columns they choose, provided that it is not already full. A disc once dropped then occupies the lowest vacant square in that column. The game is won by the player who first completes an unbroken line, vertical, horizontal or diagonal, of four of his own coloured discs. If neither player can achieve this goal then the game is a draw. The number of possible positions in Connect-4 is approximately 10^{14}, one five-millionth of the number for Checkers. The game was solved, at about the same time but working independently, by Victor Allis in the Netherlands and James Allen in the U.S.A. With correct play the game is a win for the first player, who should start by dropping a disc in the central column. Against all other opening moves the second player can draw with correct play.

---

[23] At that time two more openings were almost solved, with 173 still to be addressed.

Nine Men's Morris, which is also known as The Mill or Merrills, has a long history in Europe and is even mentioned by Shakespeare.[24] The number of legal positions in Nine Men's Morris is estimated to be 10^{10}, and the total number of possible games is approximately 10^{50}. In October 1993 the game was solved by Ralph Gasser, who compiled a computer database for the game at the ETH in Zurich. Gasser showed that the game should end in a draw if both sides play correctly.

The ancient African game of Awari is known to exist in some 400 variants, many having their own names. Mancala, Kalah, Wari and Bao are some of the more popular names and versions. Originally played on an arrangement of small pits dug in the sand, nowadays Awari is played on a wooden board on which each player has six pits on his side of the board, with each pit containing four stones at the start of the game.[25] There is also a pit, called a *house*, at each end of the board, where the owner of the house keeps the stones he has captured.[26] The number of possible positions in Awari is approximately 10^{12}. In 2002 the game was solved by Henri Bal and John Romein at the Free University in Amsterdam. Their database took only 51 hours to compute and proved that either player can force a draw from the starting position.

Go—The Most Difficult Game of All

A game of Go starts with an empty board. Each player has an effectively unlimited supply of stones, one player taking the black stones, the other taking white. The basic object of the game is to use one's stones to form territories by surrounding vacant areas of the board. It is also possible to capture the opponent's stones by surrounding them.

Go is widely regarded amongst games experts as the most difficult classic game of all to learn to play well. Similarly, within the computer games field, Go is known to be the most difficult game to program to a

[24] *A Midsummer Night's Dream* (Act 2, Scene 1), "The nine men's morris is filled up with mud."

[25] The number of pits varies from game to game, as does the number of stones in each pit at the start of the game. The version described here, with six pits on each side and four stones in each pit at the start, is one of the most popular forms of the game.

[26] To make a move a player chooses one of his own pits (not an empty one) and removes all the stones from that pit, "sowing" them counter-clockwise in the remaining pits (and skipping the original pit if there are more than 11 stones to sow). If the last stone is sown into an enemy pit that contains two or three stones after sowing, these two or three stones are captured and put in the moving player's house. In this case, if the second last pit is also an enemy pit that contains two or three stones, they are captured as well, and this process is repeated clockwise. The player who captures the most stones, wins the game.

high level. The reasons relate mainly to the enormous size of the game tree. The normal Go board has 19x19 lines,[27] with play taking place on the intersections, so at the the the start of the game there are 361 possible moves that can be made by the first player, compared to 20 at Chess.

Why Is Programming Go So Difficult?

By comparing Chess and Go we can see how immense the Go tree becomes after only a few half-moves. In Chess, with an average of approximately 37 moves to consider, the number of leaf positions on a tree of depth 20 is approximately 2.8×10^{31}. In Go, with an average of approximately 240 moves to consider in a position, the tree for a 20-ply search would encompass approximately 4×10^{47} leaf positions. In other words, a Go tree of depth 20 will have approximately 14,000,000,000,000,000 times as many leaf positions as a Chess tree of the same depth.[28] Furthermore, a winning strategy in the game of Go requires a more long-range insight than a successful strategy in Chess, and it is therefore most unlikely that even if a Go program could search to 20-ply, it would be able to play the game at the level of the strongest human players.

40 Years of Computer Go—How Much Progress?

The first writings on programming the game of Go appeared in the early-to-mid 1960s, but because the game is so much more difficult than Chess to program, it did not attract much serious interest from programmers until 1984, when a program called Nemesis took part in the Eastern U.S.A. Go Championships (for humans) and was rated at 20-Kyu.[29]

---

[27] Beginners can learn the basics on a 9 x 9 board, a format that has been gaining in popularity in recent years as a game in its own right.

[28] Another way to put the size of the Go programming problem into perspective is to compare the number of possible Go positions, estimated at 10^{170}, with the number of possible Chess positions, estimated at 10^{44}, and that for Checkers (10^{20}). While solving the game of Checkers may well be possible in the first decade of the twenty-first century, through the creation of a database containing every possible Checkers position, to do the same for Chess would require the power and memory sizes of today's computers to increase by a factor of 1,000,000,000,000,000,000,000,000, even without allowing for the fact that more memory is required to store a Chess position than a Checkers one. And for the next jump, from a complete database for Chess to one for Go, add another 126 zeros to appreciate the magnitude of the problem.

[29] The Go ranking system requires some explanation, to say the least, and this explanation is subject to variations between some countries. The ranks start at about 30-Kyu (a complete beginner) and go down to 1-Kyu as a player gets stronger. After 1-Kyu, a player becomes 1-Dan, and from there the ranks ascend to 9-Dan. A few Asian countries (China, Japan, Korea, Taiwan) have professional ranks as well as amateur ranks. The professional ranks in Japan and Korea start at 1-Dan and go up to 9-Dan. But these ranks are not on the same scale as the amateur Dans. A 1-Dan professional is roughly the same strength as a 7-Dan amateur.

In 1986 an annual computer Go tournament was started called the Ing Computer Go Congress, sponsored by the Ing Foundation.[30] Just as the computer Chess tournaments of the 1970s and 1980s did much to help advance the strength of the best programs, so the Ing tournaments started along the same path in computer Go. For some years now there have been a number of regularly held computer Go tournaments, including a World Championship.

The strength of the best Go programs has improved slowly since 1984. Mick Reiss, who has long been one of the world's leading Go programmers, estimates that the best programs in 2004 were playing at 7-Kyu and that in recent years the rate of improvement had been around half a Kyu per year on average.[31] If progress continues to be made at the rate of half a Kyu per year, the leading programs will reach 7-Dan amateur level and 1-2 Dan professional around the year 2030 and are likely to be ready to challenge the Garry Kasparov of the Go world, with a reasonable chance of success, around the year 2035.[32]

Given the magnitude of the computation problem, and the impossibility of compiling a complete database of Go positions, it is hardly surprising that Go programming has not yet progressed to the point where strong human players have anything to fear. But the magnitude of these particular problems is not the only important factor in the relative lack of progress in computer Go. Other factors include the limited size of the few openings databases that have been created and the great difficulty experienced in developing an accurate evaluation function.

Go is the best example from the field of games that demonstrates the difficulty of incorporating human expertise in an evaluation function.[33]

---

[30] Ing Chang-Ki was a Taiwanese industrialist and lifelong Go enthusiast whose foundation supported computer Go up to the year 2000.

[31] (Mick Reiss, private communication.) These estimates appear to be fairly consistent with the improvement from the Nemesis program's 20-Kyu in 1984 to the current best of 7-Kyu in 2004, a rise of 13 Kyu over 20 years, an average of 0.65 Kyu per year.

[32] It is said that the difference between two adjacent professional Dan grades is as third as much as that between two adjacent amateur Dan grades, in which case the progress from 1-Dan professional to 9-Dan is likely to take a further five years or thereabouts.

[33] In Chess it is relatively straightforward for a strong player to explain to a weaker one (or to a computer programmer) the salient features in a position and why one side or the other has an advantage, but the same appears to be untrue in Go. The science of knowledge engineering—extracting knowledge from human experts in a form in which it can be programmed—appears to have made virtually no progress in Go programming. An expert Go player can look at a position and know, intuitively, whether the formation on the board favours Black or White, and why. But as for extracting the knowledge that allows him to do this—thus far nothing has been achieved. Perhaps the most difficult concept to explain in Go is "good shape". It is relatively easy to explain to a human

Both methods of achieving this, knowledge engineering and supervised training, have been found to be extremely difficult challenges that are loaded with pitfalls. Although there has been some success in using these techniques, there has always been a huge gap between the positional judgment of the strongest human Go players and the judgement encapsulated in the heuristic evaluation functions in Go programs.

Playing Metagames—Programs that Learn to Play from the Rules

In 1985 Aubrey de Grey investigated a new concept in games programming as his BSc project in Computer Science at Cambridge University. De Grey's idea was to write a program that would learn to play a game despite being given no more than the rules.

De Grey split the task of automating the whole process into the automatic generation of the features for an evaluation function and the automatic discovery of the best weightings for those features. He devised a method for generating the features, and this was implemented and refined, to the stage where it could produce a suitable feature list for a few games.[34] De Grey also devised an original method for optimizing the weightings in the evaluation function, based on processing the game trees to a depth of only two-ply. His method consisted of refining or rejecting various constraints on the values of the weightings, so that the constraints became progressively tighter, resulting in each weighting having a smaller range and therefore becoming more accurate.

Unfortunately for the world of games programming, Cambridge undergraduate project reports are not normally published, so de Grey's work has not accrued the credit it deserves. But within a few years the same idea was being investigated by other researchers, who developed programs that could generate evaluation features automatically from the rules of the games rather than have the features available to the program

---

player or a to computer programmer that if you crowd your stones together (bad shape) you lose out by having them control less territory, while if you spread them wide apart (also bad shape) you make it easier for your opponent to adopt a divide-and-conquer strategy. So good shape is somewhere in the middle ground between the two. But this is totally inadequate as a definition or specification. Francis Roads explains the problem of defining good shape thus: "A game of go is a living organism. You can never fully understand one part or aspect of it without taking the whole into account—and the whole is usually beyond human understanding." [2]

[34]The games de Grey chose were Checkers (Draughts), Reversi (Othello^{TM}), Qubic (4 x 4 x 4 Tic-Tac-Toe), the "L" Game, Chinese Checkers and Kensington.

from the outset. De Grey's concept is an example of Artificial Intelligence *par excellence*—programs that determine by themselves, with no outside help, which aspects of a game should be codified into the features of an evaluation function. When applied to Chess such programs recognize the importance of material as an evaluation feature, counting the number of Chess pieces of each type. When applied to the game of Reversi, de Grey's program produced features which measure different aspects of positions that are correlated with mobility, recognized by strong players as being one of the most important features of the game. These methods do not generate the weights for the features—instead they serve as the input to systems that learn their weights either from the experience of playing games or by observing experts playing the game.

Another of the early systems for automatically devising evaluation functions was Tom Fawcett's ZENITH, developed as part of his PhD research at the University of Massachusetts at Amherst. When tested on the game of Reversi, ZENITH was able to create most of the features that are generally well-accepted by human expert players. It also designed a useful feature that no-one had previously thought of!

Apparently inspired by de Grey's work, another Cambridge student, Barney Pell, developed de Grey's remarkable concept into the somewhat different idea of the "Metagame", on which he wrote

> The idea is to write programs which take as input the rules of a set of *new* games within a pre-specified class, generated by a program which is publicly available. The programs compete against each other in many matches on each new game, and they can then be evaluated based on their overall performance and improvement through experience. [3]

The principal difference between de Grey's original concept and Pell's work is that, while de Grey's program would automatically devise its own features for its evaluation function, and then learn the appropriate weightings for those features based on its experience when playing games, Pell's approach was to endow his program with "advisors",[35] each of which encapsulated "a piece of advice about why some aspect of a position may be favourable or unfavourable to one of the players". Then, based on a collection of such advisors, Pell's program would analyze the

---

[35] For example, one advisor was called dynamic-mobility—the number of squares to which a piece can move directly from its current square on the current board position. Another advisor, static-mobility, counted the number of squares to which a piece could move on an otherwise empty board.

rules of the game to determine which of the advisors was appropriate for that game and the weights for each of them.

Pell's interest lay mainly in creating a program that could analyse games from within a particular class, and he later wrote a program called METAGAMER which played a class of games that includes Chess, Chinese Chess, Shogi (Japanese Chess), and Checkers.

> The strategic analysis performed by the program relates a set of general knowledge sources to the details of the particular game. Among other properties, this analysis determines the relative value of the different pieces in a given game. Although METAGAMER does not learn from experience, the values resulting from its analysis are qualitatively similar to values used by experts on known games, and are sufficient to produce competitive performance the first time the program actually plays each game it is given. This appears to be the first program to have derived useful piece values directly from analysis of the rules of different games. [4]

Zillions of Games

Pell's approach was more restrictive than de Grey's original idea but, by focussing the program on games of a particular class, Pell made the Metagame concept ideal as the basis for a commercially available multi-games system, called Zillions of Games, developed by Jeff Mallett and Mark Loeffler. With this product, which comes with several hundred games already available, users can create their own board games or specify the rules of existing board games. Within the Zillions system a game is completely defined by a "rules" file, along with some image files for drawing the pieces and the board, and sound files for providing sound effects and music. The rules file contains instructions for the program on how to play the game, written in a specially designed language called ZRF (Zillions Rules Files). Although this means that users who wish to add their own games must first go through something of a learning curve, the Zillions product allows users considerable flexibility when defining their own games. The user can define the board layout; specify how the pieces are arranged at the start of the game; specify what moves are possible by each type of piece; create various topologies by linking various positions and directions (for example, the board can be made cylindrical); create gaps in the game board; specify special zones of the board where special rules apply; specify the winning conditions; etc. Once the rules of the game have been defined, the Zillions software analyses these rules and

builds an evaluation function itself, and does so very quickly. Largely because of its efficient tree-searching software, the Zillions program is able to search very deeply in comparison with its human opponents, partly nullifying the negative aspects of having the program generate its own evaluation function rather than having an evaluation function based on expert human knowledge.

As the concept is further enhanced to encompass card games and other types of game, the robots of the future might be able to announce, for any game: "Thank you for teaching me the rules of Zappo. I understand the game. Now let's play Zappo and I will crush you".

Games with Imperfect Information and Games with Chance

Chess, Checkers and all of the other games described earlier in this chapter are categorised as being games of *perfect information*. There is nothing unknown about them—all of the player's moves and all of his opponent's possible replies are there for everyone to see. In games of perfect information, a suitably deep tree search and/or a database will usually form the most effective method of determining the best move. In contrast, some games are characterised as presenting *imperfect information* and need a completely different approach to programming.

Bridge

Bridge and most other card games are examples of games of imperfect information because you do not normally know what cards your opponent(s) is (or are) holding. In Bridge there are four players, each of whom is dealt 13 cards at the start of a hand, so although you know which 13 cards you hold, you do not know how the remaining 39 cards are distributed.

It is not necessary to understand more than a sprinkling of the rules of Bridge in order to follow our discussion on programming the game. Bridge is played by four players, normally designated North, South, East and West, who play as two partnerships (teams)—North-South are one team and East-West the other. Each hand of Bridge (often called a *deal*) takes place in two stages, the bidding (which is like an auction) followed by the play of the cards.

In the bidding stage[36] the players make bids,[37] as in an auction, for the right to determine how many *tricks*[38] need to be made by each side and to choose which suit is trumps[39]. Once the bidding has ended, which happens when three successive players say "pass" or "no bid", the play of the cards begins.

How Computers Play the Cards in Bridge

It is generally recognised amongst strong Bridge players that the play of the cards is easier than the bidding. Of the relatively small number of Bridge programs that have been written to date, most rely on some sort of planning for the play of the cards. The 1997 World Computer Bridge Champion program, Bridge Baron, employed a form of planning called Hierarchical Task Network planning. This form of planning is analogous to the goal → subgoal approach used by Newell, Shaw and Simon in their General Problem Solver program GPS,[40] planning the solution to a problem by repeatedly breaking down tasks within the problem into smaller and smaller sub-tasks, until all of the sub-tasks can be carried out successfully. The level of play achieved by this approach in Bridge has been anything but specacular, prompting many-times World Bridge Champion Bob Hamman to say of the best Bridge programs of the mid-1990s that "They would have to improve to be hopeless." [5]

In a 1989 conference paper provocatively entitled "The Million Pound Bridge Program" I proposed a different approach, using a tech-

---

[36]There are two principal goals for each of the partnerships during the bidding stage. Firstly, each *bid* is designed to convey information to the bidder's partner about the cards in the bidder's hand. There are various bidding systems and conventions that have been developed to aid this process, analogous to special purposes languages in which coded messages may be used. A second main goal of the bidding is for the partners to agree on how many *tricks* they will try to make. As in a traditional auction, the bidding process continues until no player wishes to make a higher bid, whereupon the highest bid determines the number of tricks to be made and the trump suit (if any). This final bid is called the *contract*.

[37]Most bids consist of two parts, for example "one diamond" or "two spades" or "three no-trumps". The number part of a bid indicates how many tricks in excess of six the partnership will try to make, with the *trump* suit being whatever suit is named in the second part of the bid. (Trump cards take precedence over other cards when determining which side has won the trick.) So a bid of "one diamond" means that the partnership will try to make at least seven tricks with diamonds as trumps; a bid of "two spades" means trying to make at least eight tricks with spades as trumps; and a bid of "three no-trumps" means trying to make at least nine tricks when there is no trump suit.

[38]A trick consists of four cards, one played from each of North, South, East and West.

[39]A card in the trump suit supercedes all cards in the other tree suits.

[40]See the section "The General Problem Solver" in Chapter 6.

nique known as Monte Carlo simulations for the playing phase. My reason for being provocative was that World Champion Zia Mahmood had been quoted in a British newspaper, the *Daily Express*, as offering a prize of one million pounds for a program that could defeat him and his partner at Bridge. My paper suggested how I thought this could be achieved, using Monte Carlo methods for the play of the cards. Matt Ginsberg was the first Bridge programmer to adopt this approach. His program GIB, which was introduced in the mid-1990s, was a winner of the World Computer Bridge Championship. During the 1998 (human) World Bridge Championships in France, GIB was invited to compete in an international tournament in which the participants attempt to solve difficult Bridge playing problems in which they can see all of the cards.[41] GIB competed alongside 34 of the world's top human players and at the halfway stage it led the tournament, but dropped back on the second day to finish in a highly respectable twelfth place.

In card games, the idea of Monte Carlo sampling is based on guessing in which hands the hidden cards might lie. The hope is that by examining a sufficiently large sample of situations that conform to the constraints of the deal, a playing decision that works well in a large proportion of the sample can be found. The way that Monte Carlo methods are used in Bridge play is to generate a set of deals that are all consistent with the bidding and with any play that has taken place thus far in the hand. For example, if one of the program's opponents had made a first bid of "one heart", and if the program's opponents are employing a bidding system in which an opening bid of one heart means "I have at least five hearts in my hand", then Monte Carlo sampling would generate only deals in which that particular player holds at least five cards in the heart suit. All hands dealt as part of a Monte Carlo simulation must conform to all such known constraints.

When a valid deal is generated by the Monte Carlo process the program knows the locations of all the cards in that deal. It examines each of the cards it could legally play at that point and determines how many tricks its partnership would make, taking advantage of the knowledge of where all the cards lie. After generating a large number of deals that are consistent with the information at its disposal, and after testing the result of playing each card from its current situation in each of these deals, the program simply counts how many tricks its partnership would make for

---

[41] These are called double dummy problems.

each card it could play, and then picks the card which leads to the highest average number of tricks.

As the play of a deal proceeds, certain information is revealed that increases the accuracy with which the players can estimate which cards lie in which hands and how many cards of a particular suit are held in each of the hidden hands. As more information is revealed, so a program can improve the accuracy of its probability estimates. For example, a program might estimate that the opponent on its left has a twenty percent probability of holding three or more cards in the heart suit, a thirty percent probability of the opponent holding two hearts, a twenty-five percent probability of him holding only one heart and a twenty-five percent probability of him holding no hearts at all. If, on the first occasion that Hearts are led, this particular opponent plays a card of a different suit, indicating that he holds no Hearts at all, then these probabilities are immediately adjusted to zero percent, zero percent, zero percent and one hundred percent in order to take this new information into account.

After a while Ginsberg found the Monte Carlo process too limiting, partly because it was unable to discover certain types of play that are used to elicit information about the locations of cards in the opponents' hands. Instead he developed a technique that literally solves the Bridge hand, guaranteeing optimal play.[42]

How Computers Bid in Bridge

In "The Million Pound Bridge Program" I advocated the use of a computer designed bidding system called COBRA. The system's author, Torbjörn Lindelöf, explains the development of his system thus:

> COBRA was developed with the aid of main frame computers, using a technique for studying the trick-winning ability of various cards and combinations of cards, as a function of a number of suggested properties of the whole hand. This study eventually formed the basis for a highly successful hand evaluation method, around which it was comparatively easy to build a complete bidding system. COBRA has few really innovative features, except that it performs at a level of skill which is unusual, even among international masters.

---

[42] It is not possible, of course, for a Bridge program to play perfectly without knowing for certain where all the unseen cards lie. Optimal play, based on its knowledge of the opponents' bidding, maximizes the program's performance on the play of a hand.

Its strength lies in a strong relationship between bidding rules and hand evaluation as the auction proceeds. The human user of COBRA is not required to take any action whatsoever during the auction except to "calculate" the next bid, just as a computer will perform this menial task. [6]

Ginsberg points out that while attempts at computer bidding, such as COBRA, appeared to have been able to perform at expert level during the early 1980s, the technique and style of bidding in top-class Bridge has advanced considerably since then, and machine performance using such systems "is no longer likley to be competitive." [5] Partly because of this, Ginsberg pioneered the use of Monte Carlo methods for bidding in Bridge.

The Monte Carlo approach to bidding is to generate a large number of deals, all consistent with the bidding that has taken place in the hand thus far and with the cards held by the program player which is considering its next bid. For each of the program's plausible bids, the simulation process hypothesizes how the auction might continue after that bid is made, considering all plausible bids by all four players. At the end of each hypothesized bidding sequence the program knows the final contract and it knows, from the meanings of the bids in each of these bidding sequences, certain things about how the cards lie, for example how many aces and kings are held by certain players. Knowing this information allows the program to determine, from a Monte Carlo simulation of the play of the cards, the likely result of the play if that particular bidding sequence is followed. So for each hypothesized bidding sequence the program has an estimate of the number of tricks its partnership will take, and hence of the score that its partnership will achieve. The program then chooses to make the bid for which the hypothesized follow-up bidding sequences indicate the highest average scores for its partnership.

Backgammon—A Game with Chance

Backgammon is a game with perfect information but it also has a chance element—the roll of the dice that determines what moves a player may make. This chance element creates its own problems for the programmer. One of these problems is the large number of permutations of rolls of the dice and the number of legal moves for each of the possible rolls, that a program would need to include in a game tree. The roll of two dice

creates 21 distinct possibilities:

1-1, 1-2, 1-3, 1-4, 1-5, 1-6, 2-2, 2-3, 2-4, 2-5, 2-6, 3-3, 3-4, 3-5, 3-6, 4-4, 4-5, 4-6, 5-5, 5-6 and 6-6.

For each of these rolls a program must examine every legal move that can be made by its men, and there may be many of them. The average number of legal plays for each roll of the dice in Backgammon is approximately 20, so with 21 possible rolls the average number of possible roll-and-play combinations from a position is approximately 420 (i.e., 20 x 21). Comparing this figure to the typical numbers of moves possible in Checkers (which lies between eight and ten) and in Chess (an average of 37) gives some indication of the magnitude of the problem if the game were to be programmed using a deep tree-search. Since this number of combinations is higher even than the maximum number of legal moves in Go (361), it is not difficult to understand why Backgammon programmers tend to avoid the idea of using a deep tree-search. And because of the chance element created by the dice, it has been found that beyond a few half-moves, search in Backgammon is of no benefit.

The earliest Backgammon program of note, called BKG, was developed by Hans Berliner at Carnegie Mellon University with about one man-year of effort, starting in mid-1974. Berliner is a former World Champion at Correspondence Chess and has also been a true giant in the world of strategy games programming. What was remarkable about Berliner's Backgammon program was not so much that that it was able to play at a very high level, but that its strength was due largely to the sophistication of the program's evaluation function with no look-ahead at all.

On 15 July 1979, Luigi Vila, who had just won the World Backgammon Championship, played a match in Monte Carlo against BKG version 9.8, for the first player to reach seven points. The cash stakes were a winner-take-all purse of $5,000 but the event turned out to be of far greater import for the human race than was suggested by the size of the cash prize. The program won by a score of 7-1, thereby becoming the world's first man-made entity to defeat a reigning human World Champion at a recognized intellectual activity.

Post-mortem analysis of the games showed that Vila made far fewer mistakes and that the computer was lucky with the numbers that were rolled on the dice. Berliner was rather maganimous in victory:

This event was not a match for the World Championship. My program did defeat the current World Champion, but it was an exhibition match not involving the title. Further, a match to 7 points is not considered very conclusive in Backgammon. A good intermediate player would probably have a 1/3 chance of winning such a match against a world-class player, whereas in a match to 25 points his chances would be considerably diminished. At the time of the match the bookmakers in Monte Carlo were quoting odds of 3 to 2 if you wanted to bet on the machine, and 1 to 2 if you wanted to bet on Vila. Thus the bookmakers apparently thought the program to be very slightly better than a good intermediate player...

...Further, the conditions of play may have worked somewhat against Vila taking the program seriously at first... encased inside a Backgammon playing robot

...the robot gave a semi-comic impression during the event, rolling around controlled by a remote radio link and occasionally speaking and bumping into things.

...It should also be pointed out that BKG9.8 had somewhat the better of the dice rolling. However, its dice were being rolled by a human assistant, not by itself. [7]

Neurogammon

For about a decade after Berliner's success, little was heard from the world of computer Backgammon. But then, from IBM, came a program called Neurogammon, developed by Gerald Tesauro. Neural networks had become the Artificial Intelligence community's learning method of choice,[43] so it was hardly surprising when Tesauro and others started applying neural networks to game playing.

Neurogammon was taught only the starting position for the game and a few evaluation features that incorporated the most important Backgammon concepts employed by expert human players. The program considered the game to be divided into six phases and employed a different neural network for each phase. For example, the networks responsible for making moves throughout most phases of the game were trained on a set of positions from 400 games in which Tesauro, himself a strong Backgammon player, took both sides. These networks were instructed

[43] See the section "Artificial Neural Networks" in Chapter 6 for a simplified explanation of how they work.

that each of the moves selected by Tesauro was stronger than all the other legal moves available in the position. The network employed to decide when to double[44] when to accept a proffered double and when to resign in the face of a double—was trained on a separate set of about 3,000 positions.

TD-Gammon

In 1989, in London, Neurogammon won the Backgammon gold medal at the first Computer Olympiad, in competition against three commercially available programs and two non-commercial entries. Tesauro's next Backgammon program was called TD-Gammon.[45] While Neurogammon had learned from the games of human experts, TD-Gammon learned by playing against itself, without the aid of any supervision provided by an intelligent "teacher". The result was that TD-Gammon greatly surpassed all previous computer programs in the level of its play, employing a learning method based on the approach of Richard Sutton, an extension of Samuel's work on Checkers.

While developing TD-Gammon Tesauro was rather surprised to find that a substantial amount of learning took place, even though his program started with zero knowledge about how to play the game well. During the first few thousand training games, the program's neural networks learned a number of elementary strategies and tactics. More sophisticated concepts emerged later, after several tens of thousands of training games. And as the size of the networks and the amount of training experience increased, substantial improvements in performance were observed. Without being given the benefit of any outside expertise, an early version of TD-Gammon was able to play at approximately the same level as Neurogammon. Furthermore, it appeared capable of automatically discovering features that could be employed to enhance the evaluation function—an achievement in what was, in the early 1990s, still a new field of research.[46]

Once the Temporal Difference neural networks were able to play as well as Neurogammon, despite having been primed with virtually no

---

[44]One of the most interesting aspects of Backgammon is the doubling cube, whose faces are numbered 2, 4, 8, 16, 32 and 64. At the start of a game the players are competing for one point in the score table, but during the game the players may agree to double the *stakes*, and redouble... and so on. Deciding when to offer to double the stakes, and when to accept or reject such an offer, is a major factor in distinguishing stronger Backgammon players from weaker ones.

[45]TD = Temporal Difference.

[46]See the section "Playing Metagames" earlier in this chapter.

Backgammon knowledge, Tesauro added Neurogammon's hand-designed evaluation features to the TD networks. The result was that TD-Gammon greatly surpassed Neurogammon and all other previous computer programs. This was demonstrated in numerous test games played by TD-Gammon against several world-class human grandmasters, including Bill Robertie and Paul Magriel, both noted authors and highly respected former World Champions. A later version of TD-Gammon (version 2.1) even came very close to defeating Bill Robertie in a 40-game test session. Robertie actually trailed the entire session, and only in the very last game was he able to pull ahead for an extremely narrow one-point victory. Robertie thought that in at least a few cases during the session, the program had come up with some genuinely novel strategies that actually improved on the way top humans usually play.

TD-Gammon's success, and the perception that it had reached the level of human world class players, had an enormous influence on the development of future Backgammon programs. In the subsequent decade two commercially available programs were launched, both employing neural networks and both playing at or above the level of TD-Gammon. First came Jellyfish, written in Norway by Fredrik Dahl, which was launched in 1997, and then Snowie, written in Switzerland by Olivier Egger. These programs are widely regarded as being on a par with or perhaps even stronger than the world's best human players.[47]

Poker—Imperfect and Misleading Information, and Chance

As with most card games, Poker is a game of imperfect information because some or all of the opponents' cards are hidden. Poker also has additional levels of complexity that do not pertain to any of the other games we have discussed thus far—risk management (the various possible betting strategies and their consequences), opponent modelling (identifying patterns in each opponent's strategy and exploiting this knowledge), deception (bluffing and varying one's style of play), and dealing with unreliable information (taking into account your opponent's deceptive plays). These extra dimensions, much like decision-making problems in the real world, make the programming of Poker an extremely challenging task, yet the past few years have seen the development of a Poker program that performs at the level of very strong human players.

---

[47] As of January 2005 there had not yet been a showdown in a match of a statistically significant length between any of these leading programs and the reigning human World Champion.

The earliest research into computer Poker was conducted by Nicholas Findler, using Five-Card Draw Poker[48] as his model. This is the form of Poker that was popular in the Wild West and played in many Western movies, though its popularity has been severely diminished with the advent of the various forms of stud Poker such as Seven-Card Stud, Omaha and Texas Hold'em.

Findler's Poker research spanned almost two decades and was intended as a study into decision-making under circumstances of uncertainty and risk, other examples of which are the formulation of economic policies, business management, political campaigning and military strategy. His approach was based on computer learning and employed a collection of simple heuristics combined with the results of some Monte Carlo simulation experiments. Although Findler was able to demonstrate that delicate judgemental decisions can be automated, his program was given only very limited opportunities for testing against experienced human players.[49]

The University of Alberta Poker Project

Computer Poker made no serious progress until 1991, when Darse Billings planned to write a Poker-playing program for his master's degree at the University of Alberta. But although he wrote many of the components he did not build a complete Poker-playing program. Instead, his MSc dissertation was a review of the literature, with little information about how to implement his own ideas.

On completing his degree Billings quit university life to become a professional Poker player but, as time passed, Billings' interest in playing Poker for a living day-in-day-out waned, while at the same time his enthusiasm for building a real program grew. So in September 1999 Billings agreed to return to the university as a full-time PhD student, enabling the games group to start upgrading a Poker project that had already started, with the ambitious goal of creating the world's best Poker player. In addition to Billings and Jonathan Schaeffer, the Poker group

---

[48] In Draw Poker, following the deal of five cards to each player there is a round of betting, after which all those players remaining in the pot may change any number of their cards, replacing them with fresh cards drawn from the deck in an attempt to improve their hands. There is another round of betting and then the player holding the best hand wins all the money in the pot.

[49] In the late 1970s "*Omni* magazine invited a few very good (excellent?) Poker players plus celebrities like Isaac Asimov to New York City to play against some of our programs and against other human players. Guess which player won most of the time? One of the learning strategies...." (Nicholas Findler, personal communication.)

included, at various times, Michael Bowling, Neil Burch, Aaron Davison, Rob Holte, Dennis Papp, Lourdes Peña, Terence Schauenberg and Duane Szafron. The version of Poker they programmed is Texas Hold'em, which has long supplanted Five-Card Draw Poker in popularity. This is the form of the game played in the annual World Series of Poker, where anyone with $10,000 in stake money may enter and the first prize has risen steadily over the years, now typically exceeding $5 million. Texas hold'em has also become a popular spectator sport on TV.[50]

There are different betting structures for the game. At the World Series in Las Vegas they play "no limit", which means just that. The player whose turn it is to bet may wager up to his entire remaining bankroll. The University of Alberta group based their research on the "limit raise" betting structure, where the amount of betting is strictly controlled on each round.

The Alberta group's first Poker program, developed between 1997 and 2000, was called Loki. It incorporated their first attempt at opponent modelling and the Monte Carlo sampling technique described in the earlier section on Bridge, and was tested on the IRC Poker site. Loki's progeny, called Poki, incorporated an improved opponent modelling system and various other improvements. Both programs played at the strength of average casino players but neither was world class.

Measuring the Strength of a Hand. Some versions of the University of Alberta program employ an evaluation function to measure the strength of the program's two-card hand (the hole cards) in relation to the other hands, even though the other two-card hands are unknown.[51] This measure is based not only on the program's two cards and the communal cards dealt thus far, but also on other factors such as the number of play-

---

[50] In Texas Hold'em each player is dealt two cards (his *hole* cards) and there will eventually be five communal cards dealt face up on the table. After the two cards are dealt to each player there follows the first round of betting. Then comes the *flop*—three communal cards are dealt face up, followed by the second round of betting. Next there is another communal card, called the *turn*, also dealt face up, followed by another round of betting. And for those players still in the pot after all this betting there is a fifth and final communal card, called the *river*, also dealt face up, followed by the final round of betting. All the players who are still in the pot then show their two hidden cards and whoever has the best five-card Poker hand from the seven cards (using any combination of his two hole cards and the five communal cards) is the winner of the pot. If there is a tie, the pot is split equally.

[51] This evaluation is calculated by looking at all possible future outcomes (the remaining communal cards that have yet to appear) and computing the percentage of times the program's hand will improve (called the positive potential) and the percentage of times the program's hand will fall behind those of its opponents (called the negative potential).

ers still in the pot (the more opponents there are, the less likely it is that the program holds the strongest hand), and what betting has taken place thus far (if the betting has suggested that one or more of the opponents have strong hole cards then the program downgrades its assessment of its own hand strength).

In other versions of the program, betting decisions are computed in a different way, using Monte Carlo simulations. At a given point in the hand, a simulation is conducted to the end of the hand in much the same way as Ginsberg did in Bridge. One thousand simulations are run, both for the scenario where the program's initial betting decision is a call, and for when it is a raise. (A fold decision is easy to compute—it has an expected value of zero.) The program then chooses the fold, call or raise decision[52] that has the highest expected value.

In order to increase the accuracy of evaluation of its own hand, the program assessed the likelihood of each of an opponent's possible hands being played to the current point in the pot. In effect, the program asked the question: "Given the way that the betting has gone thus far, what is the probability, for each of the various possible holdings, that my opponent does hold that hand?" This idea had been suggested by Findler as a way of placing a program in its opponents' seats at the table and reversing its reasoning processes by inferring from the opponents' actions their status in the game. The stronger the opponents' hands appear to be, the less strong Poki believed its own to be. Of course, the opponents can bluff, but when assessing the strength of an opponent's hand on the basis of his bets, Poki takes into account whatever is known about the frequency with which that particular opponent bluffs.[53]

Another important feature in Poker is the potential of a hand, which is not the same thing as its strength because the arrival of new communal cards can change everything. Hand potential is another important evaluation feature—one that assesses the probability of a hand improving or being overtaken when additional communal cards arrive on the table. For example, if a player holds the following two-card hand ♣ 2, ♣ 4, and if the communal cards showing thus far are ♣ 7, ♣ 9, ♢ Q, then at the moment his hand is low in strength (not even a pair), but it offers excellent potential for improvement to a flush (five cards of

---

[52]If the program can check, it is treated as equivalent to being able to call. If it can bet, it is treated as equivalent to being able to raise.

[53]It is of course only possible to discover that an opponent has bluffed when he stays in the pot until the showdown, after all the betting is finished, and is then forced to show his cards.

the same suit and quite a strong hand). Only one more club is needed in the two remaining communal cards, a probability of almost forty percent.

Poki's Betting Strategy. Betting strategy, in other words determining whether to fold, call, check, bet or raise in any given situation, is an exercise in risk management.[54] Poki employs different betting strategies before the flop and after the flop. Pre-flop play in Hold'em has been extensively studied in the Poker literature, attempting to explain the correct play in terms that can be easily understood by classifying all the initial two-card pre-flop combinations into a number of categories. For each class of hands a suggested betting strategy is given, based on the category, the number of players, one's position in the betting order at the table and the betting styles of the opponents.

After the flop Poki's basic betting strategy consists of three steps. First it computes its effective hand strength, which is then converted into three probability measures, after taking into account various factors such as the program's position in the betting order. These three probabilities indicate the frequency with which the program should fold, check/call (which are treated as being the same), or bet/raise (also treated as being the same). The third stage is to generate a random number between 0 and 1 and use it to determine which of the three betting decisions should be made. For example, if the probability of folding is 0.2, that of checking/calling is 0.5 and that of betting/raising is 0.3, then the program could simply make its betting decisions as follows: if the random number is less than 0.2 then fold, if it lies between 0.2 and 0.7 then check or call as appropriate, and if it is more than 0.7 then bet or raise as appropriate.

Towards World Champion. Poki quickly became the best-known non-commercial Texas hold'em program. In 1997 it started playing in games without money on various Internet Poker sites and achieved an impressive winning record against generally weak opposition, though the program's own weaknesses could easily be exploited by strong players. The program was also tested in heads-up matches[55] against human opponents, including many experienced players, and again it achieved a winning record overall. Some of the human players wrongly attributed

---

[54] As part of each betting decision a player should consider the ratio of the amount of money he will need to wager in order to remain in contention, to how much he stands to win if he is successful, and how this ratio compares with the program's estimate of the probability of it winning the pot.

[55] "Heads-up" means a two-player game, the program and only one opponent.

intelligence to the program where none was present. After losing a 1,000-hand match, one experienced player commented "The bot has me figured out now", suggesting that the program had developed an accurate opponent model for him, when in fact the version of Poki he was playing against did no opponent modelling at all.

Eventually Billings and his colleagues decided they needed a test session for their program in a long match against a world-class player, partly because the matches it had played hitherto were not long enough for the results to be statistically reliable and partly because they wanted to learn from the comments of a top player. They had a volunteer in Gautam Rao, who is recognized as one of the best players in the world and is exceptional at two-player Hold'em. Like many top-flight players, he has a dynamic ultra-aggressive style of betting. Rao played more than 7,000 hands over the course of several days in January 2003 against a version of the program called PSOpti.

Rao was able to utilize his knowledge that the program did not do any opponent modelling, allowing him to systematically probe for weaknesses in the program's strategy without any fear of being punished for playing in a methodical and highly predictable manner. Although he won the match, Rao's victory was not by a wide enough margin to be statistically conclusive, and he was impressed: "You have a very strong program. Once you add opponent modelling to it, it will kill everyone."

Creating a Model of the Opponent. In most strategy games there is usually little to be gained by modifying one's own decision making in order to address any stylistic quirks or known strategies employed by the opponent. The former World Chess Champion Bobby Fischer was very clear on this issue: "I don't believe in psychology, I believe in good moves." In contrast, not only does opponent modelling have enormous value in Poker, it will often be the distinguishing factor between players at different skill levels. The ability by a player to modify his decisions based on an accurate model of his opponents, may have a greater impact than any other factor on his success.

Creating a model of an opponent at the Poker table necessitates making appropriate inferences from certain observations and then applying them in practice. Unpredictable play makes it difficult for one's opponents to form an accurate model of one's own strategy, so not only must a strong Poker program be sensitive to changes in strategy by each opponent, it should also vary its own playing strategy over time, attempting

to induce its opponents to make mistakes based on an incorrect model of the program's play.

The original modelling system employed by Poki was based partly on a set of numerical weights, one weight for each of the 1,326 possible combinations of hole cards. A particular set of weights indicated the probability that the opponent would have played the current pot up to the present moment, if he held this particular combination. These weights were updated every time the opponent made a betting decision, so the model for each opponent was always up to date, taking into account the opponent's play in the most recent hands. For example, a raise increases the weights for the strongest hands likely to be held by the opponent and decreases the weights for the weaker hands.

This method of modelling was found to be too simplistic because the model did not take into account some of the most important details that affect the decisions of strong players, such as the number of opponents still in the pot, the size of the pot and the program's position in the betting order. An improved method also treated each opponent individually, building one model per opponent. This made a significant difference to the program's performance as it had hitherto treated all opponents in the same manner, even though each player has their own style of play.

Because of the need for a successful Poker player to learn about his opponents as the game progresses, the Alberta group investigated the use of neural networks designed to predict the next betting action by each opponent, based on a full history of that opponent's play during the previous few hundred pots. The data on which the neural networks were trained consisted of actual hands played by particular opponents. The results of this investigation were that the actions of real opponents could routinely be predicted with an accuracy of eighty percent and, in some cases, as high as ninety percent[56]. This research also identified two highly useful features for prediction that were added to the opponent model.

---

[56]The accuracy was not quite as good when competing against very strong players.

– 4 –

How Computers Recognize

Visual Recognition—How Computers See

Human vision is an amazingly complex process that involves acquiring and processing massive amounts of visual and other information. Our eyes and brain collaborate, to pick up the visual information we are seeking and to discard redundant and unwanted information. As you read these words on the printed page your eyes are focused on the words. You can see a lot of other information on the page, but for the moment your brain ignores this extraneous information to some extent, allowing you to concentrate on the word of the moment and pass on to the next one, rather than becoming confused by taking in every word that your peripheral vision can see. This collaboration between eye and brain is a computational process that has evolved over millions of years and is one of the marvels of nature that we have come to take for granted. Those of us who can see rarely stop to think about how fortunate we are—the gift of sight is an assumption.

The human eye, as shown in Figure 27, is made up of several layers, of which the innermost is the retina. The eye has muscles that control the shape of the lens, thus enabling the eye to focus. And, like a camera,

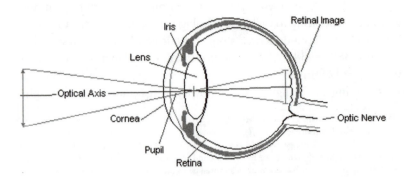

Figure 27. Cross section of the human eye (from the web site "Computer Vision" by James West)

the size of the pupil can be controlled, corresponding to the "F" setting on a camera, in order to control the amount of light that enters the eye.

The lens of the eye inverts the image of whatever the eye sees, and this inverted image impinges on the retina, where it is detected by millions of receptors known as *rods* and *cones*. The rods pick up low levels of light and are connected to the optic nerve, while the cones are more active at higher levels of light and enable us to perceive colour.

Research into computer vision, also known as machine vision, began in the 1940s when television was in its infancy. The idea of connecting a TV camera to a computer led to the invention of a crude image scanner during the 1950s, and later, in 1960, to the first recognition program for hand-written characters, based on an automated learning mechanism called perceptrons.[1] Computers see by means of cameras that replicate the functions of the human eye. The type of camera most commonly employed in computer vision systems nowadays is called a CCD camera, for Charge-Coupled Device, an electronic device that converts light into electrons and then counts these electrons to measure the levels of light in the image.[2] Light enters a CCD camera through the lens and is projected onto a CCD chip, which takes the role of the retina in the human eye. The CCD chip typically consists of one million or more individual light detectors (pixels) that are arranged in a matrix pattern, rather like a gigantic chessboard in which, instead of eight rows and eight columns, there might be 1,000 or more of each. These are the numbers that are referred to when talking about the resolution of a TV screen, a computer monitor or a similar electronic display.

The CCD chip and its associated electronic circuitry count the number of electrons within each pixel. Each number is stored, so the whole image can then be represented by a series of numbers, with one number usually corresponding to each pixel. In Figure 28, for example, the image of the fish's eye has a dark area along the left hand edge, corresponding to the 40, 38, 38, 38, 38 in the leftmost column of digits. Note that at the lower end of this column the numbers are slightly lower (36, 35, 35), corresponding to the lighter shade on the image, and in the top right-

---

[1] See the section "Artificial Neural Networks" in Chapter 6.

[2] A charge-coupled device is a computer chip containing an array of linked (or coupled) electronic capacitors. When light impinges on the surface of a CCD, some of the electrons in the surface gain energy from the light and become free to move around, congregating in small, image-forming units called *pixels* on the surface. These electrons accumulate in the capacitors on each pixel, where they can be counted by an electronic circuit. The count of electrons in a pixel indicates how much light has impinged on that pixel.

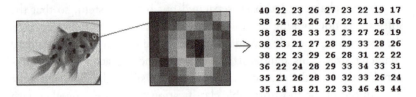

```
40 22 23 26 27 23 22 19 17
38 24 23 26 27 22 21 18 16
38 28 28 33 23 23 27 26 19
38 23 21 27 28 29 33 28 26
38 22 23 29 26 28 31 22 22
36 22 24 28 29 33 34 33 31
35 21 26 28 30 32 33 26 24
35 14 18 21 22 33 46 43 44
```

Figure 28. A fish, showing the image of its eye and the numerical representation of that image (from the web site "Computer Vision" by James West)

hand corner of the image there is an even lighter shade, corresponding to the number 17. Note also that some pixels have higher numbers than one might expect from this simplistic example—the reason being that each number here is the *code* for a different shade, and it is not always the case that a higher number means a darker shade.

In a black-and-white image the lowest possible number (usually 0) might correspond to white, with the highest number meaning black and the in-between numbers corresponding to different shades of grey. In a colour image the numbering system is more complex because the numbers not only represent how light or dark a pixel is, they also indicate the colour of the pixel (and the number of possible colours can be enormous).

Printed Character Recognition

One of the earliest vision problems to be subjected to machine recognition technology was the recognition of hand-written characters. In fact this is a more difficult task than dealing with printed characters so we shall discuss printed character recognition first.

Most early character recognition systems consisted of two main processing units—a character separator to identify each unit that needed to be recognized, and the isolated character recogniser itself. Character separation (also called segmentation) is important because printed characters can be of different sizes and can be separated from their neighbour characters by different distances, so care is often needed to avoid confusing part of one character with an adjacent character. Most printed character recognition systems operate in either a fixed spacing mode, where the sizes of the characters are known in advance and therefore the segmentation process can be very reliable, or in a variable spacing mode, where no advance information can be assumed about the sizes of the characters.

Once a character has been separated out by the system, so that the software is not looking at any extraneous data near to that character, the recognition module "normalizes" the image so that it fits a standard character size. This standardized version of the character is then compared with all of the characters stored in the system's database and a list is produced of the most probable classification candidates, together with confidence values indicating how certain the system is of its identification in each case. Each character in the database is stored as an image containing three types of pixels: black, white and grey. Different print fonts each have their own sets of stored images, in order to prevent the system from being confused by different fonts.

Handwriting Recognition

Recognising hand-written text is a much more difficult and complex process than dealing with printed characters. In the early years of character recognition there were no mass-market applications for a recognition technology that could cope with hand-written characters and therefore there was no serious demand for such systems. But with the advent of handheld PDAs[3] and other types of portable computers, instantaneous handwriting recognition has become an important field of research.

The most obvious problem in handwriting recognition lies in the huge variation in personal writing styles. And many people also exhibit a lot of variation within their own individual writing style. These variations can be related to the type of writing equipment (ballpoint pen, fountain pen, felt-tipped pen, etc.) and to the writer's situation, for example whether he is writing on his lap in a moving vehicle or sitting at a desk in his office. Such differences in style *within* the writings of any one person encourage the idea that handwriting recognition systems should be adaptive, so that the system learns the personal writing styles of its own users.

Handwritten characters vary enormously in shape, due to various factors:

1. The underlying geometrical properties of the characters vary from one writer to another, for example the relative positions and sizes of the strokes, the height-to-width ratios of the characters, and the general slant of the writing;

---

[3] Personal Digital Assistants.

2. There are some stylistic differences between the characters written by left-handed and right-handed writers;

3. There are different classes of handwriting styles according to the spacing between characters, or the lack of it: on some printed forms all the characters are written in guideline boxes; some writers separate their characters by a significant space; some run on their characters so they touch each other; while in "cursive" writing all the letters in one word are connected. Most natural handwriting styles are a mixture of discrete and cursive styles.

In the first stage of an automatic recognition process, the features of a written character are enhanced and then extracted from their images, allowing the shape of the character to be stored. In addition to the shape of a character, a recognition system might possess information about *how* a character was written—the type of pen strokes used, whether the pen retraced any strokes during the writing of a character, etc. This "dynamic" information, which is often available when the writer is using an electronic tablet or when he is writing with a stylus on a PDA screen, can also include the writing direction and the writing order of the strokes. The numbers stored for each of the pixels in the recognition image merely confirm whether or not the point of the pen visited the location corresponding to a particular pixel, but it is not always clear from this data which pixels belong to which strokes, or even what is the number of strokes in a written character.

The Parascript System

A successful natural handwriting technology was developed in the Soviet Union beginning in the late 1960s. After a long break in this research effort, interest in the problem was revived in the mid-1980s, by Israel Gelfand at the USSR Academy of Science, one of several young scientists and programmers who were keen on solving "unsolvable" problems. A group led by Stefan Pachikov attacked the problem and subsequently founded ParaGraph International, whose NHR[4] technology is employed in Parascript, a market leader in the field. The underlying idea of NHR is to use the fact that cursive handwriting is a series of movements made by a writing instrument. Each movement can be represented by one or more

---

[4]NHR is a trademark of Parascript, LLC.

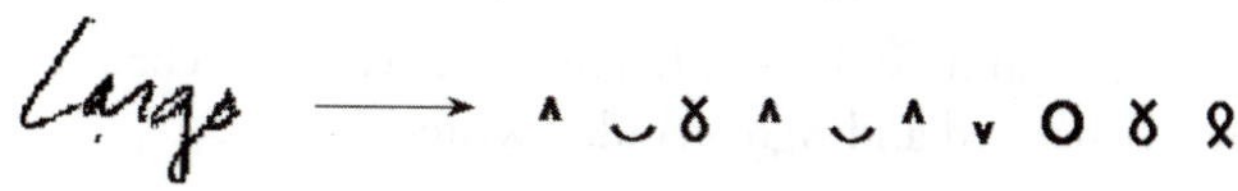

Figure 29. The handwritten word "Largo" and its representation in Parascript elements (Courtesy of Parascript®, LLC)

of eight elements that are sufficient to describe all the trajectories of the pen found in the cursive letters of the Roman alphabet (see Figure 29).

During the recognition process, NHR employs a reference vocabulary—a type of dictionary—to provide information that relates to the context of the writing. For example, the system might be trying to recognize whether the image of a particular word is "clear" or "dear". If one of these images occurred in a form where the writer was asked to specify the colour of the lenses in his spectacles, the software would look up both words in its reference vocabulary and eliminate the word "dear" as being an inappropriate response, and select "clear", which is the more appropriate word in that context.

In order to increase the accuracy of its word recognition, NHR employs two types of word recognizers, called *handwritten* and *analytical*. The handwritten word recognizer deals with the word as a single, unsegmented unit, in which the word is represented by a series of elements.

This series of elements is matched to a corresponding word in a dictionary and, as part of the matching process, the system provides a confidence level to indicate how certain it is that it has chosen the correct word from the dictionary.

The analytical word recogniser is used mainly for numerals and printed characters. It employs two classification methods—one is based on a database of symbol prototypes and the other comprises a number of artificial neural networks.[5] The combined recognition results of both types of classifier, the database classifier and the neural network, indicate the character recognized by the software and provide a confidence measure for that character.

Writer Independence and System Adaptability

Handwriting recognition systems depend on a stored database of characters and words that provide the "training" for the system. Some systems, called writer-independent systems, are trained on writing collected from

---

[5] See the section "Artificial Neural Networks" in Chapter 6.

different people, none of whom is the person whose writing the system will be attempting to recognize. By storing data from many writers with many different cursive styles, the system is able to compare any person's writing with a range of stored samples. The more writers employed in the data collection process, the more accurate will be the system.

Writer-dependent systems can be even more accurate than writer-independent systems because they are tasked only with recognizing the writing of certain people. It is the writing of those people, the "training database", which forms the data against which the recognition sample is compared. The training database in a writer-dependent system is normally large enough to contain samples of all the important variations and peculiarities of the styles of the small group of writers for which it is intended, it is not so large that it attempts to model every possible variation and peculiarity present across the whole range of natural handwriting styles. It is because these systems are not so ambitious that they offer a greater accuracy of recognition—the database against which a character or a word is being compared will be smaller than the corresponding database for a writer-independent system, and hence there is less scope for error in the recognition process. But it is possible to have the best of both worlds, by combining a writer-independent system with writer-dependent data in order to "learn" the writing of a particular user and thereby to improve the performance of a system with time.

Recognizing the Edges and Shapes in Images

Much of the early work on computer vision was devoted to the problems of finding lines and edges in images, and using this information to identify shapes. The pioneer in this field was Lawrence Roberts who, in 1965, was the first to attempt to solve the task of automatically recognizing three-dimensional objects. Two years earlier, while a graduate student at MIT, Roberts had already published an algorithm for generating a two-dimensional perspective view of a three-dimensional object. His algorithm reduced the visual information in a scene to distinct solid objects with straight line edges, which were then further reduced to flat surfaces that were also defined by straight lines. The locations of the endpoints of each line were stored in the computer together with data relating to specific viewing locations, the direction of sight from a viewing location to the object, and the position of a flat surface on which the program was required to project the two-dimensional image. Based on this

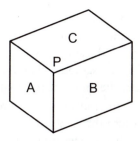

Figure 30. Line drawing of an image of a rectangular solid

information, Roberts' algorithm could generate an image of an object, in perspective, point by point.

Roberts' program detected edges by noting where the intensity of light in an image changed significantly between one pixel and an adjacent pixel. When these change points in an image formed a straight line, that line was assumed by Roberts' program to be an edge. In this way his program was able to convert images into line drawings.

Roberts worked with models of four straight-edged shapes: a cube, another rectangular (but not cube-shaped) solid, a wedge with two triangular faces, and a hexagonal prism. When the program encountered a shape in an image, it first made an intelligent guess as to which of these four shapes was the most likely. For example, in Figure 30 the point P connects the three polygons A, B and C, and a search of the models stored in the program reveals that a cube has three quadrilaterals meeting at a point, so the cube is one possibility for the shape of this object.

Having identified a possible shape as the source of the image in a line drawing, the program then "mentally" rotated that shape so it could be viewed from every angle, and compared the various views with the line drawings it had created from the image. When a good match was found, the program noted which shape it had come from, as well as the size of the shape, which angle it had been viewed from and where in the image the shape occurred. Roberts' program was also able to recognize an object made up by joining two of the shapes that it knew about. As parts of a line drawing were recognized by the program, for example a triangular part (suggesting a wedge shape), the program would eliminate from its "sight" those parts it had already recognized and then it would attempt to recognize whatever remained in the drawing.

Roberts' work was so fundamental to the study of computer vision that most of the elements of his 1965 program can be found in modern

vision programs: the detection of edges, the creation of line drawings using those edges, employing models of various shapes and mentally rotating them, and comparing the different views of models with the created line drawings.

Roberts' approach was taken further by Adolfo Guzman in 1968. Guzman's program, which he called SEE, detected junctions of lines, points where two or more lines meet, rather than the lines themselves, and could analyze scenes without the use of pre-stored models of objects. The program could "look at" a line drawing of a scene and identify all the objects in the drawing, even if they were not all completely visible. The goal of the SEE program was to divide a scene into images of distinct three-dimensional bodies so that a program such as Roberts' could be used to identify each of the objects.

Guzman's program employed heuristic rules about the junctions of lines on the surface of an object and what type of regions on the surface converged at each type of junction. This approach allowed the program to determine when two regions in a line drawing both formed part of the same object. After analysing the regions adjacent to a junction, the program would be able to suggest heuristics that were appropriate for deciding to which object a particular region belonged.

Figure 31 is one of eight types of junction employed in SEE, in which two of the regions are linked. Guzman called this type of junction an "arrow".

Guzman represented each type of object by a collection of links, indicating which regions are adjacent wherever a junction occurs in a line drawing. For example, in the case of the arch shown in Figure 32, the regions R1 and R2 are linked at two junctions, as are R1 and R3, whereas regions R2 and R3 are linked at only one junction. The program's representation of the linking relationships is shown in the lower part of the

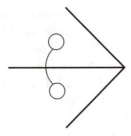

Figure 31. An arrow showing the two regions of an object that are linked

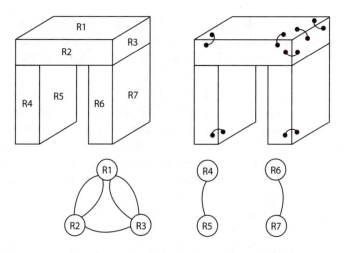

Figure 32. An arch, segmented into its component parts, with links shown between its various regions (after page 136 of *Artificial Intelligence: An Introductory Course* edited by Alan Bundy, Edinburgh University Press, Edinburgh, 1978) (Reprinted with permission by Adolfo Guzman-Arenas)

figure. Whenever a link joins two regions, this suggests that the regions are part of the same object, and the type of link suggests what types of object the program should consider.

The recognition problem in the real world is very much more complex and difficult to solve than is the case for objects represented as black-and-white line drawings. The general problem is that the computer is presented with an arbitrary scene, composed of arbitrary surfaces, each of an arbitrary colour and illuminated in an arbitrary way. With so much variation possible it is extremely difficult to reconstruct the scene correctly from a single image taken from only one perspective. One of the problems in doing so is that different spatial arrangements of an object can give rise to the same two-dimensional image. Another problem is that the appearance of an object is influenced by the material of which its surface is made, not to mention the question of clouds or haze obscuring part of a view and the angle of a light source creating or otherwise affecting shadows. And to add to all these difficulties, no matter what perspective is shown, some of the geometrical properties of a scene will be lost—parallel lines can turn into convergent lines, angles can change, and lines that are equal in length can appear to be of different lengths. All

of these factors and more make it difficult for a computer to determine which lines belong to a single object.

Some Applications of Computer Vision

One popular technique in computer vision is to compare new images of a given scene with old images, ignoring anything that has not changed and announcing when something was different between the earlier and the later images. This task is of great importance, particularly in the realm of security where noticing when a car starts to move or when a person suddenly changes direction can provide clues about possibly illicit intentions. Some advanced vision systems rely on having more than one image to view at the same time, in the way that humans benefit from having two eyes. Stereo vision systems employ two cameras that are positioned some distance apart, as with human eyes. If a common feature in an object can be identified in the images from both cameras, then the position of that feature can easily be determined by means of triangulation—calculations based on the distance between the cameras and on the angles of sight from each camera to the feature.

Being able to see something is of little benefit if your brain cannot make use of the visual information. You might be driving along a road, seeing a car in front of you, but if your brain does not notice that the image of that car is getting larger and larger at a very fast rate, then you are likely to have a disaster on your hands. Modern computer vision systems are often used in conjunction with software that employs other aspects of AI technology, allowing, for example, for the development of cars that are able to steer themselves safely along a road, or of systems that recognize and interpret facial expressions. Even a "simple" task such as determining that the black squares on a chess board are just that, part of the surface and not holes in the board, requires a sophisticated vision system.

Experience and general knowledge also play their part in enabling us (and computers) to perform recognition tasks. For example, if a small part of a scene is being hidden from view, such as when part of the road on which you are driving is obscured by snow, your experience and general knowledge tell you that, most likely, there is roadway beneath the snow. By using such experience and knowledge Artificial Intelligence programs can fill in the gaps, the missing parts of an image, to

make a picture complete. Modern vision systems are therefore usually only part of a larger system in which other AI technologies play their part.

Human Face Recognition

Your face is your most unique physical aspect, noticed by anyone who recognizes you. Different people may have different heights, weights and shapes, but there are millions of tall and millions of short people in the world, millions who are fat and millions who are thin. So faces work much better as distinguishing aspects than bodily shape or size. One of the most important uses nowadays of computer vision systems is face recognition, automatically picking someone's face out of a crowd, extracting that face from the rest of the scene, measuring the various features of that face and comparing the measurements with those of the faces in a database of stored images.

A powerful face detection system, developed at Carnegie Mellon University, examines regions of an image in blocks of 20 pixels × 20 pixels. In order to detect faces anywhere in the original image the 20 × 20 block is applied all over the image. The output from this stage of the process is a collection of these 20 × 20 "windows", each of which is accompanied by an estimate ranging from −1 (definitely no face present) to +1 (definitely a face present), signifying the system's opinion as to whether or not a face is there. Each window in this collection is then passed through a neural network recognition system that examines smaller blocks of pixels in order to attempt to recognize specific facial features. There are two smaller blocks designed to detect features such as a single eye, or a nose, or the corner of a mouth, and there are six stripe-shaped blocks that help the system to detect features such as a mouth or a pair of eyes.

Having confirmed the presence of a face in an image, the system then measures its features. Everyone's face has certain distinguishable features, for example peaks and troughs. There are about 80 of these features on a human face, including the distance between the eyes, the width of the nose and the depth of the eye sockets. After measuring these features the combination of measurements is used by the system to create an almost unique numerical code that represents a particular face. This code is called a faceprint. Usually between 14 and 22 of the 80 features in a faceprint will be sufficient to complete the recognition process. In the

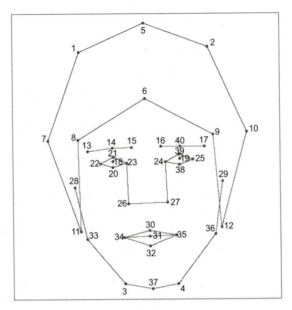

Figure 33. A wire frame model based on 40 features of a face (Figure 1 of "Finding Face Features" by Ian Craw, David Tock, and Alan Bennett, *Second European Conference on Computer Vision*, Lecture Notes in Computer Science 588, Springer-Verlag, Berlin, 1992, 92–96) (Courtesy of Ian Craw, David Tock and Alan Bennett)

case of a system developed at the University of Aberdeen,[6] 40 feature points were used (Figure 33).

When such a system is attached to a video surveillance system, the recognition software searches for faces using a low-resolution image[7] of the scene, then it switches to a high-resolution search when a head-like shape has been spotted. If there is a face in view it is detected within a fraction of a second. Once a face has been detected, the system determines the position, size and pose of the head. The image of the head is then scaled up or down in size and rotated, so that it can be registered in the system's memory in the same size and pose employed for the faces in the system's database. The system translates the facial data from every image into an almost-unique numerical code, which allows for an easy comparison of the face currently under examination with the faces in the database. Finally, the matching process identifies those stored images of faces that most closely match the face under examination. The

---

[6]This system was developed during the 1990s by Ian Craw and his team.

[7]An image with a relatively small number of pixels.

most successful face recognition system at the time of writing can match faceprints at a rate of 60 million per minute. As comparisons are made, the system assigns a value to the match on a scale of one to ten. If a score is above a predetermined threshold, the system's operator is shown the two images in order to confirm that the recognition system has found an accurate match.

The primary users of face recognition systems are security and law enforcement agencies, because faces that are "captured" in crowd scenes and elsewhere can be compared to those in a database of criminal mug shots. One of the most important security applications, particularly in the aftermath of 11 September 2001, is the use of face recognition technology as a major weapon in the war against terrorism. In 2001 Keflavik Airport in Iceland became the first airport in the world to announce its use of the technology to screen passengers.

An innovative use of this technology has been the Mexican government's idea of weeding out duplicate voter registrations. It was a popular practice for voters in Mexico to register several times under different names in order to be able to cast several votes at elections. Using face recognition technology, Mexican officials can search through facial images in their database of registered voters in order to identify duplicates. The technology was used in the 2000 Mexican presidential election with considerable success, and was later introduced into local elections.

Another potential application for face recognition technology lies in ATM and cheque cashing security. Software is able to quickly verify a customer's face, helping to protect customers against identity theft and fraudulent transactions.

Veggie Vision

Jon Connell and his team at IBM's Thomas J. Watson Research Center have developed Veggie Vision, a computer system that can recognize each of 150 or so different fruits and vegetables. By using a Veggie Vision scanner a checkout assistant at a supermarket can speed up their customers' waiting times and reduce the errors caused when a human mistakes one type of produce for another. In fact, Veggie Vision is even better than that. It can be used without a human assistant.

The Veggie Vision system makes good use of colour information. But although there are many obvious differences in colour between different items of produce, there are also many colour similarities, for example

a preponderance of different shades of green, such as green with blue and green with yellow, so the principal challenge for the IBM team was to automatically draw a fine distinction between similar colours, and to make such a system reliable.

While it is not difficult to distinguish the colours of various fruits and vegetables under laboratory conditions, a supermarket environment creates problems such as ambient lighting, which might confuse a colour detection process. A scanning system was devised to reduce this problem. The fruit or vegetable is placed on the glass surface of a scale that has been built into the checkout counter. An inexpensive CCD camera is located under the glass, pointing upwards at the produce, which is illuminated from below by a fluorescent tube. Two images of the produce are taken, one with the light on and the other with it off. The images of the produce can be separated out from its background by analyzing the differences between the two images.

Once the image of the produce is captured, its colour is analyzed into hue,[8] saturation[9] and intensity,[10] which together form a colour "signature". By combining this colour information with data on the shape, size and skin texture of the produce, Veggie Vision is able to recognize what fruit or vegetable it is looking at and to price it accordingly.

Recognition in Three Dimensions

Because light takes a finite and measurable amount of time to travel between two points, if we know how long light takes to make a journey we can calculate the distance it travels. The light illuminating each individual pixel in an electronic image sensor comes from the different features in the scene being viewed. By determining the amount of time that light takes to reach each pixel from a feature in the scene, it is possible to calculate the exact distance from that feature to the pixel. This enables a computer system to build up a three-dimensional relief map of the surfaces in the scene.

The Palo Alto company Canesta describes in a patent document several of its inventions for timing how long light takes to travel from a scene to the surface of the company's sensor chips. The basis of this technology

---

[8] A hue is a particular gradation of colour—a shade or tint.

[9] Saturation is vividness of hue; the degree of difference from a grey of the same lightness or brightness. The less grey there is in a colour, the more saturated it is.

[10] Intensity is the strength of a colour, especially the degree to which it lacks its complementary colour.

is similar to radar, where the distance to remote objects is calculated by measuring the time it takes an electronic burst of radio waves to make the round trip from a transmitter to a reflective object, such as a metal airplane, and back again. In the Canesta chips a burst of light, for example light from an LED[11] or a laser light, is transmitted instead of radio waves, and the time taken for it to return to the chip after being reflected is measured.

Life-Saving Technology for Motorists

At the Hebrew University in Jerusalem, Amnon Shashua has developed a computer chip called EyeQ that acts like a silent co-driver in your car, one that is not a back seat pest but which instead can save your life and those of as your passengers, as well as those in the oncoming path of your car. The EyeQ chip operates in collaboration with a video camera mounted on the dashboard of a vehicle. This camera sends information on what it sees to a computer containing the EyeQ chip, which can distinguish between ordinary, non-threatening observations seen by the camera, such as stationary objects at the side of the road, and imminent hazards, such as a pedestrian suddenly crossing the road in front of the car.

Figure 34 illustrates a typical example of scene interpretation: EyeQ detects vehicles (shown with a solid black line around them), pedestrians (both stationary and moving, shown with a dotted line around them) and lane markings (drawn with a white line). The numbers below the bounding boxes are the range to the target, which is computed despite the fact that the system is monocular.

In dangerous situations this information can be used by instruments linked to the computer, to sound warnings or even take automatic corrective steps involving the operation of the vehicle. Yes, this technology will one day lead to cars that drive themselves in traffic.

Among the possible applications of the EyeQ chip are a warning signal for drivers who stray out of their lanes, an automatic cruise control to regulate the speed of the car depending on traffic movement and other factors, and the automatic tightening of seat belts and extra pressure on the brake pedal in the event of an imminent crash.

---

[11] LED: Light Emitting Diode—a source of light, usually red, commonly used in electronic products.

Figure 34. Scene interpretation by the EyeQ chip (Courtesy of Amnon Shashua[12])

Lip Reading

As more of the technologies within the field of AI become mature, so there will be an increase in intelligent systems that rely on more than one AI technology. An example of the power of hybrid AI systems is the use of lip reading technology, based on computer vision, to increase the accuracy of speech recognition systems. The technology of automatic speech recognition has been developing steadily, albeit slowly, for more than 50 years, to the point where the best systems work reasonably well for tasks such as office dictation. But speech recognition is not effective in noisy environments, and the technology may stumble for other reasons.[13] Teaching computers to read lips is one way to increase the accuracy of automatic speech recognition, to help them understand the difference between "bat" and "pat", for instance.

Chalapathy Neti, a senior researcher at IBM's Thomas J. Watson Research Center in Yorktown Heights, is using face recognition and similar technology to boost the performance of speech recognition, in the same

---

[12] Professor Shashua is chairman of the School of Engineering and Computer Science at the Hebrew University of Jerusalem.

[13] See the section "Speech Recognition" later in this chapter for a mention of some of the factors that can affect the accuracy of that technology.

way that humans use a fusion of audio and visual perception in deciding what is being said. In the IBM project the computer and camera locate the person who is speaking, by searching for skin-coloured pixels, for example, and then using statistical models to detect any object in that area which resembles a face. Then, with the speaker's face in view, vision algorithms focus on the mouth region, estimating the location of many features of the speaker's mouth, including the corners and center of the lips.

If the camera looked solely at the mouth, only some 12 to 14 sounds could be distinguished visually, for example the difference between the "explosive" sound of a "p" at the start of a word and its close relative "b". So the visual region scanned by the cameras is enlarged to include many types of movements, such as jaw movements and movements of the lower cheek, as well as movements of the tongue and teeth. By combining these visual features with the audio recognition data, it has proved possible to increase the accuracy of speech recognition systems.

Although the initial results of the IBM research were promising, a studio is an ideal environment and often far removed from the conditions experienced in the real world. Many camera-based systems that work well in the controlled conditions of a laboratory fail when they are tested in situations where the lighting is uneven or the speaker is facing away from the camera. One method of combating such problems is to use an audiovisual headset, with a tiny camera mounted on a boom, enabling the mouth region to be monitored constantly, independent of any movement of the head. IBM is also exploring the use of infrared illuminators for the mouth region to provide a constant level of lighting.

Another solution to the problem of changing video conditions is a feedback system that changes its confidence levels as it combines audio and visual features, making its decisions using an evaluation function similar to those described for game playing programs,[14] on the basis of the relative weights of the two sources of information. When a speaker faces away from the microphone, the system's confidence in the lip reading and other visual cues becomes zero—the system simply ignores the visual information and relies on what it hears. When the visual information is strong, it is included. The goal of the IBM system is to do better than when relying on audio or video information alone. At worst, the system is as good as audio alone. At best, it is much better.

---

[14] See the section "How Computers Evaluate a Chess Position" in Chapter 3.

Similar research has led to promising results elsewhere. At Intel, for example, a group led by Ara Nefian has developed a system that could identify four out of five words in noisy environments, and the results were as good for Chinese as for English.

A Television Slave

One vision recognition technology that could find a wide audience is Sharp's HiMpact Sports, which applies a set of algorithms that understand what is happening in a game of Baseball, American Football or Soccer, and can extract all of the exciting moments from a game, thereby creating a summary of the video highlights without missing a single important moment in the game. The technology automatically indexes the summary so that the viewer can see at a glance the whole list of highlights. It can also provide annotated summaries of the action. The resulting summary permits sports fans to watch a game, play-by-play, with a text commentary that will tell them what to watch for, which players will be featured and the outcome of the play.

The software recognizes an event in a sport in all its permutations. For example, it can distinguish different types of base hits in Baseball (an infield grounder, or a line drive into centre field, or a bunt). The summary allows the viewer to choose to watch all of the home runs scored or all of a favourite player's hits during a game. This is achieved by using a Hidden Markov Model (HMM), a technology that has previously been successful in learning how to recognize spoken voices.[15] Just as an HMM is "taught" words by training it on samples of different people speaking the same word, Sharp's sports HMM was trained on video clips that it categorized into different training sets, one set corresponding to a "play" and one to a "not a play". Thousands of video clips in these training sets were flagged according to whether they constituted a play, and then the HMM learned these clips, adjusting its internal parameters until it could accurately classify those clips into plays and non-plays. Once it had learned the two categories, the HMM was set so it could be used to classify individual frames from a video as either part of a play or not.

Assisting Breast Cancer Diagnosis

One of the most important tools in the diagnosis of breast cancer is the mammogram, an x-ray of the breast. A shadow on the x-ray might mean

---

[15] See the section "Speech Recognition" later in this chapter.

danger or it could be a false alarm. Even highly trained radiologists sometimes miss a tell-tale sign that would necessitate further investigation of the breast. A computerized vision system can now assist in the mammogram analysis process at Innovis Health in Fargo, North Dakota, where radiologists have accepted the system as a screening tool, and the technology is becoming standard practice for many medical centres.

The computer produces an image, onscreen or on a printout, with markings to call attention to suspicious areas of the breast. Usually a spot will turn out to be benign. But studies show that computer-aided detection has found a significant number of cancers missed by radiologists. The goal is to help doctors do a better job of catching cancer early, which is often crucial to a patient's chances of survival. The five-year survival rate for breast cancers spotted at an early stage is 96 percent, compared with 20 percent for cancers detected so late that they have that have spread.

The radiologists who use the system do not expect computers to replace them. Ultimately it is still up to an experienced radiologist to decide on the best course of action once the computer system has done its job. The computer prompts doctors to take more second looks, which means calling the patient back for another mammogram, this time targeted at the suspicious area. The computer is especially good at identifying tiny deposits of calcium, which sometimes are cancerous but are usually harmless. Such deposits can be difficult to see for a human radiologist, but the computer imaging system excels at identifying them.

Speech Recognition

The "ears" of a computer are microphones, devices that contain some sort of diaphragm that vibrates in concert with any audible sound. The vibrations of the diaphragm are converted into electrical signals, which in turn can be displayed as a waveform on a screen or measured electronically. A speech waveform has an irregular shape and no two are the same. Figure 9 in Chapter 1 (page 33) shows a typical speech waveform.

In essence the speech recognition problem is one of recognizing waveforms such as this one. If everyone's voice produced the same waveform for the same word then there would be no problem at all, but this is far from what actually happens. Not only do different people say the same word in different ways, with different accents, different stresses, different

amounts of background noise, etc., but also the same speaker will often say the same word at different pitches, speaking quickly on one occasion and more slowly on another, sometimes with the effects of a sore throat or a cold, all these variations and many more creating changes in how the same spoken word is perceived on different occasions. So automatic speech recognition is most certainly not an easy task, and explaining in simple language how the technology of speech recognition works is almost as difficult as the technology is complex. Of necessity, therefore, the following explanation is a drastic simplification of how computers recognize speech[16] and it describes only one of the ways in which the recognition process works.

Let us imagine that each part of the particular word a system is trying to recognize is always spoken at exactly the same speed. For example, if we split up the word "elephant" into four parts: "el", "e", "ph" and "ant", under these idealised conditions the "el" would always be spoken in exactly the same amount of time, the following "e" would also always be spoken in exactly the same amount of time, and so on. By comparing the waveform for each segment of the sound in "elephant" with a database of stored segments, it would then be possible to identify each of the sounds, and to string them together to recreate the whole word. In practice such segments are in the region of one-fifth to one-quarter of a second on average—too long for the comparison process to be effective. Instead speech sounds are normally divided into much smaller segments, typically one-hundredth to one-fiftieth of a second in duration. By making the comparison with a larger number of shorter segments rather than with a smaller number of longer segments, the recognition process becomes much more accurate.

All speech is made up of strings of speech sounds, called allophones, and each allophone is represented by a *phoneme* consisting of one or more letters or symbols. Thus the sound of the letter "a" in "father" is represented by the phoneme "aa", the sound of the "u" in "cut" is represented by the phoneme "ah", and the sound of the "oo" in "book" is represented by the phoneme "uh". Most automatic speech recognition systems work on either a word recognition basis or a phoneme recognition basis. If a speech system can correctly recognize all of the phonemes in a word, and in the correct order, then it has recognized the whole word.

---

[16]For a more detailed yet eminently readable account of speech recognition technologies, the reader is referred to Robert Rodman's book *Computer Speech Technology* (Artech House, Boston, 1999).

One of the many problems in the science of automatic speech recognition is that the same word, and even segments of the same word, will often be spoken at different speeds, even if it is the same person speaking them. So a numerical technique called Dynamic Time Warping has been devised, which has the effect of stretching and compressing segments of the speech sound in a word, in order to make the waveform of the word easier to match with a stored waveform. In essence, the effect of Dynamic Time Warping is to stretch those segments of a speech waveform that are shorter than their stored templates, and to compresses those segments of the waveform that are longer than their stored templates.[17]

A pure matching process, by itself, will often be good enough to recognize isolated words with a high degree of accuracy, even when the software is running on a micro-processor with relatively little computing power. But some additional intelligence can be applied to the task, taking into account a knowledge of the context in which a speech segment appears. Within a word this contextual information can be applied to improve the accuracy of recognition of individual segments of the word, using a technique called Hidden Markov Models or HMMs. Here is how these models work.

Consider the word "tomato":

1. Let us assume that the probability[18] of a system recognizing the first sound in the word, the phoneme "t", is 1.

2. But assume that the system it is not certain whether the next sound is the phoneme "ah" (for which it has a probability of 0.4), or "ow", for which its probability is 0.6.

3. The system is 100 percent confident that the sound after the "ah" or "ow" is an "m", i.e. the probability of an "m" is 1.

4. But again, it is not sure what follows the "m", it might be an "ey" sound (a probability of 0.5), or it could be "aa" (also with a probability of 0.5).

5. Then there is another "t", about which the system is 100 percent certain (so the probability is 1).

---

[17]This description of the effect of Dynamic Time Warping is not *precisely* how the process works, but provides an easy-to-understand explanation.

[18]A probability of 1 represents a 100 percent certainty. To convert from a percentage certainty to a probability, simply express the percentage as a fraction or a decimal; for example, a 60 percent certainty corresponds to a probability of 60/100, i.e., 0.6.

6. And finally the system believes that the word ends with either an "a" sound (a probability of only 0.1) or with an "ow" (a probability of 0.9).

The HMM for this particular recognition task is shown diagrammatically as in Figure 35.

The individual probability estimates in the HMM allow the software to calculate the overall probability of each of the possible combinations of sounds:

$$
\begin{aligned}
\text{t ah m ey t a} &= 1 \times 0.4 \times 1 \times 0.5 \times 1 \times 0.1 = 0.02 \\
\text{t ah m ey t ow} &= 1 \times 0.4 \times 1 \times 0.5 \times 1 \times 0.9 = 0.18 \\
\text{t ah m aa t a} &= 1 \times 0.4 \times 1 \times 0.5 \times 1 \times 0.1 = 0.02 \\
\text{t ah m aa t ow} &= 1 \times 0.4 \times 1 \times 0.5 \times 1 \times 0.9 = 0.18 \\
\text{t ow m ey t a} &= 1 \times 0.6 \times 1 \times 0.5 \times 1 \times 0.1 = 0.03 \\
\text{t ow m ey t ow} &= 1 \times 0.6 \times 1 \times 0.5 \times 1 \times 0.9 = 0.27 \\
\text{t ow m aa t a} &= 1 \times 0.6 \times 1 \times 0.5 \times 1 \times 0.1 = 0.03 \\
\text{t ow m aa t ow} &= 1 \times 0.6 \times 1 \times 0.5 \times 1 \times 0.9 = 0.27
\end{aligned}
$$

So the recognition system would pick two phoneme strings as being equally likely: "t ow m ey t ow" and "t ow m aa t ow", each with a probability of being correct of 0.27.

Contextual information can also be used on a whole word basis. Consider for example the phrase "house of representatives". If a speech recognition system were to be reasonably certain that it recognized the words "of" and "representatives", but was unsure as to whether the first word in the phrase was "house", "louse", "mouse", "nouse" or "rouse", it could simply look through a database of phrases to see which occurs most often and by how much. Try it for yourself with your favourite search engine.

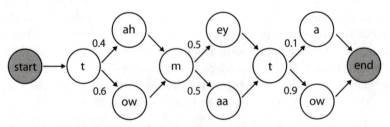

Figure 35. The Hidden Markov Model for the word "tomato" (Courtesy of James Matthews)

The number of hits I got using Google were

House of representatives	4, 180, 000
Louse of representatives	28
Rouse of representatives	27
Mouse of representatives	12
Nouse of representatives	2

giving a very high degree of confidence that "house" is the correct first word. This, of course, is very similar to the process we humans employ when recognizing a whole phrase when we are not completely certain about one of the words. It is often possible for us to make intelligent (and correct) guesses based on the context.

Since the 1960s, the science of speech recognition has been the subject of a significant research effort, yet even as recently as the early 1990s the best speech recognition systems would typically make errors on fifteen percent or more of the words they were set to recognize in a dictation task. Currently error rates as low as one to two percent are being reported for the same task. If true, this is a remarkable improvement for continuous speech recognition, a task considerably more difficult than the recognition of isolated words or phrases.

Taste Recognition

In 2001 Henrique Mattoso and Antonio Riul, working at the Brazilian Agricultural Research Corporation in São Carlos, received first prize in a government-sponsored inventors' competition. Their invention was an electronic tongue that can recognize the taste of water, wine, coffee and other beverages and assess their quality.

The tongue incorporates electronic detectors that each contain four different chemical sensors. These sensors comprise very thin films made of conductive plastic, together with some particles of the element ruthenium, all of which are deposited onto gold electrodes hooked up to an electrical circuit. When they are placed in a liquid that contains flavoured substances, each thin film absorbs the dissolved substances, thereby altering the electrical capacitance of each of the electrodes in a manner that can be measured. The end result is that different tastes are converted into different electrical capacitances, and by measuring these changes in

capacitance the electronic tongue takes advantage of the fact that the plastic films are sensitive to the substances responsible for various tastes.

Each sensor responds differently to different tastes, acting as a kind of human taste receptor. A composite detector that incorporates all four types of sensor therefore produces an electronic "fingerprint" of the taste. The electronic tongue can correctly distinguish between different types of coffee, different brands of orange juice and different types of milk. In some tests it has outperformed most humans when discriminating between various brands of mineral water, between samples of a particular type of wine taken from different wineries, and even between the same type of wine from the same winery but from different years. It easily distinguishes among the five basic human taste patterns: sweet, salty, bitter, sour and umami,[19] even in concentrations that cannot be perceived by humans. The electronic tongue can also sense low levels of impurities in water and can spot sugar and salt in concentrations that are far too low for human detection.

Instead of using plastic sensors scientists in Texas have taken a different approach to taste recognition. John McDevitt and his team at the University of Texas in Austin have developed a silicon chip with microscopic wells etched into its surface, each well acting as a tiny taste-bud. Each well holds a minute plastic bead made of polyethylene glycol and polystyrene, coated with sensitive chemicals that change colour according to the taste of the molecules that touch them. Each bead is equipped with a sensor for a particular class of chemicals, corresponding to a particular taste sensation. The colour changes are detected by placing the silicon chip in front of a light source and measuring the red, green and blue patterns. Each of the different colour patterns corresponds to a different taste.

The large number of chemically sensitive micro-sensors on the chip serve as artificial taste buds. They can be prepared a billion at a time, allowing for the creation of a sufficient supply for the entire world's needs for specific substances, such as the AIDS virus, anthrax spores or cholesterol. In one field study, in Botswana, conducted by researchers from Harvard and the University of Texas, tongues employing the technology of these chips were used successfully to detect HIV in villagers, with the results being processed in minutes, not days or weeks, and without having to send blood samples to another location for testing.

---

[19]The taste of monosodium glutamate.

Smell Recognition

Smells are caused by chemical molecules that are small enough and light enough to become vapour. A smell might be just one type of molecule or it could be a mixture of many different types. As an example, more than 600 different types of molecule waft into your nose when you smell the aroma of fresh coffee.

Humans detect smells by taking into their noses samples of air. Any dust particles are swept out of the way by a mucus layer on the inner surfaces of the nostrils together with a forest of sticky hairs. The filtered air then passes on to a collection of about five million smell sensor cells that comprise what is called the nasal epithelium.[20]

The design of most electronic noses is based on that of the human nose. There is a device, replacing our nostrils, for sampling the air, and an array of chemical sensors to mimic the function of the epithelium. One of the most common types of sensor for this purpose is called a conductivity sensor, often made of tin dioxide, which exhibits a change in resistance when exposed to molecules of various substances.

The two main components of an electronic nose are the sensing system and the automated pattern recognition system. The sensing system sometimes consists of several elements, each of which measures a different property of the chemical causing the smell, while others are made of a single sensing device that produces several measurements for each chemical. Each chemical vapour savoured by the sensor system produces its own "signature", a pattern that is characteristic of the vapour. By presenting many different chemicals to the sensor system, a database of signatures can be built up and employed to train a pattern recognition system, such as an artificial neural network, so that it produces a unique classification of each chemical. In this way the identification of a smell can be automated.

There are many important uses for our sense of smell, outside of the pleasures of wining and dining. Doctors can diagnose certain diseases from their smell alone; for example, some electronic noses are being used to aid the rapid diagnosis of lung cancer, others to diagnose gastrointestinal problems by monitoring a patient's breath. People can detect the smell of smoke and thereby warn of a fire in the home. Environmental applications of electronic noses include the detection of oil leaks and toxic wastes, the identification of household odours, and the quality of

---

[20]Dogs have 100 million of these cells, hence the greater sensitivity of their sense of smell.

air. These are just a few of the purposes of developing an electronic nose. One application for this technology that has attracted considerable attention in recent years has been the replacement of sniffer dogs, which are expensive and time consuming to train, by electronic devices that sniff out explosives as a counter-terrorism measure. A similar application is their use to locate land-mines, of which there are estimated to be more than 110 million around the world.

The Recognition of Creative Style

Every creator in the fields of art, music and literature has his or her own individual style, and those creators who are well-known have their styles widely discussed and dissected. Show an art enthusiast a painting by Van Gogh and it will almost certainly be recognized as such, even though the enthusiast might never before have seen that particular painting or its subject. Play a Chopin nocturne to a classical music aficionado and the response will most probably be something like "Oh, that's Chopin". So distinctive are the styles of many of the best practitioners of their art form, that they are easily recognisable.

Artistic Style

When the correct identification of the creator of an artistic work is a matter of some importance, as it is in the art world, for example, when a painting by a "master" comes up for sale, even the most expert of human "experts" may be proved wrong. There are countless cases, many of them quite amusing, involving art forgers who could paint a "genuine Rembrandt" or some other old master to order. One of the most famous examples involves forgeries by Henricus van Meegeren. After the end of World War II, the Dutch authorities started to hunt for citizens who had collaborated with the Nazis during the Occupation. It was discovered that a painting by Vermeer, *The Woman Taken in Adultery*, had been sold to Hermann Göring, and the trail led back to van Meeregen, who was arrested and charged with collaborating with the enemy. After six weeks in jail he announced, to everyone's amazement and disbelief, that the painting was not a Vermeer but his own work, a forgery. He also claimed to have painted five other so-called Vermeers, as well as two paintings that he had passed off as being by another Dutch artist, Pieter de Hooghs. In order to substantiate his claims van Meeregen said that, while under

police guard, he would paint another "Vermeer". He was duly provided with the necessary paints and brushes, and he proved his case well before the new painting was completed.

There is an ongoing debate within the art world on the authenticity of van Gogh's *Sunflowers*, bought for just under $40 million by the Japanese insurance company Yasuda Fire and Marine Insurance at Christie's in London in 1987. Van Gogh created several paintings of sunflowers and some art experts have challenged the authenticity of the Yasuda version as a genuine van Gogh, although it was generally accepted as an original van Gogh for several decades. The current view is that the painting is, most likely, genuine, but the lack of absolute certainty over a painting of this value is ample justification for employing Artificial Intelligence techniques to identify the painters of works of art.

In 1998 a team at the University of Bremen trained a computer to identify the drawings of the nineteenth-century French artist Delacroix, which it managed to do with 87 percent accuracy. Instead of employing techniques that had previously been used on drawings, such as analyzing the length, thickness and curvature of lines, the computer scanned the images of the drawings and simply designated each pixel as either black or white. The system then analyzed the ratio of black-to-white within each of the drawings, and across the entire set of drawings. This method was cheap and simple, and the ratio proved to be a sufficient measure to achieve an accuracy of 87 percent in identifying Delacroix drawings. But while this is relatively high for a new technology, it is not good enough for someone thinking of buying a work of art for hundreds of thousands of dollars or more.

Since 2003 a project called Authentic has been under way at the University of Maastricht with the support of the Van Gogh Museum in Amsterdam. One of the system's first crude tests was to discriminate between paintings by van Gogh, Cézanne, and Gauguin, which it achieved with 95 percent accuracy. The Authentic system detects certain elementary features in paintings and employs neural networks to identify their painters. The choice of the features employed by these neural networks was guided by discussions with an art expert, Frank Boom, and by some of the extensive literature on art. The main general characteristics known to be indicative of an artist's style are colour, brushwork, and texture.[21]

---

[21] Because a painter's brushwork yields a specific texture, brushwork and texture are treated within the system as one characteristic.

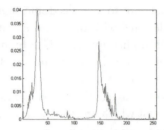 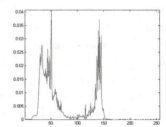

Figure 36. Hue histograms of similarly coloured van Gogh paintings (Courtesy of Eric Postma and Jaap van den Herik)

In order to analyse an artist's use of colour, two different types of colour histogram have been investigated. A colour histogram is simply a count of the relative frequencies of pixels of different colours, for example an RGB histogram is based on the three primary colours: red, green and blue. But raw RGB values are unsuitable for representing the perceptual appearance of colour, not surprising in view of the fact that humans perceive colour in terms of its hue, its saturation and its intensity, rather than in terms of its three RGB components. Figure 36 shows the hue histograms of two van Gogh paintings, *Wheat Field under Threatening Skies* and *The Church at Auvers-sur-Oise*, and the two pairs of coinciding peaks in these histograms suggest a clear link between the two paintings.

Early results with Authentic have been encouraging. The Maastricht group's initial quest for good features yielded sufficient of them to enable the successful classification of 60 impressionist paintings,[22] and many more features remain to be tested. A group at the National University of Singapore has found, for example, that the colouring of skin patches is a useful identifying feature, in particular the transition from a light colouration to a darker one when the arms and legs of the human subjects are partly in the light and partly in the shade.

The initial conclusion of the Maastricht researchers is that within fifty years the visual signature of painters will be as recognizable as those of any handwritten signature.

In a similar project at Dartmouth College, New Hampshire, Hany Farid and his team are using statistical techniques on the characteristic strokes of an artist's brush, pen or pencil, to produce an "electronic signature" of each artist's style. Details of the direction, curvature, pressure

---

[22]These comprise ten paintings by each of six artists: Claude Monet, Vincent van Gogh, Paul Cézanne, Alfred Sisley, Camille Pissarro and Georges Seurat.

and depth of brush or pen strokes contribute to this electronic signature, features that cannot normally be detected with the naked eye and are virtually impossible to forge. Farid's team has applied this technique to detecting art forgeries. In collaboration with the Metropolitan Museum of Art in New York they have analyzed 13 drawings that at one time or another have been attributed to the sixteenth-century Flemish artist Pieter Bruegel the Elder, only eight of which are now believed by art experts to be genuine, and have found that their software is able to distinguish perfectly between authentic drawings by Breughel and these eight known forgeries which nevertheless fooled the experts.

Literary Style

The literary world has long argued over the question of authorship of certain written works. Did Christopher Marlowe really write some of the plays attributed to William Shakespeare? Who was the author of the sensational but anonymously written political novel *Primary Colors*? These and other questions of literary attribution have been the target of the science of stylometry, the measurement of style.

Intuitively, we might expect one of the most tell-tale signals of an author's style to be the words that occur most rarely in their writings. In fact the most highly regarded statistical approach to the problem is to study the frequency of occurrence of the most common words, such as "and", "for", "with" and "to". The reason is that people's subconscious use of very common words provides an accurate literary "fingerprint" of their writing, precisely because the thought processes generating these words *is* subconscious. Someone trying to imitate the style of a famous author could fairly easily discover how often a few particularly rare words were employed in that author's writings, and emulate their frequency, but it is much more difficult to emulate the frequency patterns of the more commonly occurring words.

Stylometry has been around for more than a century but it was the advent of the computer that made possible the statistical analysis of large corpora of literary texts. The earliest convincing demonstration of the power of stylometry came in 1964 with the publication of the landmark research of Frederick Mosteller and David Wallace, who employed the technology to investigate the authorship of 12 "Federalist Papers", political essays published in New York newspapers in 1787 and 1788. These essays were all signed "Publius", but it has long been known that each of

them was written by one of two different people: Alexander Hamilton, a Federalist leader, and James Madison, the fourth president of the U.S.A. Because of this uncertainty of authorship the essays became known as the "disputed papers".[23]

Mosteller and Wallace measured the frequencies of the so-called function words—prepositions, conjunctions and both definite and indefinite articles—in the known writings of Hamilton and Madison. For example, they found that the word "upon" appeared 3.24 times per 1,000 words in the writings of Hamilton but only 0.23 times per 1,000 in those of Madison. On the basis of these frequency measures Mosteller and Wallace concluded that all 12 of the disputed papers were written by Madison, a conclusion that agreed with the vast majority of those history scholars who had discussed the papers' authorships.

More recently, stylometry has benefited from the application of Artificial Intelligence techniques. Neural networks have the ability to recognize the underlying organization of a set of data and therefore can be trained to learn to differentiate between the writing styles of different authors, i.e., the frequencies of use of individual words and pairs of closely occurring words. Having learned their styles, a neural network system can then attribute authorship to works that do not come from its training set. In 1993, Robert Matthews of the University of Aston in Birmingham and Thomas Merriam, a Shakespearean scholar, created a neural network that could distinguish between the plays of Shakespeare and those of Marlowe. They trained their network on examples of writings that were, without dispute, by Shakespeare, and a similar number that were by Marlowe. During the training process, whenever their network made a wrong guess as to authorship, parts of the network were adjusted.[24] By the end of the training period the network could accurately distinguish between the two playwrights.

Another popular learning technology in the field of AI, genetic algorithms,[25] is also employed in stylometry. In 1995 David Holmes and Richard Forsyth of the University of Luton in England, applied genetic algorithms to the question of the authorship of the disputed Federalist papers. They started by creating a set of rules for determining

---

[23] The "disputed papers" have been employed as a testbed for many research projects in stylometry.

[24] It was the strengths of the links in the network that were adjusted, see the section "Artificial Neural Networks" in Chapter 6.

[25] See the section "Genetic Algorithms" in Chapter 6.

authorship. An example of such a rule might be

> If the word "but" appears more than 1.7 times per 1,000 words then the text is by James Madison.

A single rule by itself might be very bad at discriminating between the possible authors so Holmes and Forsyth created 100 rules and tested each one of them on known texts written by Madison and Hamilton, arriving at a "fitness" score for each rule on the basis of how many of the known texts it attributed correctly. The 50 "least fit" rules were then removed from the set, small mutations were introduced into the surviving 50 rules in order to mimic evolution, and 50 new rules were added to the set, restoring the total to 100. This process was repeated for 256 "generations", at which point the evolved rules attributed all of the texts correctly. When the evolved rules were then applied to the 12 disputed Federalist papers they correctly attributed all 12 of them to Madison, conforming exactly to the results of the analysis by Mosteller and Wallace and to the established opinions of history scholars. What was remarkable about the evolved rules created by the genetic algorithm is that they relied on the statistics for only eight words.

Musical Style

Although stylometry has been applied to written texts for more than a century, it is only very recently that any similar form of analysis has been attempted in music. A method akin to that of literary stylometry has been employed by Dutch researchers Peter van Kranenburg and Eric Backer. They trained their system using a set of 106 pieces of music by Johann Sebastian Bach, 53 by Handel, as well as works by Telemann, Haydn and Mozart. For each composition in the training set, the system computed the values of 20 features, most of which are properties of counterpoint.[26] Other features included the stability of the duration of successive time intervals between two changes in the music, and the fraction of bars that begin with a dissonant sonority.[27] Van Kranenburg and Backer experimented with different sets of features, with results ranging from 64 percent success up to 94 percent. This area of research is still, at the time of writing, very much in its infancy. But with the increasing

---

[26]Counterpoint is the technique of combining two or more melodies in such a way that they establish a harmonic relationship.

[27]A dissonant sonority is an inharmonius sound.

availability of music in computer readable form, such as MIDI[28] files, it seems likely that research in music attribution systems will soon be making faster progress.

If a computer can recognize the style of a particular composer, might it also be able to distinguish between the playing styles of different musical performers? This is a far more difficult task because it means distinguishing between performances in which, in theory, all the notes and their durations are the same.[29] Some highly experienced lovers of classical music *are* able to distinguish the playing of one world-class violinist from another, similarly with pianists and singers. But these are the exceptions. The vast majority of concert-goers do not even notice most of the time when a soloist plays a wrong note, and they are unable to distinguish between the tones and playing styles of the leading soloists.

Research in Austria has shown that it is indeed possible for a machine to learn to distinguish and recognize famous soloists based on their style of playing. Gerhard Widmer and Patrick Zanon have identified several features of expressive timing and musical dynamics and used these features to train six different learning mechanisms. The training data comprised 12 selected movements from Mozart piano sonatas, each played by five world class pianists: Daniel Barenboim, Glenn Gould, Maria João Pires, András Schiff and Mitsuko Uchida, and Roland Batik, a Viennese pianist who is highly skilled though not in the same class as the other five. The recognition rates of the best of the six learning algorithms ranged from 66 percent for Uchida up to 75 percent for Batik. Even more remarkable was the system's performance when tested on the music of Chopin. Although it had been trained on CDs of the various pianists playing Mozart, the software was able to score as high as 68 percent when the performers were playing music by Chopin, whose composing style is quite different.

---

[28] Musical Instrument Digital Interface—A standard for representing musical information in a digital format.

[29] After all, the performers are playing the music as it was written, in theory.

– 5 –

Creative Computers

Creativity is sometimes taken to be an inexplicable aspect of human activity.

—Bruce Buchanan [1]

What is creativity? More than fifty definitions from different sources are listed by Calvin Taylor, but no matter how we choose to define it, creativity is an essential component of intelligence. Any discussion of Turing's question "Can machines think?" should therefore include a consideration of whether or not machines can be creative. And, in parallel with Turing's view that a machine should be regarded as intelligent if it gives the appearance of being intelligent, we may similarly accept that, in order to be regarded as creative, a machine need not simulate the human thought processes that underlie creativity but should merely have to produce results of at least the same quality as those produced by human creativity.

Along with Turing's question and the doubts expressed by many people at the notion that computers can have emotions, the subject of computer creativity gives rise to a host of quasi-philosophical doubts and arguments that started as long ago as 1843, when Ada Lovelace wrote in her account of Charles Babbage's proposed Analytical Engine:

> It is desirable to guard against the possibility of exaggerated ideas that might arise as to the powers of the Analytical Engine. The Analytical Engine has no pretensions whatever to originate anything. It can do (only) whatever we know how to order it to perform. [2]

Although Lovelace was arguing against any notion of creativity in the case of one specific machine, the Analytical Engine, she was by implication railing against the concept of creativity in all machines that act strictly in accordance with how we program them to perform. Fast-forward more than 100 years, to 1953, and we find a very similar dogma from Lord

Bowden in his classic tome *Faster Than Thought*:[1]

> Aesthetic judgments, which usually involve the consideration of the overall effects of sound or shapes, will be outside the proper province of computing machines for many years to come. [3]

So far so good. Bowden qualified his assertion by the simple expedient of including the words "for many years to come". But his next sentence is an example of an unnecessarily pessimistic forecast that is not uncommon in relation to some of the goals of AI —a forecast that usually expresses itself in the proclamation that "such and such will never happen".

> It seems most improbable that a machine will ever be able to give an answer to a general question of the type: "Is this picture likely to have been painted by Vermeer, or could van Meegeren have done it?" It will be recalled that this question was answered confidently (though incorrectly) by the art critics over a period of several years.

As we saw in the previous chapter, this particular case of unwarranted pessimism was put to the test exactly 50 years after Bowden wrote these words, when a team at the University of Maastricht started work on a promising project called Authentic, to distinguish genuine paintings by famous artists from fakes.[2]

The empirical proof that computers *can* be creative, for which several more tangible examples are presented later in this chapter, does much to refute the "computers can't think" attitude of some AI sceptics. But before we explore several examples of creativity by computers, there is one specific aspect of the Ada Lovelace argument against machine creativity that requires some comment. Her words, "It can do (only) whatever we know how to order it to perform", words that have been echoed countless times since she wrote them, do not present such a clear-cut argument as might first appear. The problem with this argument lies in the notion of randomness. Just about every computer program operating in a creative field utilizes some form of randomness in its decision making. One

---

[1] A ground-breaking compilation of contributions from 24 computing experts, including Alan Turing (who wrote the chapter "Digital Computers Applied to Games"). In his preface Lord Bowden paid tribute to Ada Lovelace, writing that "Her ideas are so modern that they have become of great topical interest once again,..." and he includes her paper on Babbage's analytical engine as an appendix to the book.

[2] It should also be recalled that six years before work started on the Maastricht project, another, more dramatic example of pessimistic forecasting was permanently eradicated, when Deep Blue disproved the dictum that "a computer will never beat the World Chess Champion".

of the earliest examples of the use of randomness in the creative arts is Mozart's dice game,[3] in which two dice are rolled to determine which of 11 possible bars of music should come next in a composition. In this composition process, when the computer simulates the rolling of the dice it is, in a sense, only doing what "we order it to perform". But that particular "order" is for the computer to make a random choice from the 11 possible totals of the numbers on the two dice. Does this random choice, strictly speaking, fall within the bounds of what Lovelace meant when she said, "order it to perform"? I believe not. It seems to me that she was thinking in terms of an algorithmic process that has no random element, a process in which every single step is predictable and the whole process is deterministic. But when we introduce randomness the process becomes no less algorithmic. What changes is the nature of the deterministic process so that we can not, with any certainty, determine the outcome. With Mozart's dice game, which creates musical compositions that are 16 bars in length, there are $11 \times 11 \times 11 \times \ldots$ different pieces that can be created, 11^{16} in all. We could write out each of these pieces of music, and we can argue with certainty that the process is sufficiently deterministic that the outcome will always be the creation of one of these pieces, but because of the randomness of the roll of the dice we never know which it will be.

This example tests the thinking behind the notion that a computer can only do what we order it to do. Yes, when a computer program chooses a number at random it is because the programmer has ordered it to do so. But it is because of that very command, the order to do something at random rather than in a way that allows for no choice, that the Lovelace dictum falls down. By instructing the computer to employ randomness we are making it creative, in a similar way to Steve Thaler's use of "nudges" in his Creativity Machine.[4] The use of randomness breeds creativity because the very process of creativity requires that some decisions be taken for no particular reason, i.e., at random, whether it is the choice of colour or the direction of a brushstroke in a painting, or the choice of words in a poem. In fact poetry, using random selection, was one of the first art forms to be explored by computer scientists interested in artificial creativity.

---

[3] See also the section "Mozart's Dice Game" in Chapter 2.
[4] See the section "The Creativity Machine" in Chapter 6.

How Computers Write Poetry

The first efforts at creating poetry with computers appears to have been in Stuttgart in 1959, where Theo Lutz' *Stochatische Texte* were created on a Zuze Z22 computer and published in the journal *Augenblick*. The first computer poems in English to attract attention were published in 1962 in *Horizon*, a glossy culture magazine, under the pen name Auto-Beatnik.[5] Other examples of Auto-Beatnik's poems have appeared on the Internet dated as early as 1960, suggesting that the project was almost contemporaneous with Lutz' work. One of its 1960 poems is called "Insects":

> All children are small and crusty
> And all pale, blind, humble waters are cleaning,
> An insect, dumb and torrid, comes out of the daddyo
> How is an insect into this fur?

The general method employed in Auto-Beatnik was the foundation for much of the subsequent research on computer-created poetry. This method employs a framework consisting of a line, or even a number of lines, arranged in a definite metrical pattern with recognized grammatical structures. There is a dictionary of words that are acceptable in the various slots in these frameworks. And there are simple heuristics to help make the created work poetry-like, for example choosing words that rhyme in certain slots. Auto-Beatnik's poetry generation was based on 128 different sentence structures in which the various slots were filled from a vocabulary of only 3,500 words, yet the program's poetry was considered sufficiently remarkable for the early 1960s to make it the subject of a *Time Magazine* article,[6] in which two of its poems were presented, including this one:

> All girls sob like slow snows.
> Near a conch, that girl won't weep.
> Stumble, moan, go, this girl might sail
> On the desk.
> This girl is dumb and soft.

In analysing this poem Chris Funkhouser comments that the program is able to emulate a strain of poetry influenced by the Beat Generation, and

---

[5] The Auto-Beatnik programmer, who worked in the Librascope Division of General Precision, Inc., in Glendale, California, preferred the anonymity of the pseudonym R. M. Worthy ("Ahm Worthy").

[6] 5 November 1962.

that "The first line certainly echoes the style or tenor of Jack Kerouac's poetry, especially in that it recognizes the suffering of all organisms (Kerouac was known for his use of Buddhist themes). A girl close to nature 'won't weep' but the one with the desk is 'dumb and soft'." [4]

The same method, using frameworks with slots but without the need for rhyming heuristics, was employed in a haiku generator developed by Margaret Masterman and Robin McKinnon-Wood at the Cambridge Language Research Unit in England, and some of its resulting haiku were exhibited at the Cybernetic Serendipity exhibition in London in 1968. Haiku are non-rhyming poems that traditionally invoke an aspect of nature or the seasons, one of the most popular forms of Japanese poetry, with three lines totalling 17 syllables (split five, seven, five). The Cambridge haiku generator consisted of various schemas in which the program filled the slots with words chosen from lists, with each list containing at least eight and at most 23 words. One such schema is the following:

All SLOT 1 **in the** SLOT 2

I SLOT 3 SLOT 4 SLOT 5 **in the** SLOT 6

SLOT 7 **the** SLOT 8 **has** SLOT 9

and the various lists of available words include

SLOT 1: White, Blue, Red, Black, Grey, Green, Brown, Bright, Pure, Curved, Crowned, Starred

SLOT 3: See, Trace, Glimpse, Flash, Smell, Taste, Hear, Seize

SLOT 7: Bang, Hush, Swish, Pffftt, Whizz, Flick, Shoo, Grrr, Whirr, Look, Crash.

Among the haiku created using this simple method, based on the above schema and lists of words, were

All green in the leaves
I smell dark pools in the trees
crash the moon has fled

and

All white in the buds
I flash snow peaks in the spring
bank the sun has fogged

Researchers in the field of computer poetry made no significant conceptual breakthrough until Ray Kurzweil started his Cybernetic Poet project
in the mid-1980s. His program was based on a device called Markov
chains, named after the Russian mathematician Andrei Markov (1856–
1922). A Markov chain models the behaviour of a sequence of events,
whether they be successive words in a poem, successive notes in a piece
of music or any other type of event. If it knows how often one particular event follows another, a computer system can simulate a sequence in
which events follow each other with the prescribed frequency.[7]

Kurzweil's program modelled the writing styles of 20 classical and 16
contemporary poets. The program, known as the RKCP,[8] incorporates
a software module to avoid plagiarising the original poets on which the
Markov chain data has been based, by rejecting sequences of four or more
successive words that appear in any of the works of one of the original
poets. And RKCP's Markov modelling is not only based on sequences of
words found in the original poems, it is also based on the human poets'
word structures, their rhythm patterns and the overall structures of their
poems.

When generating a poem the program employs some heuristics designed to give it creative intelligence. For example, when deciding which
word to write next, the program first determines a set of goals for that
word. One of these goals attempts to relate the next word to one or more
of the words already in the poem. Another goal is designed to choose a
word that is appropriate for its exact place in the poem. Yet another goal
encourages the selection of a word suitable for the particular human poet
whose style the program is trying to emulate.

The following poem, "A Wounded Deer Leaps Highest", was written by Ray Kurzweil's Cybernetic Poet after reading poems by Wendy
Dennis[9] and Emily Dickinson.

> A wounded deer leaps highest,
> I've heard the daffodil
> I've heard the flag to-day
> I've heard the hunter tell;
> 'Tis but the ecstasy of death,
> And then the brake is almost done,

[7] For a more detailed description of how Markov chains work, see the section "Markov Chains in Music" later in this chapter.
[8] Ray Kurzweil's Cybernetic Poet.
[9] A poet employed by Ray Kurzweil.

And sunrise grows so near
sunrise grows so near
That we can touch the despair and
frenzied hope of all the ages. [5]

How Computers Write Fiction

Amongst the first efforts to generate English text by computer was the work of Victor Yngve at MIT's Research Laboratory of Electronics in the early 1960s. His technique was similar to that employed in the slot filling haiku program described in the previous section, but because prose writing has more constraints imposed upon it, it was necessary for Yngve to base his sentence generator on a small set of grammatical rules for structuring English phrases. One type of rule was:

SENTENCE is a SUBJECT followed by a PREDICATE

where SUBJECT could, for example, be a noun, a noun phrase, a noun clause or a pronoun.

Yngve used as the basis for his grammar the simple, straightforward language of a children's book comprising 161 sentences, and he started writing a set of rules that were sufficient for describing all of these sentences. But he soon realised that it would be extremely difficult to write the whole rule set without any prior testing and he therefore decided to focus only on rules that were necessary to describe the first ten sentences in the book, which were

Engineer Small has a little train.

The engine is black and shiny.

He keeps it oiled and polished.

Engineer Small is proud of his little engine.

The engine has a bell and a whistle.

It has a sand-dome.

It has a headlight and a smokestack.

It has four big driving wheels.

It has a firebox under its boiler.

When the water in the boiler is heated, it makes steam.

From these ten sentences, which are made up from a vocabulary of only 38 words, Yngve derived 77 rules, from which it would be possible to generate all ten of the sentences and a huge number of other sentences as well, by randomly choosing nouns to fill the NOUN slots in his rules, randomly choosing verbs to fill the VERB slots, and so on. One type of sentence that can be produced by this grammar is

The engine is black, oiled,... and shiny.

where the "..." could sensibly be left blank or filled in by one or more of the other adjectives in the vocabulary: little, polished, big. Yngve calculated that, even with a tight constraint on the number of adjectives that could be used in succession in this way, there were more than 10^{20} different sentences that could be generated from the 38-word vocabulary, every one of them conforming to the 77 rules of his simple grammar.

The randomness in Yngve's approach was not suitable for Sheldon Klein, whose work at the University of Wisconsin on novel and story writing in the late 1960s required a sentence generator that could produce more meaningful text. Klein's program was based on a type of grammar called a *dependency grammar* in which each rule ensured that most of the words in a phrase were associated with (i.e., depended on) the *head* of the phrase (the most important word). For example, in "The man rides a bicycle" the word "bicycle" depends on "rides", while in the sentence "A bicycle is a vehicle with wheels" the word "vehicle" depends on "is", and the word "is" depends on "bicycle". So it would be acceptable, within a grammar containing these dependencies, to generate "The man rides a vehicle with wheels", which certainly represents an improvement on the output typical of Yngve's program.

But after Klein's work on using dependency grammars, rather little progress was made in this field over the subsequent 30 years, and it was not until the late 1990s that text generation programs were able to produce passages of well-structured, meaningful text extending to several paragraphs.

TALE-SPIN

The earliest attempt at computerized story-writing was James Meehan's research for his PhD thesis at Yale University, published in 1976. Meehan's program, TALE-SPIN, was the result of his attempts to investigate

what kinds of knowledge are needed by a computer program in order to generate stories.

TALE-SPIN generated stories about the lives of simple woodland creatures, in which a character was given a goal and then the program developed a plan to enable the character to achieve that goal. Thus the program simulated a small world of characters whose motivations were generated by the real-life problems they had to solve. Meehan employed known techniques for automatic problem solving[10] and combined these techniques with information about how social relationships and personal characteristics affect the way that people (and hence the characters in TALE-SPIN's stories) persuade each other to do things.

Meehan also had to develop a text generation module suitable for story writing, which entailed keeping track of what had happened in the story thus far. If, for example, one of its characters was a bear that had left his cave earlier in the story, the program would write "Joe Bear returned to his cave" rather than simply "Joe Bear went to his cave". Whenever an event occurred, all of its consequences were computed, and when a goal was stated, for example, "John wanted to visit Mary", a corresponding problem-solving module was called—in this case it would solve the task of how John travels to visit Mary, possibly by creating sub-goals such as John getting to the railway station and John catching a train to the town where Mary lives.

Here is part of one of TALE-SPIN's best stories from the mid-1970s, entitled "Hunger".

> Once upon a time John Bear lived in a cave. John knew that John was in his cave. There was a beehive in a maple tree. Tom Bee knew that the beehive was in the maple tree. Tom was in his beehive. Tom knew that Tom was in his beehive. There was some honey in Tom's beehive. Tom knew that the honey was in Tom's beehive. Tom had the honey. Tom knew that Tom had the honey. There was a nest in a cherry tree.... [6]

Not exactly the stuff of which Pulitzer prizes are made, but Meehan's work did contribute to the concept of *story grammars*. The idea behind a story grammar is that, just as every grammatically correct sentence in our language has a structure based on the syntactic rules of the language, so a formal structure can be developed for story writing. This formal structure usually starts with a setting (such as John being at home), followed

---

[10] See the section "Problem Solving" in Chapter 6.

by a series of episodes, each of which consists of an event (for example, John deciding to go visit Mary) and one or more reactions to that event (which might include John going to the railway station). In each episode the setting changes from one state to another, which creates the progression of the story. And in parallel to this formal structure, which is expressed in the "rules" for story writing, there is a set of semantic rules which ensure that, as a story develops, the content of the various elements of the story hold together in a meaningful and sensible way.

MINSTREL

By the early 1980s it was observed that basing stories on a highly formal grammatical approach imposed strict limitations on the mechanisms that allow a story to "work". As a result, more emphasis in story writing programs was placed on modelling the reasoning processes of the characters. More importance was also attached to the plot of a story, for which the work of George Polti in the early part of the twentieth century became a foundation stone in some programs. Polti was a French literary critic who categorised all plotlines, whether they be from French drama, classical literature or any other genre, as falling within one of 36 primary categories, for example deliverance (which might be the rescue of a character), getting revenge for a crime or some other dastardly deed, or falling prey to cruelty or misfortune.[11] A similar categorisation was developed by Vladimir Propp, who wrote an influential analysis of the structural characteristics of Russian folk tales, which he divided into 31 generic narrative units.

In the late 1980s Scott Turner developed a program called MINSTREL for his PhD at UCLA. Turner's program was designed to produce 200-word stories about King Arthur and his knights of the round table, and was modelled on various planning tasks in the storytelling process, including the application of story themes (in the same vein as Polti and Propp's work) and dramatic techniques and devices such as suspense. Turner developed a single mechanism that enabled MINSTREL to select the theme for a story, to maintain consistency during the story, to instil drama into the narrative and to present the story in a linguistically acceptable form. And he endowed the program with "imagination" by creating variations on previously known story episodes.

---

[11] The enduring nature of Polti's work can be seen almost a century later in the plots of just about all of today's movies.

When one of the characters in a MINSTREL story encountered a problem, the program solved the problem using a different approach to the goal→subgoal method employed by Meehan. Instead MINSTREL would find similar problem situations from its memory and apply the solutions used in the past to the current problem. In this way the program was able to discover new, creative solutions, by first finding a related problem that it was able to solve, then solving the slightly different but related problem, and finally adapting the solution from the related problem to enable it to solve the original one. For example, when MINSTREL was asked to create a story about a knight who commits suicide, there were no stories in its memory about a "knight who purposefully kills self", so the program modified "kills" to "injures", and it modified "purposely" to "accidentally", because it knew that it already had a story about a knight who accidentally injured himself while killing a troll. Once this earlier story had been recalled, it was modified to create a scenario in which a knight purposefully kills himself by losing a fight with a troll, thereby developing a newly-invented story fragment.

MINSTREL represented a definite conceptual advance on Meehan's work of a decade earlier, but the stories it generated were only marginally more interesting.

STORYBOOK

The mid-late 1990s saw a resurgence of interest in automatic story-writing, principally because of significant improvements that had been achieved by computer programs in the quality of their writing styles. Whereas most of the earlier efforts in story writing had focussed on employing story grammars to design acceptable plots, more recent work addressed writing quality, and to good effect.

One of the more successful of these projects was STORYBOOK, developed by Charles Callaway and James Lester at North Carolina State University. STORYBOOK requires a considerable amount of knowledge about the subject domain of the stories it is asked to generate, and it also needs a narrative model, so setting up the system to produce a story involves a significant time commitment, specifying the necessary information about the structure of the story and the world in which it takes place.

Callaway and Lester chose fairy-tales set in the world of "Little Red Riding Hood" as the domain in which to develop and test STORY-

BOOK, and they created a simple *narrative planner* capable of generating stories for which the synopses had already been specified. So STORYBOOK starts its story-writing process with a narrative plan, which is a logical representation of the characters, objects, descriptions and actions in a story. The narrative plan charts the details—the who, what, when, where, why and how of a story, along with the order of events. The story generation process then consists of making these details conform to the synopsis.

BRUTUS

While STORYBOOK clearly represented progress relative to the programs of the late 1980s, the most advanced storytelling machine at the time of writing is BRUTUS, developed by Selmer Bringsjord at Rensselaer Polytechnic Institute and a small team of collaborators.[12] BRUTUS has been endowed with a database of information about the world and the language of academia, and it can create stories that are intriguing and somewhat mysterious, with betrayal being their common underlying theme (hence the program's name). Bringsjord and Ferrucci were able to devise a formal mathematical representation of betrayal, which is defined in terms of the beliefs, goals and actions of two characters, the betrayer and the betrayed. Other themes that BRUTUS "understands" include evil, self-deception and voyeurism.

In addition to its knowledge of themes such as betrayal, BRUTUS also possesses more general knowledge of the domain of its stories (academia), relating to the various aspects and topics of academic life—knowledge about such topics as professors, students, classes, theses, etc. And BRUTUS is endowed with various types of linguistic knowledge, such as word forms, the rules of English syntax, and how to structure paragraphs. The program also has knowledge of two types of grammar. One type is story grammars,[13] which includes knowledge of the principles of storytelling, such as how to cause the reader to empathize with the characters. The other type of grammar is *literary grammars*, which are used to capture the logical structure and style of creative prose.

BRUTUS' story generation process employs all this knowledge and more, in a four-stage process. First the program sets the stage for its

---

[12] David Ferrucci at IBM's Thomas J. Watson Research Center, Marie Meteer at Bolt, Beranek and Newman, and David Porush, a colleague at Rensselaer.

[13] See the earlier section on TALE-SPIN.

story, in part by exploiting its knowledge of a theme. Next it generates the plot for the story, using the characters identified in the "stage setting" phase, as well as its knowledge of the domain and the specified theme, to generate a scenario of what action takes place and how the various characters react to that action. This scenario in turn generates a list of consequences, each of which is inferred from the program's knowledge of the "rules" that govern what consequences result from each possible type of action. For example, if character A hits character B, one result will be that character B becomes upset. These consequences are generated as a sequence of events, and since BRUTUS knows the order in which these events happen, the program is able to structure the story with events happening in the correct order. Next the structure of the story is expanded on the basis of the program's story grammars, with a list of appropriate sentence types being generated along the way. Finally the program's linguistic and literary knowledge are employed for the generation of the language, including the choice of appropriate vocabulary and phrasing.

The following example of BRUTUS' writing is entitled simply "Self-Betrayal."

> Dave Striver loved the university—at least most of the time. Every now and then, without warning, a wave of well, it was true: a wave of hatred rose up and flowed like molten blood through every cell in his body. This hatred would be directed at the ghostly gatekeepers. But most of the time Striver loved—the ivy-covered clocktowers, the ancient and sturdy brick, and the sun-splashed verdant greens and eager youth who learned alongside him. He also loved the fact that the university is free of the stark unforgiving trials of the business world—only this isn't a fact: academia has its own tests, and some are as merciless as any in the marketplace. A prime example is the dissertation defense: to earn the PhD, to become a doctor, one must pass an oral examination on one's dissertation.

> Dave wanted desperately to be a doctor. He had been working toward this end through six years of graduate school. In the end, he needed the signatures of three people on the first page of his dissertation, the priceless inscriptions which, together, would certify that he had passed his defense. One of the signatures had to come from Professor Hart.

> Well before the defense, Striver gave Hart a penultimate copy of his thesis. Hart read it and told Striver that it was absolutely first-rate,

and that he would gladly sign it at the defense. They shook hands in Hart's book-lined office. Hart's eyes were bright and trustful, and his bearing paternal. "See you at 3 p.m. on the tenth, then, Dave!" Hart said.

At the defense, Dave eloquently summarized Chapter 3 of his dissertation. His plan had been to do the same for Chapter 4, and then wrap things up, but now he wasn't sure. The pallid faces before him seemed suddenly nauseating. What was he doing?

One of these pallid automata had an arm raised. "What?" Striver snapped. Striver watched ghosts look at each other. A pause. Then Professor Teer spoke: "I'm puzzled as to why you prefer not to use the well-known alpha-beta minimax algorithm for your search?" Why had he thought so earnestly about inane questions like this in the past? Stiver said nothing. His nausea grew. Contempt, fiery and uncontrollable, rose up.

"Dave?" Professor Hart prodded, softly. God, they were pitiful. Pitiful, pallid, and puny.

"Dave, did you hear the question?"

Later, Striver sat alone in his apartment. What in God's name had he done? [7]

Bringsjord believes that in order to extend his team's work to produce truly compelling stories, the program will have to be developed to understand the "inner lives" of its characters. To do this, he argues, it will need to have awareness of a kind that, he believes, no machine will ever possess, to

> ...think experientially about a trip to Europe as a kid, remember what it was like to be in Paris on a sunny day with an older brother, smash a drive down a fairway, feel a lover's touch, ski on the edge,.... [7]

Here Bringsjord is echoing sentiments enunciated by Douglas Hofstadter,[14] but to me the assumption that a machine will never be able to think in this way seems unnecessarily pessimistic. While Chopin and Mozart were undoubtedly inspired by such awareness, David Cope's program EMI[15] does not need to have this type of awareness in order to compose music in the style of Chopin or Mozart, so why should such awareness be essential for literary creativity?

---

[14] See the section "David Cope" later in this chapter.
[15] See the section "David Cope" later in this chapter.

How Computers Write Non-Fiction

A different application for text generation is the writing of non-fiction. The Natural Language Processing Group at Columbia University's Department of Computer Science is one of the leading research establishments in this field, having developed systems that take information from several source documents and then generate a summary containing the most important information from the source material.

One of the tasks in which the Columbia group has been successful is producing biographical summaries of people described in the news, using a system called Bio-Gen that was developed in collaboration with the Mitre Corporation. Each of the source documents is analysed to identify certain readily identifiable linguistic constructions, for example,

> Presidential candidate John Kerry
>
> Kerry, the presidential candidate
>
> and   Senator Kerry who is running for president this Fall

all of which capture and reinforce Kerry's presidential aspirations. In this way several different types of descriptive information can be captured about the people named in the source documents, such as their ages, their professions and perhaps the roles that they played in past events. The software also identifies whole sentences in which someone is the subject of the sentence, providing additional well-structured information about that person. The result of this process is that, for each named person in the source documents, the software has compiled a set of well-described facts.

After pruning out erroneous and duplicated descriptions of a person, the system merges similar but not identical descriptions, for example, "Chairman of the Budget Committee" and "Budget Committee Chairman" are recognised as meaning the same thing. The system then needs to decide which of the descriptive facts about a person are the most important to include in its biographical summary, a process based on the frequency with which a fact or description is found in the source material. When fed with several documents about President George W. Bush, for example, the description "President" is likely to be the one that appears most often and so the system will assume that his presidency it is the most important fact known about him. The system is also able to perform a similar task on relative clauses in the source material. It counts

how often the main verb in a relative clause is strongly associated with a particular person, compared to how often that verb occurs in the whole corpus of source material. Verbs that occur very frequently across all text and are associated with several different people, such as "get", "like", "intend" and "think", are deemed to be promiscuous and are ignored, while those verbs that are associated with fewer people are deemed to be more significant as descriptors of those people.

The Columbia University group has also developed a much more general summarization system called Newsblaster that automatically tracks the news of the day. Every night the software crawls around several news sites on the Internet, downloads articles, groups them together into clusters of articles about the same topic and produces a summary article on each topic. The end result is a summary of the news generated entirely by a computer program, complete with links to source articles on other Web sites.[16] The following article appeared on the Newsblaster site on the day after the final TV debate in the 2004 presidential election campaign. The program not only wrote the article, it also wrote the headline and determined that the summary, which was based on 51 source articles, related to U.S. news.

Bush defends his presidency, Kerry makes case for change in final debate
(U.S., 51 articles)

Analysts say the debate's importance had grown after strong performances by Senator John Kerry in the first two debates were followed by rising support for him in opinion polls. Among registered voters who watched the debate, 42 percent called Kerry the winner, 41 percent said US President George W Bush won and 14 percent called it a tie. Kerry said Wednesday night that Bush bears responsibility for a misguided war in Iraq, lost jobs at home and mounting millions without health care. Kerry said on Wednesday Bush had not done enough to protect America from another attack while the president labeled his Democrat rival's approach to preventing terrorism as "dangerous". There is a frantic feel to tonight's final presidential debate as both campaigns see the confrontation as their best remaining chance to take control of a race that enters its final three weeks in a dead heat. Moran closed saying that Bush campaign officials admit Kerry has an advantage on health care but believe a strong showing tomorrow night would introduce voters

---

[16]Newsblaster is available at http://www.cs.columbia.edu/nlp/newsblaster/.

to a relatively unknown Bush health plan . Of about 2,500 women polled, 49 percent said they'd rather date the president, while about one-third of singles said they would prefer Democratic challenger John Kerry.

How Computers Compose Music

The composition of music by computer has attracted the interest of many researchers since the earliest days of AI, possibly because music is the world's most popular art form. Computer music now has an extensive literature of its own, somewhat independent of the rest of the AI community, including academic journals that focus on the subject. With the advent of the personal computer, thousands of enthusiasts have tried their hand at developing programs to compose music of one sort or another.

The automated composition of music has its roots in several inventions that used dice and other methods of chance to randomly select the notes, bars and even the rhythms of a piece of music. Examples of such techniques are the writings of Athanasius Kircher and Wolfgang Amadeus Mozart described earlier.[17]

Markov Chains in Music

Algorithmic methods of music composition have always been popular with computer music enthusiasts, partly because they are relatively easy to program. One such method is based on the use of Markov chains. In the February 1956 issue of *Scientific American*, Richard Pinkerton published an article on Markov chains, entitled "Information Theory and Melody", in which he described how a statistical analysis of nursery tunes enabled him to create a system of melody, rhythm and harmony, itself capable of composing tunes in the same genre. His system modelled the basic qualities of symmetry and regularity found in the melodies of the tunes he used to set up his database.

Pinkerton selected 39 tunes from a children's song book and counted the frequency with which each note appeared[18]—there are eight possible notes altogether, including the rest (or pause). From the frequencies of

---

[17] See the section "Creativity" in Chapter 2.

[18] To simplify matters Pinkerton converted all the tunes to the key of C and he treated all the notes as though they were in a single octave, so that middle C and the next C above it were both counted simply as C.

occurrence of the notes, Pinkerton calculated the probability of occurrence for each:

$$\text{Rest} = 0.297$$
$$C = 0.163$$
$$D = 0.112$$
$$E = 0.132$$
$$F = 0.066$$
$$G = 0.149$$
$$A = 0.045$$
$$B = 0.036$$

The crudest form of music composition based on this data would simply
select each note based entirely on the frequencies, with no other constraints. But in composing music, the occurrence of a particular note
will normally exert a strong influence on the choice for the next note, for
example it is far more common to hear a series of notes going up or down
a scale in small steps, such as D, E, F or B, A, G, than it is to hear, for example, D, A, E, B. So Pinkerton counted the number of times that each
note is paired with a particular successor note, and from these counts he
calculated the relative frequencies, shown in Figure 37.[19]

Suppose for example that a particular note in a tune is an F. Pinkerton's data indicates that the probability of the next note being a rest is
0.15, the probability of it being a C is 0.00, the probability of it being a
D is 0.14, etc. In order to compose a tune using the data in this table,
Pinkerton simply chose the first note of the tune at random and then
chose each successive note (or rest) by reference to the frequency data in
the table.

Pinkerton discovered that snatches and phrases of his newly composed sequences sounded quite tuneful, but that the random appearance
of rests, which would sometimes occur in the first note of a bar, would
create an unnatural effect. Since a rest can *never* occur at the beginning
of a bar, it proved necessary for Pinkerton to create tables with frequencies for the rests that more closely resembled where they do occur in
a bar, with different tables catering for different rhythms. This technique improved the system's performance noticeably. Further possible
enhancements include using similar methods for composing harmony,

[19] From page 80 of "Information Theory and Melody" by Richard Pinkerton, *Scientific American*,
vol. 194, no. 2, February 1956, pp. 77-86.

	Rest	C	D	E	F	G	A	B
Rest	0.38	0.17	0.10	0.10	0.06	0.13	0.03	0.02
C	0.36	0.23	0.13	0.07	0.02	0.10	0.03	0.07
D	0.26	0.20	0.21	0.19	0.03	0.06	0.01	0.05
E	0.22	0.15	0.18	0.16	0.16	0.12	0.01	0.00
F	0.15	0.00	0.14	0.35	0.14	0.20	0.01	0.01
G	0.29	0.14	0.00	0.16	0.06	0.26	0.08	0.00
A	0.17	0.05	0.07	0.00	0.02	0.36	0.15	0.17
B	0.18	0.30	0.12	0.01	0.01	0.08	0.21	0.08

Figure 37. Pinkerton's first order Markov chain probabilities

when a group of notes are sounded simultaneously. The eighteenth-century French composer Jean-Philippe Rameau carried out a similar statistical analysis of harmony as the basis for much of his extensive writings on the subject.

The Markov chains employed by Pinkerton are called *first order* Markov chains because they rely only on the one previous note to select which row of the table should be examined when selecting the next note. First order chains are the most appropriate ones to use when the amount of source data is insubstantial. But when there is a large amount of source data available as there is today,[20] then it is possible to delve more deeply into the source material and create Markov chain databases that are even closer to the original style. If, for example, we had data for how often each of the sequences of notes C, rest; C, C; C, D; C, E; C, F; C, G; C, A; and C, B; is followed by a rest, by an A, by a B, etc., then whenever there is a two-note sequence starting with a C, we would know the frequency of occurrence of the note after that two-note sequence. Such data creates what is called a *second order* Markov chain. And the same method can go even deeper, giving rise to a third order Markov chain in which sequences of three successive notes each have associated with them a list of the frequencies for the fourth note being a rest, an A, a B,.... By using a third or fourth order Markov chain, based on all the compositions of a well-known composer, it is possible to create new works in which that composer's style is recognisable. Imagine being able to generate literally thousands of tunes in the style of Lennon-McCartney, or anyone else you wish.

---

[20] For example, enormous corpora of written English and vast numbers of songs and other musical works available in MIDI and other digital formats.

Rule-Based Music Systems

A month or so after the publication of Pinkerton's article, but clearly inspired more by Mozart's dice game and similar methods, two American mathematicians, Martin Klein and Douglas Bolitho, devised a program running on a Datatron computer that composed music. Their program used random decisions combined with a simple set of rules devised from an analysis of hit songs that had reached the U.S.A. "top ten". Klein and Bolitho discovered a surprising similarity between the musical patterns of these top ten songs, and observed that the similarity was "... so great, that it sounds as if the same piece of music had been written over and over." [8] They distilled these similarities into three rules:

1. There are between 35 and 60 different notes in a popular song.

2. A popular song has the following pattern: part A, which runs eight measures[21] and contains about 18 to 25 notes, part A repeated, part B, which contains between 17 and 35 notes; part A, again repeated.

3. If five notes move successively in an upward direction, the sixth note is downward and vice versa.

These first three rules were augmented by three more, set down by Mozart and designed to assist in the writing of melodies:

4. Never skip more than six notes between successive notes.[22]

5. The first note in part A is ordinarily not the second, fourth or flatted fifth note in a scale.

6. Notes with flats are followed by the note one tone down, notes with sharps are followed by the note one tone up.

The Datatron computer's musical education consisted entirely of these rules. In order to generate music the program would select a note at random, verify that the note conformed to all of the rules (discarding any notes that did not and generating new attempts in their place), and then generate the next note. Despite its simplicity the program was able to

---

[21] A measure in music is a unit having a given number of beats. Measures are separated on the staff by vertical lines called bars, and the term "bar" has become synonymous with measure.

[22] Here, when Mozart says "skip" he means to include the current note as the first one skipped.

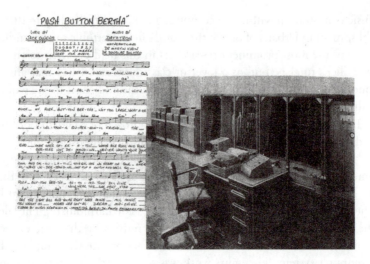

Figure 38. The music of "Push Button Bertha" and the music's composer, a Datatron computer, from "Syncopation by Automation" by Martin Klein (*Radio Electronics*, June 1957, pages 36–38))

compose an example of song music deemed good enough, by the nationally known song writer Jack Owens, for him to write the lyrics. Owens called the song "Push Button Bertha" (see Figure 38) and less than one week after he completed it there were five recordings on the market.[23] This novel success created a problem of its own—when an attempt was made to claim copyright in the name of the computer, the U.S. Library of Congress refused to issue a copyright certificate because they had never before been confronted with a piece of music written by a machine.

These early efforts highlight the two most important techniques adopted in music composition programs for the next 30 years, namely

1. The random selection of notes, each selection being constrained by a set of rules.

and

2. Markov chains.

The first of these approaches was adopted by Lejaren Hiller and Leonard Isaacson at the University of Illinois in Urbana, whose composition *Illiac Suite* generated a little media publicity and stimulated interest in

---

[23]It was broadcast for the first time on the KABC-TV station in Los Angeles on 15 July 1956.

music composition within the computing community, leading to the establishment in Urbana of one of the world's first electronic music studios. Many of the most prominent researchers in this field cut their teeth under Hiller's tutelage but, as a lover of classical music, I feel that almost none of their output would find favour with the audiences in the world's leading concert halls.[24]

In 1984 Kemal Ebcioğlu extended the approach of Klein and Bolitho by developing a successful rule-based system for composing chorales in the style of Johann Sebastian Bach. Ebcioğlu's program was based on some 350 rules, rather more than the six used to create "Push Button Bertha" almost 30 years earlier, and these rules were augmented by Schenkerian analysis[25] on Bach's procedures for harmonizing chorales. The results were far nearer to the level of musicality that one expects to hear in the concert hall than was anything previously composed by a computer system. Ebcioğlu's work was a landmark in computer composition, but his computing interests moved away from music almost immediately thereafter. Fortunately the baton of stylistic imitation had already been taken up by David Cope, whose work has dominated the field since 1987.

David Cope

On 5 April 1997 a most unusual performance took place as part of the University of California's "April in Santa Cruz" music festival. The highlight of the concert was a performance of Mozart's 42nd symphony, a work that incorporates all the glorious features that music lovers expect from a Mozart symphony, especially his later ones such as its predecessor, the 41st, which Mozart called "Jupiter" and which he composed in 1788. What was so remarkable about the Santa Cruz concert was that Mozart's 42nd symphony was composed more than 200 years after the Jupiter. Its composer was a program called EMI,[26] running on an Apple

---

[24]This is, of course, a somewhat subjective opinion. But although I have heard only a very small proportion of the work of what might be called the "Hiller school of computer music composition", none of it appealed to me in the slightest, and I cannot imagine anything similar appealing to a wide audience of regular classical music concert goers.

[25]The Polish-born music theorist Heinrich Schenker was a pupil of Anton Bruckner. Schenkerian analysis explains how music is made up of a series of common melodic fragments and operates by reducing a piece of music from the detail of its surface to a few simple fragments that lie far below its surface, resulting in what Schenker called a background—a few simple progressions that span and define the entire work.

[26]Experiments in Musical Intelligence.

MacIntosh computer, and developed by David Cope, a professor in the UCSC Music Department.

Cope's inspiration for the program came in 1981 when, as a composer in his own right, he found himself floundering in a period of "composer's block". So he decided to try to write a program that could emulate his own compositional style, with the idea that the program could act as his collaborator. He started by designing a system that could take, as its input, many works by a classical composer, and from them extract the characteristics that typify the particular composer's style. EMI was the eventual result, and turned out not only to be able to compose a "Mozart" symphony that attracted the critical acclaim of music aficionados, but also to be able to compose works in the styles of Brahms, Rachmaninov, J. S. Bach, Beethoven, Chopin and even the ragtime composer Scott Joplin. Cope was so enthralled by his program's success in writing Mozart's 42nd that he declared, "There's no expert in the world who could, without knowing its sources, say for certain that it's not Mozart." [9]

EMI has also made a nice job of increasing the repertoire of Chopin mazurkas. Douglas Hofstadter, a cognitive science professor and the Pulitzer Prize-winning author of *Gödel, Escher, Bach*, is also a gifted pianist and a passionate lover of Chopin's music. In his book, published in 1979, Hofstadter speculated on whether an artificially intelligent computer would ever be able to compose uplifting music, and concluded that, in order to create music as mesmerizing as that of the famous composers, a program would have to learn what it feels like to be alive. It...

> ...would have to wander around the world on its own, fighting its way through the maze of life and feeling every moment of it. It would have to understand the joy and loneliness of a chilly night wind, the longing for a cherished hand. [9]

In this belief Hofstadter overlooks an important point. It is not necessary for a computer program to *learn* what it feels like to be alive, rather it is sufficient for the program to learn how people who *do know* what it feels like to be alive, behave when they are composing music. It is not the *experience* of life that a program must mimic, it is the *behaviour* of those who *have* experienced life.

Cope's success in programming this mimicry amazed Hofstadter when, two decades after writing the above words, he played some of EMI's "Chopin" mazurkas for the first time. Hofstadter described the experience as

...a shocking comeuppance. They sounded eerily Chopin-like to me—and I'm someone who feels sure that music is a soul-to-soul communication. I think works of art tell you something very central and deep about their creator. How could emotional music be coming out of a program that has never heard a note, never lived a moment of life? [9]

So impressed was Hofstadter at the resemblance to Chopin's style of EMI's mazurkas that he took the program to the Eastman School of Music at the University of Rochester, widely regarded as one of America's leading music schools, where he witnessed more than half of the academic staff vote for one of EMI's mazurkas as genuine Chopin, while a truly genuine but rather little-known mazurka was voted to be a computer composition.

How Does EMI Work?

Cope's basic assumption is that "The genius of great composers lies not in inventing previously unimagined music but in their ability to effectively reorder and refine what already exists". [9] This is a statement with which I disagree, as it belittles the craft of the composer in creating previously unheard melodies, phrasings and orchestrations, etc. But the topic of human genius goes well beyond the scope of this book and I am happy to accept that that this is Cope's view and the assumption underlying much of his work. In fact I suspect that Cope's "error" (as I see it) has been very much to the benefit of his research—without such a strong belief in this statement it is doubtful whether he would have achieved such phenomenal success in his design and development of EMI.

EMI's design is based on the analysis of the structure of all the music presented to the system, for example all the symphonies of Mozart or all the mazurkas of Chopin. The system extracts from this data important elements of a composer's style and then recombines some of the musical elements of that style with some of the composer's recognizable musical structures, thereby creating new compositions in the same style as the original composer. For EMI's purposes, "style" means those identifiable characteristics of a composer's music that are recognisably similar from one work to another, including the pitch and duration of the notes and their timbre.[27] The musical logic is programmed partly as strict struc-

---

[27]Timbre is the quality of a musical note which distinguishes different types of musical instrument.

tural formats, as found by the analysis process, and partly by implementing harmonic rules that determine what follows what.

EMI's algorithm operates in stages. First, the analysis process identifies the "bricks-and-mortar" of a composition, the material that has to match the same structure as one of the original compositions in order to generate new music with the same structure. EMI's analysis module can distinguish between different harmonic functions,[28] as well as taking into account their purpose and directionality (whether a melody goes up, down, in a straight line or jumps around). The analysis module also takes into account certain hierarchical relationships between musical elements and, following Schenker's method, each piece can be reduced by eliminating unnecessary details, layer by layer, a process that can be continued until the bare essence of the piece is revealed. The analysis module can also focus on certain cues in the music, such as chord changes and the patterns of melody, in order to segment a musical work into sections, then to further segment these sections into musical phrases, which in turn are segmented into individual beats and chords.

The deconstruction module takes the music that has been tagged in the analysis phase (each fragment being tagged according to its function) and stores it as little building blocks that are usually one measure long. Those building blocks are stored in *lexicons*, which are lists of all those musical elements that have been analysed as having the same functions, and which, later in the process, will be recombined in order to create new music.

Knowledgeable listeners are able to detect certain stylistic elements (called ornaments in musical parlance) that uniquely identify the specific style that the system is trying to imitate. There are three types of ornament, of which *signatures* are the most important. Signatures are stylistic patterns that are specific to different works of a certain composer, sequences of notes that occur in more than one of the composer's works. Most signatures are between four and ten notes in length and usually consist of combinations of harmony, melody and rhythm. Cope's research has shown that, besides recognizing these patterns, EMI needs also to learn to recognize where in the music they reside. They frequently appear at the end of phrases and last a certain length of time, usually two or more beats.

---

[28] A harmonic function in music is sometimes known as an *interval*, an example of which is a major second.

EMI's pattern matcher is able to detect such patterns automatically, although variations between several works by the same composer might mean that it needs to find not only patterns that are exactly alike, but also patterns that are almost the same. Humans are able to detect similar musical ideas rather easily, but the task is considerably more difficult for a computer program because rhythms, pitches, and the locations of patterns within a measure can vary without changing the composer's basic musical idea. EMI surmounts this problem by employing a method that examines the *intervals*[29] between the pitches of successive notes, rather than the actual pitches themselves. This method, converting the transitions from one note to the next into a sequence of numbers, enables EMI to recognize similar sequences of notes even though the actual pitches in the sequences may be very different.[30] When a signature pattern is detected it is stored in the *signature dictionary*, a database that provides the fragments to be used in the reconstruction stage.

Once it has prepared the functionally tagged lexicons of building blocks and a signature-dictionary with patterns of different types, EMI can use these to generate new music. This process, which is carried out by EMI's reconstruction module, consists of composing musical phrases in which building blocks are re-ordered using the same ordering logic from their functional analysis, and inserting signatures in appropriate places within a block. The success of *recombitant* music, as this is called, depends on retaining the musical logic present in the original work. Music follows progressions that our ears have learned to expect, with certain sounds happening only at the beginning, middle and end of a phrase. EMI analyses all the musical groupings from the pattern matching function to determine their role within these progressions. This involves determining what harmonic pattern the groupings correspond to, so that when the elements are mixed and matched they are placed in the right order according to their function. This allows the piece to be put together following musical "logic" rather than being randomly selected.

The system creates a musical phrase by choosing as its first bar one that has been tagged as a "first bar" because it was the start of a musical phrase in one of the original pieces of the composer's music. (This is determined by the analysis module.) Then it chooses a next bar from all

---

[29] An interval is the distance between notes on a musical scale. EMI counts these distances in half steps—the difference in pitch between adjacent keys on a piano.

[30] For example, two sequences might consist of identical notes in terms of whether they are A, B, C, D, etc., but one octave apart. Although the pitches are completely different, the intervals are the same.

those bars that are of the same functional type as the one that originally followed what is now the first bar. It also takes into account that the melody of this next bar may have a strong tendency towards a certain continuation, as might be determined by a Markov chain database, and so it looks for a next bar that fits this preference of the melody. This process continues until a full musical phrase has been generated, with a structure that adheres to the style being imitated and with a signature in the correct place within the phrase. The composer module also transposes all of the snippets into the same key and adapts the individual voices (musical instruments) to the appropriate range. This prevents melodies and harmonies from overlapping in range or from becoming too distant.

Using this method in order to recombine musical elements would be too simple an approach, without the use of signatures, to enable the system to describe an ordering of the melodic and harmonic material that conformed to the constraints of the composer's style. The result would perhaps be a piece of Mozart that is recognizable as a symphony but rather a boring one. The pattern matcher is essential in creating works that sound typical of the intended composer, because the signatures identified by the pattern matcher provide the composition with sequences of notes that fit the newly generated structure, while at the same time including recurring motifs from the composer's oeuvre, sequences of notes that typify that composer.

Ironically, although EMI sometimes produces music as sublime as Mozart's, the program cannot tell the difference between a work of genius and a badly written nursery song, so for the time being it is left to Cope to decide which pieces of EMI's music are good enough to let out of his studio. Most of it goes straight into the waste bin once Cope has heard it. The small minority of pieces that survive are those that Cope finds pleasing, a judgment that EMI can not yet make for itself. Eventually EMI will probably be endowed with an accurate evaluation function that can measure both the musical quality and the level of originality of a new work, whereupon it will be able to make the decision itself as to which works to discard and which to present to its operator.

How Computers Create Visual Art

The beginnings of computer art parallelled the early days of computers, when artists used to create their work on oscilloscope displays using *raster*

175

graphics. An oscilloscope screen comprises many lines of phosphorescent pixels (dots) that are turned on and off by an electron beam sweeping across the screen, row by row (or raster by raster to use the more technical term). The artists would create their works by varying the electrical inputs to the oscilloscope and twiddling with knobs and buttons to control the various parameters on the device. The images produced on oscilloscope screens were then photographed, creating a permanent record of the artistic creation.

One of the first computer artists to come to prominence using this technique was Ben Laposky, an American mathematician and artist whose first "oscillons" (as he called his works) were created in 1950, while in Germany, almost simultaneously, Herbert Franke also started producing works in this art form. The aesthetic appeal of the oscilloscope images lay in the various shapes and forms that can be created using the many varied functions (or graphs) from different branches of mathematics. True, these patterns are not original works of art, but they can be employed as components in new creations and many of them are sufficiently appealing to stand as works of electronic art in their own right, despite lacking originality.

In the early 1960s, although computers were available only to a few people, those interested in their artistic potential had a new toy with which to create—the digital plotter. This was a device, a kind of electronic Etch-a-Sketch® machine,[31] that moved a pen across a sheet of paper, raising and lowering the tip of the pen in order to "plot" the desired image, which was determined by a computer program. Digital plotters allowed the few computer artists of that era to develop algorithms for creating line drawings, a process that could take a plotter anything from a few minutes to several hours.

One of the pioneers in the use of digital computers in the visual arts was Michael Noll, who created his earliest computer art in 1962 while he was working at Bell Telephone Laboratories in Murray Hill, New Jersey. In 1964 his program famously employed random numbers to create, on a microfilm plotter,[32] the drawing *Computer Composition With Lines* (see

---

[31] Etch-a-Sketch is a registered trademark of the Ohio Art Company.

[32] The microfilm plotter was a device incorporating a 35mm camera pointed at the screen of a cathode ray tube (a similar type of tube to those employed in TV sets and oscilloscopes). The plotter could draw black and white pictures on the screen, composed of segments of lines, some of which were connected and some unconnected. These images were photographed by the camera, which was controlled by the plotter.

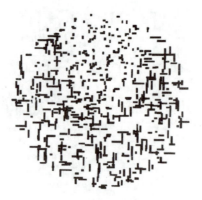

Figure 39. *Computer Composition with Lines* (1964) Copyright © A. Michael Noll 1965 (Courtesy of A. Michael Noll)

Figure 39), which mimics Piet Mondrian's painting *Composition With Lines* (see Figure 40).[33]

Noll created the vertical and horizontal bars in this image by a series of parallel segments, the centers of which were spaced closely enough that the segments slightly overlapped each other. Although Mondrian's bars were apparently placed in a very orderly manner, Noll's program was written to locate the bars randomly within a circle. Not only was the location of each bar randomly chosen, but so was the choice of whether a bar was horizontal or vertical, and their widths and lengths were also randomly chosen within specified limits. If the program decided that it wanted to put a bar inside a parabolic region at the top of the image, the length of that bar was reduced by a factor proportional to the distance of that bar from the edge of the parabola. With a little trial and error, Noll was able to set the program to create effects similar to that of Mondrian's original.

When Noll showed reproductions of both works to 100 people at Bell Labs, only 28 of them were able to correctly identify the computer generated picture, while 59 of them preferred the computer version and believed that it was by Mondrian.

Largely because of their lack of access to computers, a few putative computer artists of the 1960s designed and built their own creative machines. A few of these machines and some of their works of art were

---

[33]This painting has been described by the prominent French art critic Michel Suphor as "the most accomplished" of Mondrian's series of paintings based the horizontal-vertical theme.

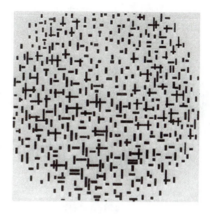

Figure 40. *Composition with Lines* (1917) by Piet Mondrian (Courtesy of Collection Kröller-Müller Museum, Otterlo, the Netherlands)

shown in London in 1968 in the Cybernetic Serendipity exhibition at the Institute of Contemporary Arts. Also on show at the exhibition were examples of computer creations in the fields of music, choreography, poetry, animated films, architecture and mobiles.

One of the inventors represented at Cybernetic Serendipity was Desmond Henry, a philosopher at the University of Manchester who also had an interest in and wrote about the logic machines devised by William Jevons.[34]

Henry had a life-long passion for all things mechanical. During the 1960s he created a series of three mechanical drawing machines (see Figure 41), powered by one or two servomotors, all constructed from analogue bombsight "computers" that had been used to calculate the accurate release of bombs from aircraft during World War II. Henry had found the design and workings of these bombsights quite inspirational, with their arrangement of gears, belts, cams, and differentials,[35] and he called the configurations made by these inner workings "peerless parabolas". The initial inspiration behind the development of his machines was to capture graphically these peerless parabolas.

Henry's drawing machines were not precision instruments and could not store information or be pre-programmed, due to the aleatoric el-

---

[34]See Chapter 1.

[35]The bombsight computer was an instrument that formed part of the cockpit of a bomber aircraft. Into this computer was fed information regarding wind-speed, altitude and wind direction, from which the computer calculated the accurate release of bombs onto its target from the moving aircraft. Such predictor systems inspired Norbert Wiener to coin the term "cybernetics" in 1948.

Figure 41. The Henry Drawing Computer (Courtesy of Elaine O'Hanrahan)

ement inherent in their operation, so Henry had only general overall control of the machines. Yet at the same time he was able to intervene, whenever he wished during its operation, to influence the course of the drawing taking shape. Both the pen and the table on which the paper rested were moveable, the pen moving in elliptical paths of various dimensions while the table moved in harmonic paths that distorted the ellipses at various points, at the same time moving the paper in a curved path.

There was also a chance element in Henry's machines due, as he freely admitted, to imperfections in the machines' construction, but this chance element, combined with the combination of paper-shifting and ellipse distortion, meant that no two drawings were ever alike and each therefore retained the virtue of infinite variety.

Figure 42. A drawing created by one of Henry's drawing machines (Courtesy of Elaine O'Hanrahan)

Henry enjoyed the surprise element of the various graphic effects generated by his machines, and liked to allow them to "do their own thing". He was so pleased with the drawings created by his machines (see Figure 42) that in the late 1960s he came to the conclusion that special purpose machines, rather than computers connected to digital plotters, would constitute the most profitable future development in computer art. And when, in 2001, at the age of 80, Henry was introduced to the world of fractals, he began referring to his own machine-produced effects as "mechanical fractals".

The Cybernetic Serendipity exhibition was a milestone in computer art. For the first time a wide public was made aware of the creative possi-

bilities of computers, not only those who visited the exhibition during its 11-week run but also those who read the many articles about the various exhibitors and their works that appeared in newspapers and magazines on both sides of the Atlantic. As a result there was a significant increase of activity in the field, which benefited as much from artists who took up an interest in computing as it did from those who arrived in the opposite direction. For several years much of the published output of computer artists was highly mathematical in nature, for example creations based on fractals or mathematical curves, but then came Harold Cohen and he changed everything.

AARON

> AARON exists; it generates objects that hold their own more than adequately, in human terms, in any gathering of similar, but human-produced, objects, and it does so with a stylistic consistency that reveals an identity as clearly as any human artist's does. It does these things, moreover, without my own intervention. I do not believe that AARON constitutes an existence proof of the power of machines to think, or to be creative, or to be self-aware, to display any of those attributes coined specifically to explain something about ourselves. It constitutes an existence proof of the power of machines to do some of the things we had assumed required thought, and which we still suppose would require thought, and creativity, and self-awareness, of a human being.
>
> If what AARON is making is not art, what is it exactly, and in what ways, other than its origin, does it differ from the "real thing?" If it is not thinking, what exactly is it doing? [10]

AARON is a program created by the English abstract artist Harold Cohen, who studied painting at the Slade School of Fine Arts in London and taught there for several years, developing an international reputation as an artist before moving to California in 1968. Cohen's development of AARON started when he was a visiting scholar at Stanford University's Artificial Intelligence Laboratory in 1973, and the work has been ongoing ever since. Cohen and AARON have had exhibitions at the Tate Gallery in London—the leading modern art museum in Britain—and in the Brooklyn Museum, the San Francisco Museum of Modern Art, the Stedelijk Museum in Amsterdam and many more of the world's major art spaces.

Cohen's ambitious goal when he started was to teach the computer the rules of artistic composition on which his own work was based. Instead his project has turned into the world's most advanced computer artist. In 1980 Cohen began to examine the scribbling behaviour of young children, focussing on the moment in the scribbling process at which the first part of a scribble migrates outwards and becomes an enclosing form for the rest of the scribble. Cohen perceived this to be the moment at which the child becomes aware that the scribble means something, and he found the geometry of enclosure, the physical relationship of the enclosing form to what is being enclosed, quite baffling. Cohen attempted without success to simulate this early human drawing, but he became convinced that the range of forms AARON could generate would be greatly enhanced if its steering strategy could be made to find its way around a pre-existing "core figure", the computer equivalent of the way that a child's initial scribble partly determines the path it later traces to enclose it. This conviction proved, in due course, to be justified. The construction of simple core figures, plus a simple strategy for tracing a path around them, yielded forms of sufficient complexity to enrich AARON's drawings. For example, AARON employs a sort of internal scribble when it draws trees. From the knowledge Cohen has given it about how trees grow, AARON grows its own tree skeleton, and then draws a line around it to create the complete image of the tree.

AARON's approach to the creation of drawings is based on the idea of beginning with something simple and building on it. AARON generates a drawing by using its knowledge about the objects that it draws and its knowledge about how to build visual representations of those objects. These two types of knowledge interact with each other continuously in complex ways and are fundamental to AARON's drawing expertise. All of its knowledge has been given to it by Cohen, but it is AARON that decides how to use this knowledge.

Some of AARON's knowledge is easy to represent, for example, how long are a person's arms and legs. Cohen simply make a list of parts, such as the left-upper-arm, the torso, etc. Each part is represented by a list of all the data points in that part, giving the position of each point in relation to the "origin" (or end-point) of the part. Thus, the origin of "left-upper-arm" is "left-shoulder", while the position of "left-elbow" is specified to be at some position relative to "left-upper-arm", such as two inches below, one inch to the left, three inches in front. And AARON knows that the "left-upper-arm" is attached to the torso, but it can only

attach it in ways that are plausible for a real body. AARON also knows some useful things about faces, for example that a face has eyes, a nose and a mouth, and that there are limits to where they can appear and what size they can be.

AARON can draw or paint anything it knows about, but in fact it knows about very little: people, potted plants and trees, simple objects such as boxes and tables, and how to decorate the background of a picture. It knows never to put a tree in a pot, or to put a person in a pot, though in some art movements such devices would be perfectly acceptable. Cohen could have made a policy decision to teach AARON about more things, different things, but instead his work has concentrated on attempting to make AARON draw and paint better, preferring quality to quantity. As part of this quest for quality, and spurred by a desire to have AARON paint a portrait of one of his friends, a poet, Cohen started to wonder whether AARON could be persuaded to produce a recognizable likeness of a particular person. For almost two months he developed AARON's knowledge of the structure of heads and faces, during which process the number of data points needed to represent a head and face grew from a few dozen to several hundred, while at the same time the structural development of software for face generation advanced considerably. Giving AARON the ability to create more detailed faces and heads than had hitherto been possible, necessitated organising the data points in AARON's representation of a face into distinct parts. Instead of having simply a set of points for a face, there was one set for the upper mouth, another for the lower mouth, a set for a beard, one for the forehead, an eyelid, the lower part of an eye, and so on, in such a way that the individual parts of a face could be scaled up or down in size and moved around the image at will, thereby enabling AARON to create heads and faces that showed a true perspective of the three-dimensional image. When Cohen completed this expansion of the program's facial knowldge, AARON knew enough to generate from it a varied population of highly individualized physical and facial types, with a range of haircuts to match.

AARON remembers everything it does in a drawing or painting, building up a very elaborate internal representation of an image as it develops. It has to remember where the lines are, partly so that each line corresponds to an appropriate part of the final creation and partly so that AARON can colour and fill in a creation without daubing one brush stroke unintentionally on top of another. In order to get everything in

a painting into perspective, AARON also needs to know what each little patch of the painting represents, what is in front of what, and so on. When it comes to colouring, for example, AARON is quite likely to find that a part has been cut up by something passing in front of it, and it has to be able to find each of the sub-parts so that it can make them all the same colour. AARON can only accomplish this if it remembers everything it has done in the painting.

AARON draws its enclosing forms, starting in the foreground, proceeding towards the background and without erasing anything already drawn, and most of its decision-making follows from this simple approach. For example, AARON is unlikely to draw a big object in the foreground, because it knows that that would obscure any small objects that it might decide to locate in the background later in the drawing process. Although the data representing a figure of a person are three-dimensional, AARON's construction of the representation of a figure is entirely two-dimensional, which is exactly how a human artist represents the outside world. In order to create a two-dimensional representation of a person, AARON was supplied with an extensive set of inference rules by means of which it could determine which part was in front of which on the basis of its knowledge of the figure. The problem was simplified by the realization that the hands move around much more than any other part of the body, and that a great deal could therefore be determined by examining the articulation of the arm. For example, AARON can make the following type of deduction:

> **if**   the left wrist is closer (in three-dimensional space) than the left elbow
>
> **and if**   the left elbow is closer (in three-dimensional space) than the left shoulder
>
> **and if**   the left wrist is to the right (in two-dimensional space) of the left shoulder
>
> **and if**   the left wrist is not higher (in two-dimensional space) than the right shoulder
>
> **then**   the left arm will obscure the torso

Such a deduction will lead AARON to the conclusion that the left arm will have to be drawn before the torso is drawn.

When Cohen started to endow AARON with the ability to colour its work, the program at first acted as though it was preoccupied with bright-

ness. As Cohen's colouring rules became more complex, AARON became more able and varied in its performance. Cohen devised a notion of colour "chords"—ways of choosing colours that worked together in various spatial relationships, and AARON was able to construct these colour chords as a way of controlling the overall colour structure of the image. Even so, the importance of brightness remained central to AARON's ability to colour; while the structure of a colour chord demanded that all its components, however selected, retain some required level of vivacity as they became lighter or darker. Cohen also gave the program knowledge about the range of colours from which it is sensible to choose when painting people, and on the basis of this knowledge AARON makes its own decisions as to which colours it uses.

When AARON became able to paint in colour Cohen decided to link the program to a painting robot. He designed and built the robot entirely by himself, even though he had no training in either electronics or engineering but instead had to learn about these subjects from robotics books and magazines. The robot starts by filling its own palette of paint pots, actually a row of bottles containing fabric dyes.[36] The robot mixes its own colours, selects the brush most suitable for the next area to be painted, and keeps track of how many inches of brushstroke it has used with each colour—this is how AARON knows when the robot is running out of paint. AARON can cope with filling arbitrarily complex shapes in a painting once it has located them, under the overall constraint that it should attempt to keep the wet edge of a shape moving forwards, as far as possible, and not leaving an edge to dry in one part of the shape while it is working in another part.

The colouring and filling-in process requires some care on AARON's part. When the program has to colour a potted plant, first it mentally outlines every patch in the image, determines whether the boundary of that patch is the edge of the leaf to which it belongs, and determines whether that patch is part of a leaf or part of the background. Next AARON assigns a number to each patch, in such a way that the same number is assigned to every patch belonging to any single leaf. And finally AARON employs a strategy, devised by Cohen, for ensuring that it does not draw attention to the shape of a leaf as an isolated event, but allows it to be seen as part of the whole object to which it belongs.

---

[36]Cohen would have preferred AARON to use oil paints but mixing and thinning the colours would have been beyond the capabilities of his robot, not to mention the robot's inability to clean the brushes adequately.

AARON and its robot take about four hours to colour a whole paint-ing, but the results are worth it, commercially as well as aesthetically (see Figure 43). Even at its first showing at the Computer Museum in Boston in 1995 the paintings AARON was generating were selling for $2,000. But eventually Cohen decided to put the robot to one side, because the attention it was attracting by its physical dexterity and cleverness was masking the creative side of AARON's talents.

Figure 43. *#041114* (the fourteenth image in November 2004), a painting by AARON (Courtesy of Harold Cohen)

– 6 –

How Computers Think

Most people think computers will never be able to think. That is, really think. Not now or ever. To be sure, most people also agree that computers can do many things that a person would have to be thinking to do. Then how could a machine seem to think but not actually think? Well, setting aside the question of what thinking actually is, I think that most of us would answer that by saying that in these cases, what the computer is doing is merely a superficial imitation of human intelligence. It has been designed to obey certain simple commands, and then it has been provided with programs composed of those commands. Because of this, the computer has to obey those commands, but without any idea of what is happening.

Indeed, when computers first appeared, most of their designers intended them for nothing only to do huge, mindless computations. That is why the things were called "computers". Yet even then, a few pioneers— especially Alan Turing—envisioned what is now called "Artificial Intelligence"—or "AI". They saw that computers might possibly go beyond arithmetic, and maybe imitate the processes that go on inside human brains.

Today, with robots everywhere in industry and movie films, most people think AI has gone much further than it has. Yet still, "computer experts" say machines will never really think. If so, how could they be so smart, and yet so dumb?

—Marvin Minsky (in 1982) [1]

This chapter describes the key fundamental processes of thought, as they are implemented by Artificial Intelligence methods: logical reasoning, general problem solving, planning, reasoning on the basis of common sense or by the use of precedents, learning, discovery, invention and acquiring and using expert knowledge to solve problems within a specific domain such as medical diagnosis. All of these processes, normally associated with human intelligence, can today be carried out by a computer system and will, as the twenty-first century unfolds, become integral aspects of the mental capabilities of robots.

What is Logic?

In Chapter 1 we saw how the science of logic was first mechanized in a number of devices, built mostly in the nineteenth century, to solve simple problems. To recap a little, the underlying foundation of logic is the question of whether a particular statement is true or false, and those nineteenth-century devices represented truth and falsity in a visual way that was easy for the user to follow. For example, in Jevons' abacus[1] the presence in a vertical rack of a wooden tile indicated that a statement corresponding to that tile was true.

The mechanical methods for solving simple logic problems were given a boost, in 1854, when the British mathematician George Boole devised Boolean algebra, a system of logic based on the three fundamental operators: **not, and** and **or**.[2] Boole was able to express his system of logic in mathematical formulae that could be manipulated and subsequently simplified, but he could hardly have imagined the great power his system would one day wield, becoming the basis of logic as it is used for computation. Because logic allows us to describe situations and knowledge in a way that can be understood and manipulated by computers, it allows us also to represent problems that we want to solve and to use its rules to enable us to find the solutions to these problems.

The language of logic, unlike spoken language, is precise and unambiguous. For example, the statement

Everyone in my family likes a cat

is ambiguous. Does it mean that everyone in my family likes a particular cat, or that everyone likes a different cat (one cat per person), and which cat for which person? In logic we would need to express our intended meaning of this statement in a more precise way, for example,

Everyone in my family likes at least one particular cat called Muffin.

This level of precision is necessary in computer programs if they are to avoid going down blind alleys in their search for the solutions to problems. And once we have this level of precision in our premises (the statements on which a logical argument depends), we can express the situation or problem in a precise and compact way for programming. For example,

---

[1] See the section "Nineteenth-Century Logic Machines" in Chapter 1.
[2] See the section "Early Logic Machines" in Chapter 1.

the statement

> **for all** men referred to as **A** and **for all** men referred to as **B**, if **A** is the brother of **B then B** is the brother of **A**

can be expressed in a more computer-like way as

> **for all** (A,B) [the brother of **A** is **B implies** that the brother of **B** is **A**][3]

More symbolically, and using the normal shorthand symbols of logic, where $\forall$ means "for all" and $\supset$ means "implies", we can express this as

$$\forall (\mathbf{A, B}) \, [\, \text{brother} (\mathbf{A,B}) \supset \text{brother} (\mathbf{B,A}) \,]$$

You should be able to see that this symbolic expression represents the original statement: "**for all** men referred to as **A** and **for all** men referred to as **B**, if **A** is the brother of **B then B** is the brother of **A**". But the symbolic form is more easily manipulated inside a computer than is the original statement.

Logical Reasoning

One of the most important problems in Artificial Intelligence is how to enable computers to draw logical conclusions from known facts, the process of reasoning. We can look at the task of proving a theorem in logic, or solving a problem that requires the use of logic, in the following way. A proof is a sequence of expressions such that every expression in the sequence is either one of the already accepted theorems or axioms, or an expression that can be obtained from one or two of the already accepted theorems or axioms by applying one of the rules of inference, such as that described in footnote 5.

The Logic Theory Machine

The first programmed deduction system was developed by Allen Newell, Herbert Simon and John Shaw at Carnegie Institute of Technology[4] and described by Newell and Simon during the 1956 Dartmouth workshop.

---

[3] Note that "X implies Y" means that "Y can be deduced from X".
[4] Later called Carnegie-Mellon University.

It was called variously the "Logic Theorist" and the "Logic Theory Machine", and was able to deduce the proofs to theorems in a form of logic called the propositional calculus, a system of logic propounded by Bertrand Russell and Alfred Whitehead in their classic three-volume series *Principia Mathematica* which was published in 1910–1913.[5]

The Logic Theory Machine was able to use the axioms and inference rules proposed by Russell and Whitehead, together with **and, not, or** and **implies** (which are called connectives), in order to operate on mathematical theorems that had already been proved, and thereby to generate new theorems. Starting with an expression that it wants to prove, i.e., a new theorem, and with the set of five axioms (which it had been told are true) and with the theorems it had already proved (which it knew to be true), the Logic Theory Machine would apply the three inference rules again and again until it generated the desired mathematical expression, the one that it wanted to prove.

If a computer were to be given an infinite amount of time and memory we could use the Logic Theory Machine to prove any provable theorem and to solve any solvable problem.[6] The process would simply be this:

1. Check to see if any of the existing expressions (including the original axioms and the already-proved theorems) is an exact match of what we want to prove. If so, the task has been completed and the program should stop and print out or display its proof.

2. If none of the existing axioms or already-proved theorems is an exact match of what we want to prove, apply each of the inference rules to all the existing expressions and to all pairs of existing expressions (but with the restriction that the program should not apply a rule twice to the same expression or to the same pair of expressions). This process will create new expressions that are known to be true, and the program can then go back to Step 1.

---

[5]The system was based on a set of five logical axioms—expressions that are always true, for example: [p is **true or** (q is **true or** r is **true**)] **implies** that [q is **true or** (p is **true or** r is **true**)]. In order to use these axioms to create new theorems from known ones, the system also employs three rules of inference, for example the rule of *detachment*, which states that if **A** is **true**, and if "**A implies B**" is **true**, then **B** is also **true**. Russell and Whitehead attempted to show that it is possible to derive *all* mathematical truths from their five axioms and three inference rules.

[6]Proving a theorem and solving a problem are very similar tasks from a logic perspective. In solving a problem the solver hypothesizes a solution and then tests it—this testing process is akin to proving a theorem. If the test is successful, the "theorem" (i.e., that the hypothesized solution works) has been proved.

Clearly this process, which Newell and Simon called the British Museum Algorithm, will eventually lead to a proof of the target theorem or to a solution of the given problem, provided one exists. This is because eventually the program will construct all possible proofs in a systematic manner but in a random order, checking each time to eliminate duplicate attempts and to see if the target theorem has been proved. The difficulty with this approach is that working in a random manner is not guaranteed to provide success. The process will generate far more useless sequences of expressions than useful ones, and for any difficult or complex theorem or problem the program is likely to run out of time or out of computer memory for storing all of the newly generated expressions. Nevertheless, the method is sufficiently powerful that one of the first 246 theorems generated and proved by the Logic Theory Machine was the same as one from Chapter 2 of *Principia Mathematica*, and it was proved by the program in only four steps, while three more of the theorems from the same source were proved in six steps, and one more was proved in eight steps. Thus, five out of the first 246 theorems to be generated and proved by the Logic Theory Machine, using this random approach, are amongst those published by Russell and Whitehead.

In order to improve the efficiency of the process the Logic Theory Machine needed a way to radically improve the order in which possible proofs of theorems and solutions to problems were generated. Three methods were developed for searching for a proof or solution, called substitution, detachment[7] and chaining.

The substitution method finds an axiom or a previously-proved theorem that can be transformed into the expression that needs to be proved. The transformation can be accomplished, for example, by a series of substitutions that are similar to those employed in proofs in high school algebra.

The detachment method replaces a problem expression by a new subproblem, that has been chosen so that if the new sub-problem is solved, its solution provides a proof for the original problem expression.

The chaining method is based on two techniques called forward chaining and backward chaining. If the problem expression is "**A** is **true implies** that **C** is **true**", forward chaining is used to search for an axiom or theorem of the form "A is true implies that B is true". If one is found then the expression "B is **true implies** that **C** is **true**" is set up as a new

---

[7] See also footnote 5.

sub-problem, because the program knows that this new sub-problem can form the second part of a chain with "**A** is **true implies** that **B** is **true**", resulting in the following proof:

> **A** is **true**
>
> But **A** is **true implies** that **B** is **true**
>
> Therefore **B** is **true**
>
> But **B** is **true implies** that **C** is **true**
>
> Therefore **C** is **true**
>
> So **A** is **true implies** that **C** is **true**

Backwards chaining works in a similar way. Given the problem of proving that "**A** is **true implies** that **C** is **true**", the backward-chaining method seeks a theorem of the form "**B** is **true implies** that **C** is **true**" and, if one is found, the expression "**A** is **true implies B** is **true**" is set up as a new sub-problem, because solving this new sub-problem, combined with the knowledge that "**B** is **true implies** that **C** is **true**", provides a solution to the original problem.

Each of these three methods of improving the efficiency of performance of the Logic Theory Machine is independent of the others. They can be used in sequence, one working on the sub-problems generated by another, resulting in the various sub-problems becoming intermediate expressions in a proof sequence.

The proof process is rather like growing a tree. Each application of each method can be thought of as a branch of the tree. At the end of each branch is the updated list of axioms, theorems and sub-problems. This whole process continues until a proof has been found, or there is no more available computer time, or the computer runs out of memory, or there are no untried problems on the sub-problem list (in which case the program will never find a proof). In the days of Newell and Simon's original work on the Logic Theory Machine, the prospect of running out of time was a real one—a simple proof of four steps required about ten seconds, while one example of five steps needed about 12 minutes. Clearly deep and complex proofs would have taken forever in 1956, but in fact the program was able to prove, within a reasonable amount of time, 38 of the first 52 theorems found in *Principia Mathematica*.

The Resolution Principle and Beyond

It is easy to see from the above description of the Logic Theory Machine that when searching for a proof of a particular theorem, the process of generating more and more intermediate theorems can easily get out of hand. The British Museum Algorithm is the extreme example of this syndrome. Researchers in this field have therefore attempted to reduce the search space by means of various heuristic techniques. One very effective rule for achieving this is the Resolution Principle devised by Alan Robinson in 1965. Put simply, this principle enables the elimination of redundant and conflicting statements.

Although Alan Robinson's Resolution Principle achieved a certain measure of success at improving the efficiency of the theorem-proving process, researchers in the late 1960s and the 1970s were unable to develop programs that could solve difficult problems or prove difficult theorems. For any non-trivial theorems the tree of intermediate expressions was enormous and little success was achieved in finding powerful heuristics to reduce the magnitude of the problem. It was also discovered that purely deductive logic was not up to the task of solving many problems. What *did* advance this particular niche within AI was the advent of computers that were much more powerful and had much larger memories than those used during the 1950s, 1960s and even the 1970s. The use of faster, bigger computers has since enabled some notable successes in automatic theorem proving, including one that has been of great interest to mathematicians.[8]

Problem Solving

Most of the early work on deduction by computer was focussed on automatic theorem proving, a highly specialized task that of itself is of little interest outside the rather rarified world of higher mathematics. But the real promise of this work, in terms of AI, a promise created by the similarities between theorem proving and problem solving, was the

---

[8] In 1996 there was an announcement from the Argonne National Laboratory in Illinois that a famous open conjecture in mathematics had been proved by a theorem proving program called EQP, developed by William McCune. The wording of the announcement from Argonne was "The Robbins problem—are all Robbins Algebras Boolean?—has been solved: Every Robbins algebra is Boolean." In itself this statement may not engender much excitement in those who do not have the benefit of a PhD in mathematics, but the fact that this conjecture had first been stated in the 1930s, but had lain unproved for 60 years, was a great triumph for AI.

automation of reasoning in the solution of commonsense problems. This was a dream of John McCarthy in the early days of AI.

The Advice Taker

McCarthy's 1959 paper, "Programs with Common Sense", presented an intriguing view of how computer programs might be made to think like human beings. The simplest program proposed by McCarthy was called the Advice Taker,[9] and was based on the idea of drawing immediate conclusions from a list of premises. The Advice Taker was a method proposed for solving problems by manipulating sentences in formal languages, such as the language of logic described above. And the Advice Taker was unusual, for those days, in that it allowed the user to improve the program's performance "in real time" (i.e., while it was working on a problem) by giving it advice in the form of true statements that it might find useful. McCarthy's idea was that, when it received a piece of advice, the program would already have generated several new and true statements from those with which it started the problem solving process. By adding a new piece of advice, the user would encourage the program to combine this new statement with each of the other true statements it already had in its memory, thereby allowing it to create even more true statements. It might at first appear that encouraging the program to grow its "tree" of true statements in this way would lead to it running out of memory space or time. But the fact that a human was providing the new advice would mean that the human would, in effect, be guiding the program towards a solution.

The ultimate objective of the Advice Taker project was to make programs that learn from their experience as effectively as humans do. In McCarthy's opinion, a system that is to evolve intelligence akin to that of humans should have at least the following features:

1. It must be possible to represent all behaviours of the system in the language of the system.

2. Interesting changes in behaviour must be expressible in a simple way.

3. All aspects of behaviour except the most routine must be improvable. In particular, the improving mechanism should itself be improvable.

---

[9]The Advice Taker was a program concept. It was never actually written.

4. The machine must have or evolve concepts of partial success because on difficult problems decisive successes or failures come too infrequently. The system should therefore be able to assess, for any statement proved by the system, how near that statement is to (or far from) a solution or proof.

5. The system must be able to create shortcuts. The learning of shortcuts is complicated by the fact that the effect of a shortcut is not usually good or bad in itself. Therefore, the mechanism within the system that selects shortcuts should be able to identify interesting or powerful shortcuts which might be useful to the system.

McCarthy illustrates the way the Advice Taker is supposed to act by means of the following example. Assume that I am seated at my desk at home and I wish to go to the airport. My car is parked at my home, so to solve the problem I can walk to my car and drive it to the airport. McCarthy's paper presents a formal statement of the premises used by the Advice Taker as the basis for its solution to this problem and explains the interpretation of each group of premises. Here I have paraphrased his explanation for the sake of simplicity. The "program" presented by McCarthy is written in the style of the programming language LISP, a language conceived by McCarthy in 1956 during the Dartmouth workshop specifically for use in Artificial Intelligence work.[10]

1. "at(x,y)" is a formal way of writing "x is at y".

Under this heading we have the premises:
at(I, desk)
Meaning: I am at my desk.

at(desk,home)
Meaning: My desk is at my home.

at(car, home)
Meaning: My car is at my home.

at(home,county)
Meaning: My home is in a particular county.

---

[10]LISP caters for computing with symbolic expressions (such as those in this example of the Advice Taker) rather than with numbers, and it allows symbolic expressions and other information to be represented in the computer's memory in the form of lists, hence the name "LISP" (for list processing language).

at(airport, county)
Meaning: The airport is in that particular county.

2. The Advice Taker also needed to know that, if A is at B and B is at C then A is at C. So if I am at my desk and my desk is at home then I am at home. In logic this relationship is expressed, using the "implies" symbol ⊃ thus:

at(x,y), at(y,z) ⊃ at(x,z)

3. There are two rules needed by the Advice Taker concerning the feasibility of walking and driving. They are:

walkable(x), at(y,x), at(z,x), at(I,y) ⊃ can(go(y,z,walking)

and

drivable(x), at(y,x), at(z,x), at(car,y), at(I,car) ⊃
$$\text{can(go(y,z,driving)}$$

Meaning: The first of these rules means that if the journey to x is walkable [indicated by **walkable(x)**], and if y is at x [indicated by **at(y,x)**], and if z is at x [indicated by **at(z,x)**], and if I am at y [indicated by **at(I,y)**], then all of these being true implies [shown by the ⊃ symbol] that I can walk from y to z [indicated by can(go(y,z,walking)]. This is easy to understand—I am at y and the journey to x is walkable so I can walk to x, and since z is at x, by walking to x I also walk to where z is located.

The second of these rules has a comparable meaning, relating to driving from y to z.

The system also needs two specific facts:

walkable(home)
Meaning: The journey to home is walkable.

drivable(county)
Meaning: Anywhere in the county is drivable.

4. Next there is a rule concerned with the properties of going:

did(go(x, y, z)) ⊃ at(I,y)
Meaning: I did go from x to y by method z implies that I am now at y.

5. And the problem to be solved is expressed by:

want(at(I,airport))
Meaning: I want to be at the airport.

The problem-solving process employed by the Advice Taker leads to the following step-by-step solution:

at(I,desk) ⊃ can(go(desk, car, walking))
Meaning: I am at my desk implies that I can go from my desk to my car by walking.

at(I,car) ⊃ can(go(home,airport,driving))
Meaning: I am at my car implies that I can go from my home to the airport by driving.

did(go(desk, car, walking)) ⊃ at(I, car)
Meaning: I did go from my desk to my car by walking which implies that I am at my car.

did(go(home, airport, driving)) ⊃ at(I, airport)
Meaning: I did go from my home to the airport by driving which implies that I am at the airport.

can_ultimately_achieve (at(I, desk), go(desk, car, walking), at(I,car))
Meaning: I am at my desk and I can go from my desk to my car by walking, implies that I can ultimately achieve being at my car.

can_ultimately_achieve (at(I,car), go(home, airport, driving), at(I,airport))
Meaning: I am at my car and I can go from my home to the airport by driving, implies that I can ultimately achieve being at the airport.

can_ultimately_achieve (at(I, desk), action(go(desk, car, walking), go(home, airport, driving)) ⊃ at(I, airport))
Meaning: I am at my desk and I can go from my desk to my car by walking, following which I can go from my home to the airport by driving, which together imply that I can ultimately achieve being at the airport (which is my goal).

This example of the Advice Taker in action demonstrates how everyday problems can be expressed in a symbolic form that can be understood by computer programs based on the rules of classical mathematical logic. And the example of the Logic Theory Machine demonstrates how computers can manipulate such logic expressions in order to prove theorems,

which is analogous to solving problems. Because mathematical logic lends itself so readily to automation, many problems can be solved by representing them and their relevant background information as logical statements (axioms) and treating the solutions of the problems as though they were the proofs of theorems based on these axioms.

Different Directions in Problem Solving

During the 1970s there was flurry of interest amongst researchers in a new type of programming called logic programming, and in a programming language called PROLOG[11] that showed some promise of becoming a panacea in the field of problem solving. It was realised that the way in which knowledge is converted into formal expressions and axioms in logic can affect the manner in which an automatic deduction process operates. And Robert Kowalski pointed out that the process of making decisions as to how to convert knowledge into logical forms is akin to the process of writing programs in a conventional programming language. This led him to the observation that many logical operations can be formalized in such a way that, when the methods of automated deductive reasoning are applied to such operations, the effect is the same as when executing these operations in computer programs. This was the thinking behind the development of logic programming and the invention of the PROLOG programming language.

The power of PROLOG for problem-solving tasks lies in the way in which knowledge can be represented in the language. For example, the PROLOG statements

$$parent(X,Y) :- mother(X,Y)$$

and

$$parent(X,Y) :- father(X,Y)$$

mean that if Y is the mother of X then Y is a parent of X and if Y is the father of X then Y is a parent of X, so the meaning of

$$grandfather(X,Z) :- parent(X,Y), father(Y,Z)$$

is that if Z is the father of Y and if Y is a parent of X, then Z is a grandfather of X. This example illustrates how the manner in which information

---

[11] For Programming in Logic.

is represented in PROLOG contributes to the ease with which the language can manipulate information in order to arrive at the solution to a problem.

The General Problem Solver

Another early attempt at automatic problem solving was a program called GPS developed by Allen Newell and Herbert Simon at Carnegie Institute of Technology during the late 1960s. GPS employed an approach called means-end analysis, which is based on the idea of breaking down the goals of a problem into sub-goals and then attempting to solve each sub-goal. If a difference is detected between the current state of a problem and a goal, then a sub-goal is created by minimizing or otherwise reducing the most important difference between the goal and the current state. For example, let us assume that the problem I want to solve is that I am hungry and I want to be full. What is the difference between my current state (being hungry) and the state I want to reach (being full)? The answer is an empty stomach. What changes the emptiness of my stomach? Eating food. So eating food becomes a sub-goal. The process then determines whether I can eat food, in order to satisfy the new sub-goal. If I do not have any food, what is the difference between what I have (no food) and what I want to have (food)? The answer is the presence of food. So achieving the presence of food becomes another sub-goal on the path to solving the original problem. And so on.

Solving problems in this way becomes an exercise in growing trees of the various goals and sub-goals, hopefully leading eventually to a solution path from the original problem state to the original goal. If the original problem is solvable, then eventually the goal state will be generated. The solution to the original problem then consists of carrying out the various actions that convert the original problem state to the goal state.

A serious difficulty in using this approach for complex problems is that, as the number of sub-goals in the search tree increases, so the number of possibilities for creating new sub-goals also increases, and unless a program has the benefit of extremely powerful heuristics for encouraging it to search for a solution in sensible ways, the growth of the search tree can get out of hand, consuming vast amounts of computer time and, in some cases, computer memory.

During the heyday of GPS in the late 1960s, the project was being viewed optimistically within the AI community as a breakthrough in the

realm of automatic problem solving. But, without infinite resources of computer time or memory, it became essential to find ways of managing the expansion and search of the goal/subgoal tree so that a solution could be found economically. This ultimately proved impossible to achieve, causing the generality of the aim of GPS to be its undoing. Given any specific problem, such as solving a Rubik Cube, heuristics could be developed that would enable a program to focus on the most promising paths through the tree and the most important goals, thereby keeping the size of the tree in trim. But it proved to be simply impossible to find such solutions for every tree. Going from the specific to the general was too difficult for GPS, which always needed to be primed with large amounts of knowledge about the specific domain in which it was being asked to solve a problem. There was no general strategy available that would automatically lead to the solution of every problem.

Planning

With a decline in the belief that GPS or some other system might lead to the creation of programs with general problem solving capabilities, the attention of many within the AI community shifted from this utopian dream to the subject of planning. Creating a plan involves determining all the small tasks that must be carried out in order to accomplish a goal, and the achievement of a suitable plan is therefore synonymous with solving a problem. If your goal is to buy a pint of milk, superficially a trivial task, there might be several small tasks involved in achieving that goal. These might include finding your car keys, finding your wallet, starting your car, driving to the grocery store, finding milk on the shelves at the store, picking up the milk, taking it to the checkout, paying for it and finally driving home. The process of planning takes into account various constraints (or rules), that affect when certain tasks in your list can or cannot happen. In this simple example, the constraints include finding your car keys and wallet before you drive to the store and, once you are there, finding milk on the shelves before you take it to the checkout.

The main benefits of planning include reducing the amount of search required to solve a problem, resolving possible conflicts between different goals, and providing a basis for enabling a program to recover from errors. One of the earliest planning systems to gain recognition within the AI community was STRIPS, developed in the early 1970s by Richard

Fikes and Nils Nilsson at Stanford Research Institute. STRIPS is a simple model for expressing planning problems that need to be solved on behalf of a hypothetical robot operating in a "world" of several rooms, with doors, some boxes and some other objects that the robot can manipulate. STRIPS represented this world by a set of facts expressed in first-order predicate calculus—the logic of Russell and Whitehead's *Principia Mathematica* described earlier in this chapter.

The STRIPS robot can implement a plan of action by means of *operators* such as going somewhere or moving something. Each operator works on the basis of preconditions, for example if the robot needs to go from A to B it must first be at A, and if it wants to move an object it must first be next to that object. The traditional form of STRIPS-based planner starts with a goal, looks at what preconditions must exist for that goal to be achieved, and then decides what action will come *last* in the planning process in order to achieve that goal. It does this by finding an action that directly leads to the goal state. It then needs to determine whether the robot will be able to take that action, so it must ensure that the preconditions of that action are satisfied. It does this by adding those preconditions to its list of what it needs to achieve.

The planning process proceeds in this fashion, looking at the facts in its new goal (i.e., the new list of what it needs to achieve), finding another action that achieves some of these facts, adding the preconditions of that action to the list of what it needs to achieve, and so on. If STRIPS encounters an action that will not achieve any of the facts that make up its goal, then it admits defeat in this particular part of the plan (for the time being at least) and backtracks, before choosing a different action. A STRIPS planner as described here could be improved in its performance by selecting actions for consideration in a hopefully intelligent order, for example with those having fewer preconditions being examined first. This heuristic is based on the simple notion that the more preconditions there are, the greater the amount of work that the planner has to do in order to achieve the goal.

Although the terminology differs, the STRIPS approach to planning is actually nothing more than another search problem. This type of planner is usually referred to as a backward search—one in which the search process starts with the goal and works back towards the original problem. The opposite approach, called forward search, starts from the original problem and plans in a "forward" direction, towards a solution. In this respect a forward planner bears considerable similarities to the

methodology of GPS, growing a tree of possibilities in the search for a solution.

In common with many other types of search problem in AI, planning gives rise to enormous tree growth, as programs are required to search vast problem spaces. During the 1990s, the technology of automatic planning proliferated, producing several powerful techniques that overcome the problems presented by vast search spaces. As planning methods grew more sophisticated, the types of problem with which planner systems were faced grew more intricate. Not only were planners being required to cope with relatively straightforward problems, those that could be solved completely on a step-by-step basis *before* the first step was actually undertaken, new types of problem were attracting interest in the world of AI, problems that involved more than one robot and problems that required certain steps to be taken before the later steps in the plan could even be considered.

Ruth Aylett and her team at the University of Salford have developed a software design for a robot that must plan continuously, as for example the robots in a soccer team.[12] In her design, the robot's goals are generated as a result of its motivations. Having autonomy in the goals it selects makes the robot's planning considerably more sophisticated— the planning process interleaves planning steps with execution steps and sometimes a planning step can involve delegating a task to a different robot.

Aylett's software architecture is designed to provide continuous planning for a robot by meeting the following requirements:

1. Planning and executing are interleaved—the robot plans a step and then examines its situation as it will be if that step is executed.

2. The planner is able to accept new goals generated by the robot at any time.

3. Time passes while the robot plans and executes, so the planner must be able to reason about time.

4. Actions are planned for many robots so the planner must be able to reason about which robots should carry out which actions.

5. It may not always be possible to achieve all current goals within the given time, so it must be possible to prioritise outstanding goals. It

---

[12]See the section "Robot Soccer" in Chapter 8.

must also be possible to remove goals that cannot be met from the planning process, along with any partial plans and constraints they required.

6. The robot has long-term motivations, which enable it to generate goals, prioritise them, and select the best plan to achieve them.

The key to the overall architecture shown in Figure 44[13] is the robot's motivations, which can be thought of, depending on the problem domain, as its long-term aims or objectives, or as its drives or emotional states. For example, a robot might be motivated by hunger (the need to recharge its batteries), by tiredness (the need to rest in order to cool down its motors), or by curiosity (the need to discover knowledge that might be relevant to a problem solving task).

A robot's motivations also have another purpose, allowing the robot to evaluate the plans it generates in order to achieve its goals, and to choose between alternative plans. For example, a plan to acquire food that involves a robot walking to the shops to buy a pizza, might conflict with its tiredness (lack of battery power) and cause it to explore an alternative plan that involves cooking something it already has at home.

An important question here is, "when exactly should a robot's motivations be updated and new goals generated?" In Aylett's design this happens only after an action has been executed, but in certain types of situations, and robot soccer is a good example, things happen so quickly that it is necessary to update a robot's motivations even before one of its actions has been completed. In soccer, when a player has decided to pass the ball to a particular teammate, immediately after he has kicked the ball the passing player will plan what he is going to do, and it is only by such anticipatory action that a player can maximize his usefulness on the soccer field. Thus a robot's actions can affect its motivations. The specification of an action also determines which robot (or robots) is (or are) intended to execute it.

The flow of the planning algorithm can be followed by reference to Figure 44. A robot first chooses the best partial plan at its disposal. When deciding whether to plan to satisfy a goal or to execute an action, the robot first selects the "best" goal from those that it needs to achieve, and then determines which actions are ready to be executed. In selecting

[13] Figure 1 of R. S Aylett, A. M. Coddington, and G. J. Petley. "Agent-Based Continuous Planning" *19th Workshop of the UK Planning and Scheduling Special Interest Group (PLANSIG 2000)* (http://mcs.open.ac.uk/plansig2000/Papers.htm).

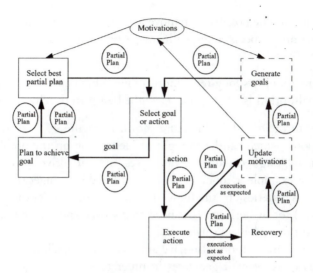

Figure 44. Ruth Aylett's overall planning architecture (Courtesy of Ruth Aylett)

the best goal the robot assesses the importance of each goal, estimates the effort required to execute each, and determines whether there is an urgent deadline for one or more of its current goals. A decision will be taken to execute an action rather than to satisfy a goal if there is an action at the top of a partial plan, an action for which all of its preconditions are currently true and one whose immediate execution is appropriate at the current time. The priority value for an action is calculated in the same way as for a goal, namely taking into account its importance, the effort required to carry it out and whether it has an urgent deadline.

If a goal has been selected, rather than an action, the planning process starts up and a new partial plan is added to the list of partial plans awaiting execution, unless the goal of this new partial plan cannot possibly be achieved. When the new partial plan has been created, the system estimates deadlines for all of its actions, based on its knowledge of how long each action will take to execute. When all of these time estimates are known, the system checks to see whether there is sufficient time in which to meet all of its goals, failing which the user is offered the opportunity of editing the plan manually in order to change the time constraints imposed on satisfying the main goal, and if the user rejects this option the current plan is rejected and the goal it was hoping to achieve is removed from the list of active goals. If on the other hand there *is* sufficient time for the plan to meet the main goal, the goal it now knows can be achieved

is removed from the current goal list because the system can be certain that that goal will be satisfied during the execution of the plan. There will then be one goal less to achieve during the remainder of the plan.

If an action was selected rather than a goal, the system determines what time it will be and what will be the state of the robot's world after this action has been executed. If the time is too late for a deadline or if the state of the robot's world is not to its liking, some part of the plan may be faulty. If the system finds this to be the case it will be because one or more goals that should already have been satisfied have not been, in which case these goals are added back in to the partial plan to which the action belonged, which is also updated by removing the executed action from its list. This re-planning process is an integral part of planning and execution, since these unsatisfied goals will require additional planning activity, and the subsequent actions that depended on them will not be executable until this additional planning has been carried out.

Once an action has been proposed for execution in a plan, the robot's motivations are updated in order to reflect the changes in its world. Also the planning system's clock is updated to reflect the amount of time taken to execute the action. The system then returns to the start of the algorithm and continues until an entire plan has been created and found to work within the allotted time.

Commonsense Reasoning

In 1984 Doug Lenat began work on a new approach in automatic reasoning, an incredibly ambitious and courageous research project aimed at creating a massive database of real-world information that could serve as the basis for commonsense reasoning. When it started, the project, called Cyc, was a very long-term, high-risk gamble, which in 1994 led to the incorporation of Cycorp, Inc., the company that markets the system. During the first 20 years of the project, many tens of millions of dollars were invested by U.S. Government funding agencies and others, and more than 1.5 million pieces of knowledge were acquired.[14] The underlying idea was to employ computer inferencing methods on the mass of knowledge being collected, in order to provide the computer with

---

[14]The figure of 1.5 million dates from early 2002. It increases as more knowledge is added but it is reduced as similar items of knowledge within the system are merged.

commonsense reasoning. Thus Cyc is a kind of expert system,[15] in which the knowledge is employed by an inference engine in order to answer a user's queries.

In April 2000 Lenat gave a demonstration of Cyc to officials from the U.S. Department of Defense. At the request of one of the officials, Lenat asked his program about anthrax, whereupon Cyc requested that he clarify whether he meant the band,[16] the bacterium, or the disease caused by the bacterium? This somewhat surprising response illustrates the breadth of Cyc's knowledge, but the program's next response was even more surprising, and at the same time uncannily prescient. When Lenat told Cyc that he meant the bacterium, and asked the program to comment on its toxicity to people, Cyc's response included the prophetic remark that one could use FedEx to mail packages containing anthrax to high-ranking government officials. Eighteen months later the dreaded bacterium started showing up in the mail.

Cyc's knowledge base consists of a vast quantity of fundamental human knowledge—facts and rules of thumb for reasoning about the objects and events of everyday life. Information is given to the program using a formal language called CycL, which is far from ideal for everyday users but which allows Cycorp's staff to impart knowledge to the program in a format that it already understands. The knowledge itself is made up of many thousands of "micro-theories", each of which is essentially a group of assertions that share a common set of assumptions. Some of these micro-theories are focused on a particular knowledge domain, some on particular levels of detail, some on particular periods in time. New knowledge is continually being added manually and in addition the system has the capability to automatically create facts from those that it already knows, and to infer new facts.

Despite its apparently vast amount of knowledge and some impressive examples of its capabilities, Cyc appears to be only scratching the surface of the sum total of human knowledge. Partly prompted by this realisation Push Singh, a PhD student of Marvin Minsky at the MIT Media Lab, started the Open Mind Common Sense Project in the fall of 2000. Singh has quantified commonsense knowledge thus:

> Common sense is a problem of great scale and diversity. How big is
> common sense? The scale of the problem has been terribly discour-

---

[15] See the section "Expert Systems" later in this chapter.
[16] The heavy metal rock group based in New York City.

aging. Those courageous few who have tried, have found that you need a tremendous amount of knowledge of a very diverse variety to understand even the simplest children's story. Many researchers regard the problem of giving computers common sense as simply too big to think about. There have been several attempts to estimate how much commonsense knowledge people have. These attempts range from experiments that demonstrate that people can only acquire new long-term memories at the rate of a few bits per second, to counting the number of words we know and estimating how much we know about each word, to estimating the brain's storage capacity by counting neurons and guessing at how many bits each neuron can store. Several of these estimates produce approximately the same number, suggesting there are of the order of hundreds of millions of pieces of commonsense knowledge. This is a huge number. [2]

From the estimates quoted by Singh it is clear that, admirable though the efforts of the Cyc project have been, 1.5 million pieces of knowledge is a mere drop in an ocean of several hundred million. Singh's fundamental goal, therefore, is to make much more manageable the task of acquiring all this knowledge in a format that a computer can understand. The real problem is that computers do not know anything about us, how we think and behave, or what we can achieve. They know nothing of the patterns of our lives, the homes, offices, schools and other places where we spend much of our time, of inter-personal relationships, of our hopes and fears, our likes and dislikes, our personalities or our emotions. So Singh concluded that we should teach them. Yes, you and I! His idea is to tap into the vast volunteer workforce of humankind that has access to the Internet.

Many of the millions of ordinary pieces of knowledge we acquire are so obvious that we take them for granted: the sun rises in the morning and sets in the evening; 12:30 p.m. is later than 12:15 p.m.; a cheetah runs faster than a man. In order to accelerate the process of teaching computers all this knowledge Singh decided to make a manual attempt to build a system with common sense, a problem he described as being "as large as any other that has been faced in computer science to date".

Singh's idea of inviting everyone to participate in his project takes its impetus from the fact that some 100 million of us have access to the Internet. This makes it possible for millions of people to collaborate on the project, to help build the mother of all databases. Anyone with

Internet access, even though they may not have any training in computer science or Artificial Intelligence, can participate in the Open Mind database project. If everyone with access to the Internet contributed just one (different) piece of knowledge, the task would be well on the way to being completed. So Singh and his colleagues built a Web site located at http://www.openmind.org/commonsense/, and in its first 15 months of life more than 8,000 volunteers contributed many hundreds of thousands of pieces of knowledge. This was still only a quarter of what was in Cyc at that time, but the knowledge in Open Mind was acquired in less than one-tenth of the time and at a tiny fraction of the cost.

Because the Open Mind project relies on the general public, knowledge is input in plain English sentences, rather than in a precise but difficult-to-use language[17] such as is employed for the Cyc project. But the manner in which Singh's project uses plain English has only been feasible in recent years, made possible by the advances in the techniques for Natural Language Processing, which are now at a level good enough to translate a large proportion of the sentences people have supplied to the system into the syntax understood by the software.

Case Based Reasoning

Possibly the main disadvantage of any form of reasoning based purely on logic, is that the process of searching for a solution to a problem does not benefit significantly from any kind of knowledge. Instead the process relies on hopefully intelligent heuristics guiding the creation of the next step in the solution process, and growing a tree of possibilities until a solution to the original problem is found.

In the early 1980s, Roger Schank at Yale University investigated the role that a memory of previously occurring situations can play in problem solving. This is how we often think when we are faced with a problem. A Chess grandmaster will often encounter a position on the board that is similar to positions with which he is familiar, and he knows from his past experience, his memory of past positions, what moves are the most promising ones. An airline pilot encountering particularly bad turbulence in an area will often know the geography and weather patterns of the area well enough to be able to decide correctly in which direction he should head or to which height. In almost every domain of problem solv-

---

[17] Difficult for anyone who is not familiar with the CycL language.

ing, past experience will often be of great assistance in finding a solution to a problem.

Schank's early work in this field led to a model for what has become known as Case Based Reasoning—the process of reasoning based on knowledge of past cases, on precedents. Case Based Reasoning operates on a knowledge-base of past cases, attempting in the first instance to find a *source case* that is relevant to a given *target case*. One problem domain ideally suited to this form of reasoning is the law, since laws are founded very much on the use of precedents, so it is hardly surprising that Schank's work in the 1980s prompted interest from those researchers who had taken an interest in how AI might be used to aid legal reasoning.

One early example of a Case Based Reasoning system was Katia Sycara's PERSUADER program, developed in 1990, which reasoned about negotiations between a workforce and management. When faced with a problem in this domain, PERSUADER searched its knowledgebase of past cases, which included agreements between parties who were in similar situations to those in the dispute currently under consideration. Those earlier agreements suggested various proposals that might be successful in the current negotiations, allowing PERSUADER to compare the various past situations with the current dispute (as regards rates of pay and other features) in order to find the closest match.

Retrieving the most similar case or cases from a program's knowledgebase is the first of four stages in Case Based Reasoning programs. The whole cycle of processes is shown in Figure 45.

After retrieving one or more previously experienced cases from its knowledge-base,[18] a program reuses its knowledge of the relevant previous case(s) by revising (i.e., amending) the earlier solution(s). Sometimes an earlier case will be so similar to the current problem that the program need make only the smallest of changes in order to be able to apply the earlier solution and thereby find a solution to the current problem, while on other occasions more drastic changes will need to be made to an earlier solution before it is reusable. Finally, whatever parts of the new case are likely to be useful for future problem solving are retained in the knowledge-base.

As shown in Figure 45, the program's knowledge-base incorporates not only knowledge that is specific to past cases, but also some knowledge of a more general nature but still pertaining to the same problem

---

[18] Also called a case-base.

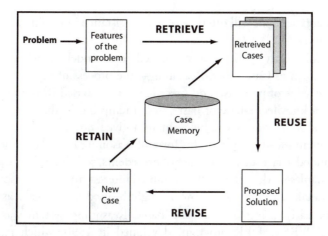

Figure 45. The Case Based Reasoning Cycle, from "Çase-Based Reasoning: Foundational Issues, Methodological Variations, and System Approaches" by A. Aadmodt and E. Plaza (*AI Communications* vol. 7, no. 1, 1994, pages 39–59) (Courtesy of Agnar Aamodt)

domain, which is often useful knowledge in the Case Based Reasoning process. For example, when a program is diagnosing a medical condition in a patient by retrieving and reusing cases of previous patients, a model of anatomy is clearly useful, together with a knowledge of relationships indicating how various pathological states are connected.

Case Based Reasoning is now the most popular form of reasoning employed within the field of AI. One research project currently showing great promise is the Sherlock Holmes program being developed by Jeroen Keppens at the Joseph Bell Centre for Forensic Statistics and Legal Reasoning at Edinburgh University. A known problem faced by investigators is that police officers often jump to premature conclusions. In general, they have a tendency to decide at a very early stage of an investigation on the most likely suspects, decisions that bias the officers' subsequent investigations. In the case of a suspicious death, for example, it is human nature for police officers to latch onto a particular theory and then attempt to confirm it, an approach that might well divert them from what actually happened.

The Sherlock Holmes program uses Case Based Reasoning to decide on the most likely cause in instances of suspicious death. The program is designed to highlight less obvious lines of inquiry that police detectives might overlook, taking an unbiased overview of all the available evidence

before speculating on what might have happened. A knowledge base within the program contains data from previous cases of various causes of death, together with evidence that can either support or contradict a particular explanation. When the investigating officers have entered data into the program, such as eyewitness accounts, medical and forensic evidence, Sherlock Holmes applies the knowledge from previous cases in order to indicate the likelihood of each of a number of possible scenarios. In this way, the program encourages the police to consider all feasible possibilities, rather than leap to an obvious and possibly false assumption.

How Computers Learn

Humans and other animals can be observed in their gradual acquisition of new responses to an old situation, thereby demonstrating an essential feature of intelligence—the ability to learn from experience. Any self-respecting robot must therefore be endowed with appropriate learning capabilities.

The study of Machine Learning dates back to 1943, and by the time of the Dartmouth workshop in 1956 the topic had already been embraced by some of the early AI researchers. Samuel's Checkers program,[19] for example, employed two different forms of learning, one of them a simple rote learning mechanism and the other, temporal difference learning, a much more sophisticated idea that is still in use today. Here we shall examine three additional types of learning mechanism: reinforcement learning, artificial neural networks and genetic algorithms.

Reinforcement Learning

The idea behind reinforcement learning is extremely crude but rather effective—reward success and punish failure. Employing this concept in a computer system appears to have been first suggested by Pierre de Latil in his groundbreaking 1953 book *La Pensée Artificielle.*[20] De Latil's idea was described using a simple game such as Tic-Tac-Toe[21] as an example. The machine would keep track of the positions it encountered and the moves it played during each game. If it lost a game it would discard all

---

[19]See the section "Checkers (Draughts)" in Chapter 3.
[20]Published in translation (by Y. M. Golla) in 1956 as *Thinking by Machine: A Study of Cybernetics.*
[21]Known in the U.K. as Noughts-and-Crosses.

the information about its moves in that particular game, but if it won the game it would retain the information and use it to reinforce the information it had previously acquired. This reinforcement process increased the weight of the moves played in winning games. As the number of games it played increased, so would the amount of data it built up for each position that it encountered in the games that it won, with the best moves building up the largest weights. Whenever it encountered a position, the machine would select the move corresponding to the largest weight of reinforcement data, and gradually the machine would acquire more and more reinforcement data for those moves that it played in the games that it won, leading it to play the best moves most often and the worst moves least often.

De Latil later proposed a modification to his algorithm. Rather than discard the information pertaining to the games that it lost, the machine should collect that data and employ it to reduce the likelihood of it repeating bad moves. Just as the good moves received reinforcement in the form of an increased weighting, so the bad moves would have their weightings reduced.

Donald Michie employed de Latil's idea in the construction of MEN-ACE,[22] a Tic-Tac-Toe playing system constructed as an assemblage of matchboxes (see Figure 46[23]). MENACE consisted of 288 matchboxes, each one corresponding to one of the 288 essentially different positions with which the first player can be confronted in a game of Tic-Tac-Toe.

Each of the boxes functions as a separate learning machine, tasked only with making a decision when its own unique position arises in a game. In each box there were a number of coloured beads, the various colours representing codes for the different locations on the board where MENACE could make a move. When a particular board position was encountered in a game, MENACE's operator would open the appropriate matchbox and shake it. Each matchbox had a V-shaped cardboard fence fixed in the front, so that when the box was tilted forward one of the beads was selected at random by being the first bead to roll into the apex of the V. The operator would then make the corresponding move in the game, leaving the box open, with its randomly selected bead visible, until the end of the game. If MENACE won the game then all of the boxes corresponding to the moves it had made in that game would

---

[22] MENACE: Matchbox Educable Noughts And Crosses Engine.

[23] This figure originally appeared on the page facing page 137 of *Machine Intelligence* 2, edited by Ella Dale and Donald Michie, Oliver and Boyd, Edinburgh, 1968.

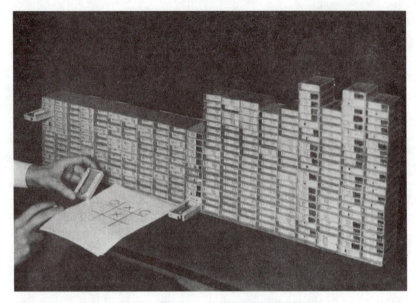

Figure 46. The original matchbox version of MENACE (Courtesy of Donald Michie)

have the chosen colour reinforced by increasing the number of beads of that colour. But if MENACE lost a game, the numbers of beads of those colours corresponding to the moves MENACE had made in that game would be reduced.

Michie describes how he tested MENACE by playing a "tournament" of 220 games against it, occupying two eight-hour sessions on successive days.

> By the end of the first 20 plays the machine was settling into a stereotyped line which ensured a draw in face of 'best strategy'. I therefore resorted to a series of theoretically unsound variations, in order to draw the machine into unfamiliar territory. Each of these paid off for a time, but after 150 plays the machine had become capable of coping with anything, in the sense that whatever variations I employed I could not get a better average result against it than a draw. In fact after this point I did much worse than this, by unwisely continuing to manoeuvre in various ways. The machine was by then exploiting unsound variations with increasing acumen, so that I would have done better to return to 'best strategy' and put up with an endless series of draws, or retire from the tournament. This I eventually did after sustaining 8 defeats in 10 successive games. At every stage, I used what tactics I judged to be the most hopeful.

It is likely, however, that my judgement was sometimes impaired by fatigue. [3]

Artificial Neural Networks

The neurons in the human brain are rather simple processors, each of which receives input signals from many other neurons, then merges the information from these input signals, and finally sends out many signals to other neurons based on this merged information. It is the network of these neurons, the huge number of neurons and their connections, that create the immense processing power of the human brain.

Artificial Neural Networks (ANNs), often called simply "neural networks" or "neural nets", are a computer-based representation of how psychologists and neurologists believe this process works in our brains. The first artificial neurons were devised in 1940 by Warren McCulloch and Walter Pitts, who were able to prove that, by linking together several simple processing units (artificial neurons) to form a network, it is possible to create in the whole network more computational power than one would expect from the sum of its units. Each unit simulates a biological neuron, receiving one or more inputs and producing an output based on the combined information from these inputs.

In 1957 Frank Rosenblatt, a psychologist working at the Cornell Aeronautical Laboratory, invented the "perceptron", which consists of a layer, or possibly more than one layer, of artificial neurons. A neuron receives one or more input signals and assigns a numerical weighting to each signal, akin to the numerical weightings employed in the evaluation functions in game playing programs.[24] By multiplying the strength of each input signal by its corresponding weighting and calculating the total of these products, an artificial neuron is able to compute a numerical score, akin to an evaluation function's score for a position in a game. If this score for an artificial neuron is above a certain threshold, the neuron is said to "fire".

Just as perceptrons are made up of one or more layers of artificial neurons, so ANNs are made up of a network of perceptrons and were originally called multi-layer perceptions. The more artificial neurons and layers there are in a network, the more connections there will be between them and the more knowledge an ANN will be able to acquire, since it is the connections that hold the knowledge.

---

[24]See the section "How Computers Evaluate a Chess Position" in Chapter 3.

Figure 47 illustrates a simple artificial neuron with three inputs, labelled i_1, i_2 and i_3. Associated with these three inputs are the three weights: w_1, w_2 and w_3 respectively. So the weighted sum of these inputs—its *score*, Σ—is

$$[w_1 \times i_1] + [w_2 \times i_2] + [w_3 \times i_3]$$

The sum total of this expression (Σ) determines the output from the artificial neuron—this total is fed into the *activation function* which decides how this particular artificial neuron should react to whatever total it is fed, and it is this reaction that determines what output is sent from *this* neuron to act as the inputs for other neurons in the network. If the total score is equal to or above a certain threshold then the artificial neuron fires, otherwise it does not. In essence, the neuron fires if it *recognizes* that a particular pattern which it is examining has some special characteristics. If the weighted sum of its inputs (i.e., its score) exceeds the threshold, then the neuron recognizes the pattern as belonging to some particular category of patterns. This is the real crux of the idea— recognizing that a particular input pattern belongs to some specific class of patterns. The task could be recognizing whether a photograph is of a man or a woman, whether a voice is that of an adult or a child, whether or not a piece of music was written by Brahms, whether a spoken word was "one", "two", "three" or another digit. Just about any classification task can be accomplished by a network of artificial neurons—an Artificial Neural Network.

In order to function accurately as a classifier an ANN needs to have its numerical weights—the strengths of its connections—set to suitable values. In an ANN, an input to one neuron will often be an output from another neuron, and the weights we have been referring to relate to

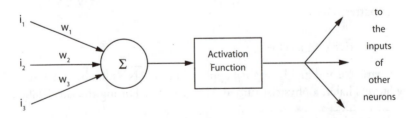

Figure 47. A simple artificial neuron

what are called the *strengths* of these connections between the neurons.[25] Setting appropriate weights is the really clever part—an ANN usually learns its weights from examples. If the task is to distinguish the spoken digits from "zero" to "nine", the ANN will normally be presented with dozens or even hundreds of samples of different people speaking those words. It is rather like saying to the ANN: "I am going to play you recordings of 500 people saying the word 'zero' so that you can learn what the word sounds like." The ANN then learns from the experience of hearing all these recordings, repeatedly adjusting its own weightings until it has learned as much as it can from the sample data.

An ANN normally starts life with random values assigned to each of its weights, so at that stage the ANN knows absolutely nothing. The network is then exposed to what is called the training set of data, such as the 500 samples of the spoken word "zero" referred to in the previous paragraph. When the first training pattern (also known as an input pattern) is presented to the network, the weights on each of the connections are adjusted by the program itself by a very small amount, in order to increase the likelihood of the network recognizing that pattern if it sees it again. Looked at in numerical terms, the changes to these weights will make it more likely that the sum totals of the various calculations at the artificial neurons will result in those totals classifying that same input pattern correctly. Once this has been done for the first input pattern in the training set, a second pattern is presented to the network and the process repeated, and then again for each of the patterns in the training set. When every pattern in the training set has been presented to the ANN, the whole process is run again and again with all of the training patterns, hundreds or thousands of times or more. The reason the weights on the connections are only adjusted a very small amount each time is to allow the ANN to learn to recognize *all* of the patterns reasonably well. If the adjustments were larger, the ANN would be able to recognize the most recently presented pattern every time, but all the other patterns never.

The Creativity Machine

A remarkable and highly original application for ANNs has been devised by Steve Thaler, a physicist turned AI-researcher. During the late 1980s

---

[25]The strengths may be thought of as measures of how important it is, in a particular classification task, that some combination of two or more different inputs are present simultaneously.

and early 1990s Thaler had the idea of investigating what would happen to an ANN if he tried to "kill" it by randomly degrading the weights connecting the artificial neurons. He trained a network, then held its input information constant and observed what happened to the output, while, one by one, he switched off the network's connections or reduced their strength at random—the computer equivalent of killing individual connections between the neurons in a human brain. Intuitively one would expect this to have the effect of completely destroying the performance of the network, but instead the stunted versions were not incapable, they merely performed differently.

The changes in the internal state of Thaler's network were interpreted as though they were caused by changes in the input information rather than by the death of some of its neurons, and the network made a stab at what the outputs might be, much as a human might guess a word that has letters missing. Thaler discovered that, as the neurons "died", the network generated memories, then fragments of memories, and finally new concepts created from fragments of memories. He also found that an ANN will respond in the same way if, instead of deleting connections, he simply changed some of their weights. His first breakthrough came on Christmas Eve in 1989, when he typed the lyrics of some of his favourite Christmas carols into an ANN. Once the network had learned these songs he started to switch off its connections. Gradually the network "began to hallucinate", creating new ideas in the process. As it was degrading, the network dreamed up new carols, each of which was created from shards of its broken memories. One of its final creations was the line "All men go to good earth in one eternal silent night." What most intrigued Thaler about the program's dying gasps was how creative the process of dying could be. This prompted the idea: "What if I don't cut the connections, but just perturb them a little?" [4]

Thaler tried "tickling" a few of the connections in the network, a process akin to giving a human a shot of adrenaline or a small electrical jolt to the brain. These tickling disturbances (called noise) caused his ANNs to generate variations on whatever the original network was trained to produce. For example, in order to generate silhouettes of car shapes, Thaler first provided the network with the positions of the salient points of a car's profile, such as the top and bottom of its windscreen. As the network was being trained, some of its nodes came to represent particular components of a car's shape, while the weights on the connections represented ways in which these various components can be

combined. Once the network reached that stage, Thaler started to change the weights on the connections at random. When he added too much disturbance, the network would throw components together randomly, producing what Thaler called "Picasso cars", bizarre designs with strange shapes and perhaps having their wheels in the air. When he added only a little disturbance, the network produced designs for the cars it has seen during training and ones that were very similar. But between those two extremes the network would suggest putting together components in new ways, that nevertheless conformed to the fundamental idea of what is a car.

The disturbances applied in an ANN can be likened to the often subconscious nudges to our brains that are at least partly responsible for creativity in humans. No-one knows exactly what process causes human creativity but clearly it must be some sort of stimulus to the brain. Thaler's idea mimics this stimulus.

On its own, the purely creative side of the Creativity Machine would produce not only interesting and successful ideas, it would also produce a lot of boring and useless variations as well. So Thaler added a second ANN to the system, this one trained on good examples and on bad ones, so that it can distinguish ideas of acceptable quality from poor ones. The first ANN creates the ideas, the second ANN performs quality control on the creations and rejects the duff ones.

In March 2002 Thaler was granted a U.S. patent, entitled "Device system for the autonomous generation of useful information."[26] The patent document describes some of the many successes claimed by Thaler, in areas as diverse as writing music, designing coffee mugs and thinking of new hard materials that are half the price of existing ones with similar constituents. For example, Thaler reports that one weekend he showed the machine short phrases, each of about ten notes taken from popular songs, but without any accompanying harmonies, whereupon the Creativity Machine generated so many new themes that, even after the second ANN had carried out its filtering process, some 11,000 tunes remained.

Another task that Thaler gave to the Creativity Machine, was the invention of new, ultra-hard materials. The ANN he designed for this purpose was trained by showing it some 200 examples of molecules, each

---

[26]U.S. patent number 6,356,884. The diagrams of the two automobiles in Figure 48 come from this patent.

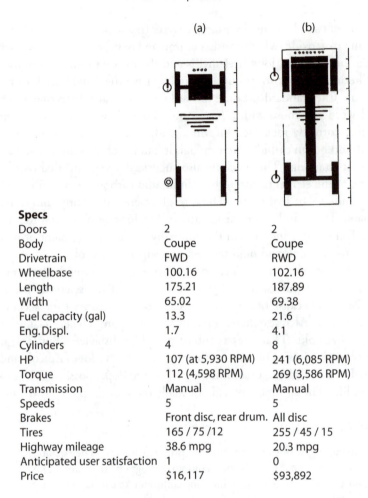

(a)   (b)

Specs

	(a)	(b)
Doors	2	2
Body	Coupe	Coupe
Drivetrain	FWD	RWD
Wheelbase	100.16	102.16
Length	175.21	187.89
Width	65.02	69.38
Fuel capacity (gal)	13.3	21.6
Eng. Displ.	1.7	4.1
Cylinders	4	8
HP	107 (at 5,930 RPM)	241 (6,085 RPM)
Torque	112 (4,598 RPM)	269 (3,586 RPM)
Transmission	Manual	Manual
Speeds	5	5
Brakes	Front disc, rear drum.	All disc
Tires	165 / 75 /12	255 / 45 / 15
Highway mileage	38.6 mpg	20.3 mpg
Anticipated user satisfaction	1	0
Price	$16,117	$93,892

Figure 48. Two automobile designs produced by the Creativity Machine: (a) a design specified to achieve at least 35 miles per gallon with a retail price of less than $25,000.00 and to have a favorable rating in terms of user satisfaction and design and (b) a design specified to be capable of accelerating from zero to 60 miles per hour in less than eight seconds and to achieve a top speed of at least 150 miles per hour

made up of two elements, for example water (made of hydrogen and oxygen) and iron oxide (which consists of iron and oxygen), in order to teach it plausible combinations of elements and their proportions. He trained the filtering network so its output correctly gave the *Moh's score* for each molecule (the standard measure of hardness), so that this second ANN could weed out those with a low score. When it was set to work the machine correctly identified known ultra-hard materials such as boron nitride and boron carbide, even though it had never seen these materials during its training. The machine also suggested several untried combinations of the elements boron, beryllium, and carbon, each doped with small amounts of hydrogen. These results were sufficiently impressive to enable Thaler to license the Creativity Machine to Advanced Refractories Technology, Inc., to aid the company in the development of new ultra-hard materials and high-temperature superconductors.

Thaler's research has a decidedly commercial focus, with the result that he has not published any technical details in the scientific literature. Perhaps because of this, his work has not yet received the acclaim of mainstream AI researchers, some of whom regard Thaler's claims as being rather bold. Thaler reacts robustly to their disbelief: "But look at Columbus. He was an outcast for the most part, with a lot of ridicule and scorn. Look at Galileo, excommunicated by the Pope, possibly about to lose his life. The community had not made the same leap they had." [4]

Genetic Algorithms

> Living organisms are consummate problem solvers. They exhibit a versatility that puts the best computer programs to shame. This observation is especially galling for computer scientists, who may spend months or years of intellectual effort on an algorithm, whereas organisms come by their abilities through the apparently undirected mechanism of evolution and natural selection. [5]

Genetic algorithms are sections of computer programs that attempt to mimic the process of natural evolution. Their goal is to evolve a solution to a problem. The algorithm starts by devising a population of possible solutions, and then subjecting each member of this population to operations akin to the processes of natural selection, including mating and mutation. As a result of these processes, and in accordance with Darwin's principle—"survival of the fittest"—the population of possible solutions gradually becomes "fitter", until, hopefully, one of the members of the

population corresponds to a satisfactory solution to the original problem.

Most organisms evolve by means of the processes of natural selection and reproduction. Natural selection determines which members of the population survive and reproduce, while reproduction mixes and recombines the genes of the parents in their offspring. The process of natural selection is simple. Each offspring of a reproductive process is subjected to one or more tests of fitness. For example, in the case of a bird, one test of fitness is recognizing that cats are predators and evading them. If a bird fails such a test then it is likely to be attacked and seriously injured or killed. During human reproduction, when a sperm and an ovum meet, pairs of matching chromosomes line up together and then cross over partway along their lengths, thereby swapping some of the genetic material of each. This mixing process propagates evolution much faster than if the genes of each offspring were merely those of a single parent.

In the early 1960s Hans Bremermann, at the University of California at Berkeley, developed the idea of using computer programs to simulate biological reproduction, carrying over some of the characteristics of parents into their offspring by reference to the genes of both parents. Bremermann's simulated mating procedure was limited in its scope, but his idea was soon augmented by the work of John Holland, who had become convinced that the recombination of groups of genes during the mating process was a critical part of evolution. By the mid-1960s Holland had developed the idea of what he called "genetic algorithms" that evolved by combining the benefits of mating and gene mutation.

At the start of the genetic algorithm process there is a population of "genomes", each of which represents a possible solution to the original problem. This starting population might be created randomly or it might be based on heuristic rules derived from the program's own knowledge of the problem domain. Genetic algorithms work iteratively, that is to say the evolution process is carried out hundreds or thousands or millions of times, depending on the complexity of the problem, until the software cannot improve the algorithm any further. Each iteration corresponds to a generation in human evolution and when each new generation is created, each member of its population is tested for its fitness according to some pre-defined criteria, using a fitness-function, akin to the evaluation functions employed in game playing programs. Each member of the new population is thus assigned a fitness score.

Many different techniques are employed for deciding which members of the population in a new generation should be selected as the basis for creating the next generation.[27] One of the most effective is Holland's original method, which he called "fitness-proportionate selection", a method that avoids the "hill-climbing" problem described in footnote 27. With Holland's method, the probability of selecting a particular member of a population as a parent is proportional to its fitness. This means that the fittest will indeed be more likely to be selected than the less fit, but statistically some of the less fit (and some of the least fit) will also be selected. The rationale is that a lesser member of a particular generation just might turn out to be on the right track for the global summit—the best of all possible solutions within the "hilly" area of candidate solutions.

Once the selection has been made to determine which of the members of the current generation are allowed to reproduce and which are to be removed from the population, new members of the population are created by two processes, both derived from genetics, called *crossover* and *mutation*. The crossover process in a genetic algorithm mimics the way that biological chromosomes cross over one another when two gametes[28] meet to form a zygote.[29] When two genomes in a genetic algorithm line up, a point along the genomes is selected at random (called the crossover point) and the portions of both genomes to one side of that point are swapped around, thereby producing two offspring. One of these offspring contains the data from one of the genomes up to the crossover point and the data of the second genome beyond that point, and the other offspring vice versa. These two offspring are then used to replace two less fit genomes that have been weeded out by the fitness-proportional selection process (or whatever method of selection is used instead). So each creation of two new genomes corresponds to the discarding of two from the previous generation, ensuring that the total population size remains constant.

---

[27] Superficially it might appear as though the evolutionary process should automatically select all of the fittest members of one generation as the gene pool for the next. To understand why the process is not that simple, consider the problem of being placed somewhere in a hilly area and told to climb to the summit of the highest hill in the area. Intuitively one might simply climb higher and higher at every opportunity, but that approach can lead to the top of what turns out to be only a local summit, from where it would be necessary to go downhill before going up again to a higher summit.

[28] A gamete is a mature sexual reproductive cell having a single set of unpaired chromosomes.

[29] A zygote is a cell resulting from the union of an ovum and a spermatozoon (two gametes).

The mutation process is designed to ensure that evolution does not gravitate prematurely to a local summit. What happens is that a small fraction of the genomes, selected at random, undergo a very slight modification (corresponding to mutation). Following the hill-climbing analogy, this part of the process can be thought of as short excursions for exploration purposes, "looking around" the hilly area for a potentially promising route to the global summit, rather than stagnating at a particular location from where it is impossible to see how to improve the next generation.

The optimal solution at the end of the evolutionary process will be the fittest member of the final generation of the population, which may be the best possible solution or it may not. Thus the evolutionary methods employed in genetic algorithms do not guarantee to find the perfect solution to a problem, or, expressed in hill-climbing terms, genetic algorithms will not always find the highest peak in the area. But genetic algorithms have been proven to be extremely effective in an enormous variety of tasks, with a huge raft of success stories to their credit. One classical problem in mathematics that has been solved with the help of genetic algorithms is the so-called Travelling Salesman Problem.[30] Another success story is a program called GenJam,[31] developed by Al Biles, that learns to play jazz solos. GenJam creates jazz improvisation "riffs" (short, rhythmic phrases) and Biles tells the program whether each riff is good or bad, thereby improving the program's fitness measure. After much training, GenJam has become a formidable improvisation partner for jazz musicians.

The most remarkable achievements of all in this field are, perhaps, those from an offshoot application that is still in its infancy, a topic called Genetic Programming. Imagine for a moment what would happen if, instead of using genetic techniques to breed better and better solutions to a problem (the algorithms), we were instead to employ the same techniques to breed better and better computer programs. What an enormous power that would give us! Instead of asking a robot to go away and find a solution to a particular example of the Travelling Salesman problem, we would be able to say it: "Here is the nature of the problem. Go away and develop a computer program to solve all such problems." A discussion of

---

[30] A salesman must visit several cities once and only once. What route enables him to accomplish this while travelling the shortest possible distance?

[31] For Genetic Jammer.

this topic, since it relates to the reproduction of the robot itself, will be found in the section "Self-Reproducing Software" in Chapter 11.

How Computers Discover and Invent

> Creativity is the ability to bring something new into existence, by seeing things in a new way. Those who have this in the greatest degree are considered geniuses. [6]

Discovery

The earliest demonstration of discovery[32] by computer was Douglas Lenat's program AM,[33] written in 1976, which could propose interesting mathematical theorems based on its own "intuition". AM was given a small number of mathematical concepts together with a set of heuristic rules for creating new concepts and deciding how interesting they were. The program worked by first creating a list of things to investigate and then trying them in the order "most interesting" first. Any new ideas that resulted from an investigation would be added to the list and the cycle repeated. In essence this process is similar to the method employed by the Logic Theory Machine—generate a new result from existing knowledge and then employ the new result together with the old knowledge to see what else can be generated.

AM started life with a small nucleus of ideas from set theory, a branch of mathematics. Amongst the mathematical concepts that the program discovered were counting, addition and multiplication, all of which appear very simple to us. But it also discovered for itself the concept of prime numbers, and then, having "invented" prime numbers, it announced what had been known since 1742 as Goldbach's Conjecture, namely: "Every number that is greater than 2 is the sum of three prime numbers,"[34] a conjecture that the program did not find particularly interesting.

---

[32] Scientific discovery and invention are both highly creative processes and might, therefore, argue for inclusion in the previous chapter. In the case of the Logic Theory Machine, for example, Newell, Shaw and Simon wrote in 1958 about the possibility that their "machine's" discoveries of new proofs in logic were creative. I make no apologies for including discovery and invention here instead—they simply seem to me to be more closely allied with some of the techniques described in the present chapter.

[33] Lenat has claimed that the name AM means nothing, but stands alone as in the biblical "I AM that I AM", though others have suggested that originally it stood for Artificial Mathematician.

[34] Remember that one is a prime number.

Having devoted its attention to prime numbers, AM then discovered the concept of maximally divisible numbers,[35] which in a sense are the opposite of prime numbers, and proceeded to propose theorems about them. One reason why this was interesting is that Lenat did not know anything about maximally divisible numbers when he wrote the program and in fact he believed the program to be on a false track when it first went in that direction!

The methodology of Lenat's program is not dissimilar to the approached used in more recent work on automating the discovery of scientific knowledge—employing heuristics to search the whole "space" of ideas and concepts within the realm under investigation. By the late 1990s there were already several programs sufficiently accomplished at scientific investigation to enable the publication, by their human collaborators, of discoveries that had been made by the programs. One example is the use of the rule induction system RL,[36] in discovering qualitative laws that can predict which chemicals are likely to have the long-term effect of causing cancer in humans exposed to them. John Aronis, Bruce Buchanan and Yongwon Lee, the program's human collaborators at the University of Pittsburgh, were taught a set of rules by the program, rules that themselves state which conditions must be present for a particular rule to be valid. These rules were found to be substantially more accurate than existing prediction schemes and became the subject of papers in the learned journals *Mutation Research* (in 1995) and *Environmental Health Perspectives* (in 1996).

MECHEM, developed by Raul Valdés-Pérez, finds explanatory hypotheses in chemistry from experimental evidence, the hypotheses explaining the sequence of steps for a given chemical reaction. The program is given information about the starting materials of a chemical reaction, any observed products of the reaction, and any products that exist only at intermediate steps in the reaction. The program also has background knowledge about catalytic chemistry expressed as constraints (i.e., rules). MECHEM is able to discover all of the simplest hypotheses that explain how the products are formed, given the known constraints. The program works by carrying out an examination of all the reaction *pathways*, as these processes are known, and starts by looking at the

---

[35] A number is maximally divisible if it has more prime factors than any lower number. For example 12 is maximally divisible because it has six divisors: 1, 2, 3, 4, 6 and 12, whereas no number less than 12 has as many divisors.

[36] For Rule Learner.

simplest pathways—those involving the fewest substances and steps—before going on to examine more complex ones. The known constraints help the program to eliminate certain pathways and to help in identifying any products that are created at intermediate stages in the reaction. The program's output is a set of the simplest pathways consistent with the background knowledge, that conform to the constraints and explain the experimental evidence.

Because the pathways are generated from scratch rather than being taken from a database of common reactions, MECHEM will quite often suggest hypotheses that are original. In addition, the program's systematic and exhaustive approach to the task is often responsible for unearthing hypotheses that have been overlooked by human scientists, who simply lack the time and resources to conduct such an exhaustive search of the problem space. Not only are the pathways proposed by MECHEM sometimes original, they are often of scientific interest because of their simplicity. By starting the search on the basis of "simplest first", the program tends to reveal pathways that minimize the number of steps in the chemical reaction and the number of chemical substances that are assumed to be created at intermediate stages in the reaction.

A more interactive approach is used in the ARROWSMITH program, developed by Don Swanson at the University of Chicago, which helps the user to interrogate a database of medical publications called MEDLINE, in order to spot new relationships between causes and effects in the world of medicine and to form and assess novel scientific hypotheses. The underlying philosophy behind the program is that information developed in one area of research can be of value in another area without anyone being aware of the fact. The user might suspect that some connection exists between, say, eating a lot of bananas and contracting gout.[37] So the user begins with a direct question concerning the connection between bananas and gout, but a conventional search of the database will provide no answer if there are no publications containing both words "bananas" and "gout", and some sort of link between them. (Thus far the search process described here has been rather like using Google.) But then the user can ask ARROWSMITH: "Can bananas, or a deficit of bananas, influence the cause or risk of gout?" Now it is possible that bananas might influence some hitherto unconsidered factor,

---

[37] This is a purely hypothetical example born of the author's total ignorance of both gout and compulsive banana eating.

which we shall call "factor X", that *is* mentioned in some publications relating to bananas that do not mention gout, and *is* mentioned as influencing gout in some publications that make no mention of bananas, but is *not* mentioned in any publications that refer to both bananas and gout. This type of relationship cannot be discovered with a conventional search of MEDLINE, but ARROWSMITH provides a straightforward solution to discovering such relationships.

Invention

One of the most remarkable scientific advances at the turn of the twenty-first century has come from a marriage of two methodologies: the theory of inventive problem solving (TRIZ[38]), first developed in the former Soviet Union, and genetic algorithms, one of the most prominent and fastest growing areas of AI.

TRIZ was devised by Genrich Altshuller while working as a clerk in a patent office. Altshuller had the idea of using the wealth of information contained in more than 200,000 patent applications, which were available at that time, as a source for finding some common rules to explain the creation of new, inventive, patentable ideas. In December 1948 Altshuller, together with his colleague and former schoolfriend Rafael Shapiro, wrote a long letter to Joseph Stalin, explaining that there was a state of chaos and ignorance in the U.S.S.R. concerning the country's approach to innovation and inventing, and saying that "There exists a theory that can help any inventor invent." The result of their efforts to improve the climate of invention for the good of their country was that they were both sentenced to 25 years of imprisonment, though they served only about four years, due to Stalin's timely death in 1953. During his imprisonment in the Vorkuta gulag, and especially in the winter of 1952–1953, Altshuller continued to work on his ideas about the creativity process in invention.

After they were released and rehabilitated, Altshuller and Shapiro wrote their first paper about TRIZ, which was published in the journal *Voprosy Psihologii* (*Problems of Psychology*) in 1956, expounding many of the basic concepts of TRIZ that he had already developed.

The underlying philosophy behind Altshuller's ideas was that problems in invention stem from contradictions or tradeoffs between two or

---

[38] The acronym of the original Russian: *Teoriya Resheniya Izobreatatelskikh Zadatch*.

more elements of the invention. For example, if we are designing a car and we want to improve its acceleration, we might decide that we need to use a larger engine, but that will increase the cost of the car. The net effect will be that by creating more of something desirable (i.e., acceleration) we are also adding more of something undesirable (i.e., cost) or perhaps we would reduce the amount of some other desirable aspect of the product. Altshuller referred to such tradeoffs as "technical contradictions". He also defined what he called "physical" or "inherent" contradictions, where we may need both more and less of something, simultaneously. For example, during a process to manufacture some sort of mixture, we might need a higher temperature in order to melt a compound more rapidly, but at the same time it could be necessary to lower the temperature in order to achieve a sufficiently homogeneous mixture. An inventor might be faced with several such contradictions, for which the solutions often come from employing a creative approach to the problems. But finding a suitably creative approach is often difficult.

Altshuller and his friends scrutinized huge numbers of patent documents in order to discover what type of contradictions had been resolved by each invention and the way this had been achieved. From this knowledge, he developed a set of 40 inventive principles and subsequently a *matrix* of contradictions. Each row of the matrix corresponded to one of 39 features that an inventor is typically attempting to improve by his invention, such as speed, weight, accuracy of measurement, etc. Each column of the matrix refers to a frequently occurring but undesirable result, such as increased cost. Each cell in the matrix indicates those principles that were most frequently described in the patent documents, in order to resolve a contradiction.

Altshuller also studied the way that various technical systems had been developed and improved over a period time. This led him to discover several trends, which he called the "laws of the evolution of technical systems", that help engineers to predict what are the most likely improvements that can be made to a given product. The most important of these laws relates to what Altshuller described as the "ideality" of a system, a qualitative ratio between the sum of all the desirable benefits of the system and its cost or other negative effects. When an inventor attempts to decide how to improve an invention, his goal is to increase its ideality, either by increasing its beneficial features or decreasing its cost or other negative effects, or possibly both. The utopian solution

for the inventor would be to have all the benefits of the product at zero cost, but that can never be achieved. A more realistic goal of inventors is therefore that successive versions of a technical design improve its ideality.

After he was released from the gulag, Altshuller continued for 20 years to develop his theories with the assistance of more than 80 engineers and scientists who helped him in the study of more than 1,500,000 patent documents.[39] In 1976 the Minsk School of TRIZ was started by Valerei Tsourikov at the Radio Electronic University. Tsourikov was a disciple of Altshuller's, and rapidly rose to become head of the Intelligent Systems Laboratory at Minsk University. In 1987 he started a project called the Invention Machine, based on TRIZ, which he took to the U.S.A. in 1991 and commercialised by setting up the Invention Machine Corporation, which numbers IBM, General Electric and Motorola among its 500 or more clients.

Not surprisingly, the Invention Machine[40] is itself the subject of a patent in the U.S.A.,[41] granted in May 1999. The title of the patent, "Computer based system for displaying in full motion linked concept components for producing selected technical results", belies the beauty of the idea and the simplicity of the technology behind it. The Invention Machine assists the user in solving technical and engineering problems, and helps engineers, scientists and inventors to better understand the products, processes, or machines they are attempting to invent and improve—a methodology known as concept engineering. Such systems serve not only to increase a designer's inventive and creative abilities in solving problems in engineering and science, but also, in the course of solving them, to induce inventors to consider new structural and functional concepts that are applicable to their design goals.

The Invention Machine comprises knowledge-based and logic-based systems that generate concepts and recommendations for solving problems at a conceptual level. More than 150 inventive rules and procedures are included in the system's knowledge base, and when the user describes the problem that needs to be solved, the system selects certain rules and presents them to the user for consideration as possible routes to a solution. What the system is really doing is employing the TRIZ approach

---

[39]The number of published patents in the U.S.S.R. had grown significantly since Altshuller had first planned the project in the late 1940s.

[40]The Invention Machine is a trademark of the Invention Machine Corporation, Inc.

[41]U.S. patent number 5,901,068.

by solving engineering contradictions, in order to reduce the natural tendency of users to make sacrifices in one feature of a design as a trade-off for achieving an improvement in some other aspect of the design. The knowledge-base includes data relating to more than 1,200 physical, geometric, chemical and other effects used in the past to solve other problems in engineering, and a selection of these effects is presented to the user of the Invention Machine for consideration as potential solutions to the current problem. The system also includes a technology evolution and prediction capability that aids the user in understanding the dynamics of the product's evolution and the most logical next generation(s) of the product or its function. This stimulates the user to forward plan and extrapolate the dynamics of the life cycle of the technology and to originate its next generation.

The Invention Machine and its spin-off systems have several interesting success stories to tell, one of which was the design of a new type of pizza box that required the reconciliation of apparently incompatible goals. When a Russian refugee engineer called Michael Valdman emigrated to the U.S.A. in 1990, he worked for a while delivering pizzas and discovered that the traditional pizza-box design suffered from two contradictions. Firstly, pizzas being delivered in boxes to locations outside the restaurant needed extra insulation, so that they would stay hot longer than they normally would in a conventional pizza box, but it was important to achieve this increase in insulation without increasing the bulk and, therefore, the cost of the box. The software suggested using a void inside the box instead of a more solid insulating substance, and recommended making the box in a spherical shape. Valdman did this by making the bottom of the box concave, thereby creating a dome of air beneath the pizza. The other contradiction was that, in order to keep a pizza hot, the pizza box must be kept closed, but in order to keep the pizza from getting soggy the water vapour given off by the pizza must be allowed to escape. The program suggested changing the shape of either the pizza or the box, in order to reduce contact between the pizza and the water vapour, as a result of which Valdman designed a box with small, moulded pyramids rising from the inside of the base so that the water vapour condenses between the pyramids and away from the pizza crust. By implementing both of the software's suggestions in his design, Valdman's container keeps the pizza hot and crispy for almost three times as long, around 45 minutes, while costing the same as a conventional pizza box. In 1995 Valdman was granted two patents for his design.

Genetic programming[42] has proved to be another route from within the field of AI to automatic invention. The founding father of genetic programming, John Koza, was able to report, in 2000, in the very first issue of the journal *Genetic Programming and Evolvable Machines*, on ten potential products that had been created by genetic software, in each case starting with nothing more than the basic parameters of a problem posed by the programmers. In one instance, Koza's team asked the program to design a device for receiving television signals, providing the software only with a few requirements as to the approximate size and performance of the desired device. The software then "invented" an antenna which turned out to be the same as the familiar ladder-shaped Yagi-Uda antenna, designed by two Japanese scientists in the 1920s and still the most common type of terrestrial TV antenna to be found on the rooftops of houses today. That Koza's software had proposed a design that was already known is of no negative consequence—in fact it proves the power of genetic software for inventing useful and commercially viable products. As of December 2004 the web site http://www.genetic-programming.com/ listed 36 instances where genetic programming has produced a *human-competitive result*.[43] These results include 15 instances where genetic programming has created something that either infringes or duplicates the functionality of a previously patented twentieth-century invention, six instances where genetic programming has done the same with respect to a twenty-first century invention, and two instances where genetic programming has created a patentable new invention. These results come from such fields as computational molecular biology, cellular automata, sorting networks, and the synthesis of designs for analog electrical circuits, controllers, and antennae.

These achievements testify to the ability of genetic programming to act as a successful invention machine, and it is this capacity for innovation that will, I believe, raise genetic programming to God-like status in the eyes of the scientific community, if not the entire civilized world. Eventually, all we will need to do in order to solve a problem is say to our robots something akin to "A new illness has just been discovered in

---

[42] See the section "Self-Reproducing Software and Genetic Programming" in Chapter 11.

[43] A human-competitive result is defined for these purposes as a result that satisfies one or more of eight criteria, including: (A) The result was patented as an invention in the past, is an improvement over a patented invention, or would qualify today as a patentable new invention; or (B) The result is equal to or better than a result that was accepted as a new scientific result at the time when it was published in a peer-reviewed scientific journal.

the Nile delta. Please find a cure for those already affected by it and a vaccination to guard against it."

Knowledge Discovery

The sum total of recorded data and human knowledge has become absolutely staggering in the Internet age. The School of Information Management and Systems at the University of California at Berkeley publish an annual report entitled *How Much Information?*, summarizing the amount of information produced since mankind began, and in particular during the year prior to that year's report. The total amount of information that existed on printed, film, magnetic or optical storage media in 2002 is estimated to be roughly 18,000,000,000,000,000,000 bytes of data (18 exabytes), of which more than one-quarter was created in 2002 alone.

With this ever-growing electronic mine of knowledge, available in a digital form that can be processed electronically, come opportunities to employ computers to unearth useful knowledge that we might not otherwise have noticed. Unsurprisingly, this discipline, which is often called Data Mining or Knowledge Discovery in Databases, is burgeoning.

The development of the necessary technology for Data Mining started in the field of cognitive psychology in the mid-late 1950s at Yale, with the work of Carl Hovland and his research student Earl Hunt, on the computer simulation of concept creation. In 1966 Hunt, together with two other cognitive psychology researchers, Janet Marin and Philip Stone, published a model of concept formation called Concept Learning System (CLS). This model was based the idea that, in order to isolate discrete concepts from cluttered data, the human mind tends to divide complex concepts into groups of smaller similar concepts that have common characteristics.

Within this field of AI research a concept is defined formally as a classification rule that divides a set of examples into two classes, such as those examples that satisfy the concept, and those that do not. As an illustration, if the CLS is set to learn how to play the Chess endgame of king and rook versus a lone king, as played by Torres y Quevedo's rule-based machine,[44] the system might begin by dividing all such positions into those in which the lone king is within two horizontal or vertical

---

[44]See the section "Torres y Quevedo's Chess Endgame Machine" in Chapter 1.

rows of the edge of the board and those positions in which it is not. Then it might sub-divide each of these classes further, according to other criteria suggested by useful Chess heuristics. This "divide and conquer" approach is employed by CLS to build *decision trees* that learn from their training examples to discriminate between different classes of examples.

In 1979 Ross Quinlan, the leading AI practitioner in Australia, adapted the CLS model of concept formation to develop a variation on its algorithm which he called ID3.[45] Quinlan made two significant modifications to the original algorithm. Firstly, rather than have the attributes proposed by human operators of the system, the attributes in ID3 were chosen by heuristics so as to be applied in an order which reflected their usefulness as discriminators, with the most useful being applied first. The other significant change was to allow only a selection of examples within a class, those that fell within a "window" specified by the system, to be used for training the decision tree—the other examples in that class were then tested on the tree and when it made a mistake the tree was modified in order to correct the mistake. Both of these changes speeded up the tree creation process considerably by enabling Quinlan's algorithm to learn from a relatively small set of training examples. Quinlan also used a combination of mathematics and logic to enable ID3 to operate on both numerical and non-numerical data, thereby making it applicable to many more classification problems than are those algorithms which can cope only with numerical data.

Quinlan tested ID3 on the Chess endgame of king and rook versus king and knight. We saw in Chapter 3 how Ken Thompson and others were building huge databases of Chess endgame positions together with the correct move for each, so that programs can look up any of these positions and know instantly which was the right move. Instead, Quinlan and Donald Michie became intrigued by the problem of using databases to *teach* a program by example, teaching it the heuristic rules for playing endgames. This idea was inspired by Torres y Quevedo's success at isolating the few rules that were necessary for his machine to play with king and rook versus a lone king. ID3 was also adopted for other employment at the chessboard by two of Michie's research students in Edinbugh, Tim Niblett and Alen Shapiro. Michie had long been interested in computer Chess and, at that time, was particularly fascinated by the idea of automatically inducing rules for playing Chess endgames, rather than

---

[45] For Itemized Dichotomizer.

teaching a program a set of such rules. Niblett and Shapiro tested ID3 on the endgame of king and pawn versus a lone king, and found that the decision trees generated by the algorithm were 100 percent accurate when validated against a complete database of such positions.

This early work of Quinlan's, and an improved algorithm which he called C4.5[46] and published in 1992, served as the foundation for much of the subsequent research into data mining, a discipline that deals with extremely large amounts of data and which usually employs some form of automated learning. For many years the commercial organisations that offered data mining services tended to make exaggerated claims. One apocryphal tale describes how data mining, applied to the contents of each of its customers' shopping baskets, alerted a major supermarket chain to the fact that sales of baby diapers and beer were highly correlated. This was assumed to be because young fathers who dropped in at its stores on their way home from work to pick up supplies of diapers, often decided to stock up on beer at the same time. The supermarket chain then, supposedly, put the two items side-by-side on its shelves, whereupon sales soared—an excellent advertisement for data mining, not allowing the truth to stand in the way of a good story.

But in recent years the claims from the data mining industry have increasingly been true. Thanks to bigger and faster computer hardware and to advances in data mining algorithms, fiction has turned into fact. A British supermarket that had just about decided to discontinue a line of expensive French cheeses that were not selling well, discovered through data mining that the few people who did buy those cheeses were among the supermarket's most profitable customers, so the cheeses acquired a fresh lease of life in order to retain the custom of these big spenders.

Expert Systems

Expert systems[47] are sophisticated computer programs that use human knowledge to solve problems normally requiring specialist expertise.

---

[46]C4.5 starts with large sets of cases belonging to known classes. The cases are scrutinized for patterns that allow the classes to be reliably discriminated. Those patterns are then expressed as models, in the form of decision trees or sets of if-then rules, that can be used to classify new cases, with the emphasis being on making the models understandable as well as accurate. The system has been applied successfully to tasks involving tens of thousands of cases described by hundreds of properties.

[47]Also known as "Knowledge-based Expert Systems" and "Intelligent Knowledge-based Systems".

These programs behave like human experts in various useful ways: they operate efficiently and effectively in a narrow problem area such as specialist medical diagnosis; they solve problems that are difficult enough to need a significant level of human expertise for their solution; they can offer intelligent advice; and they can explain the reasoning behind their decisions in such a way that a user of the system can understand their "thinking".

The domains in which expert systems have been used cover a wide field, including not only medicine but also mathematics, engineering, geology, crime, computer science, business, law, politics and defence. Examples of the type of decisions that can be made by expert systems are "From which of three possible thyroid complaints is this patient suffering?"; "Why is my car not starting?"; "How likely is it that the stock market will crash in the next month?"; "Is there anywhere on this parcel of land where there might be oil deposits?"; and "Is the judge more likely to find me guilty or not guilty?"

Expert systems usually represent their expertise and knowledge as rules within the program, often supplemented by data of various types. In a weather forecasting system, for example, the rules might include "If the pressure does not rise then it is likely to rain today", while the data in the program might include the rainfall and air pressure statistics for the relevant area over the previous 20 years. In general, each rule in an expert system represents a chunk of expert knowledge, and most systems contain hundreds of rules. For example, the MYCIN medical diagnosis system employed about 450 rules while the PROSPECTOR system for locating mineral deposits had more than 1,600. These rules are usually obtained by a process called knowledge engineering, based on interviewing human experts for periods of weeks or longer. The person who carries out these interviews and converts the resulting information into rules is called a knowledge engineer.

The philosophy of expert systems design is rather different to that of programming most other types of task. Computer programs that employ conventional decision-making logic are usually driven by algorithms (methods) designed to solve a specific problem. Such algorithms are embedded as part of the program and whenever the algorithm needs to be modified, so will the program. But knowledge-based systems work in a different way. Their methodology remains constant but what can and usually does change with time is the set of rules incorporating the system's knowledge. Rules are stored as data that can be added, removed

or changed, without the need to modify the program itself, so new information (i.e., new rules) may be easily accommodated. The ease with which the rule-base can be changed is a boon to expert systems designers and users—bring on a new expert, or have the original expert(s) add new rules, and the rule-base can be improved rather easily and quickly. Another advantage of employing a knowledge-base of rules in the system is that knowledge in this form is active, in that it can be used to infer new information from what is already known about a problem.

Expert systems are deliberately made narrow in their domain of expertise because, like human specialists, by knowing more about less, the system is able to perform with a higher level of understanding within the chosen problem domain. For example, there is no medical diagnosis expert system that encompasses the entire range of medical knowledge, but there are several systems that function in specialist areas of medicine and which are able to perform at the level of leading human experts in these fields. One of the earliest and best-known expert systems, MYCIN, was developed at Stanford University in the 1970s to diagnose certain types of blood infection, and in tests the system was able to outperform even some members of the Stanford Medical School. Even today, after three decades, MYCIN remains a classic example of how expert systems are designed and how they function. But although such a level of specialisation clearly has its advantages, it also means that the users of an expert system must take care to employ it only within the appropriate domain; otherwise the system could produce nonsense. One researcher, who used a skin disease expert system to diagnose problems with his rusty old car, was advised that the car had probably developed measles!

The Structure of an Expert System

Figure 49 shows the structure of a typical expert system program. The user interacts with the system through the user interface, which will normally ask questions via menus or using natural language. The user interface sits between what is called the inference engine and the user. It translates the system's answers from the internal representation created by the program to something the user can understand, and it passes questions from the system to the user and checks that the user's replies are valid. For example, it would query or reject a ludicrously low number as the answer to a request for your weight. The inference engine is employed to do the reasoning for the system, using both the expert

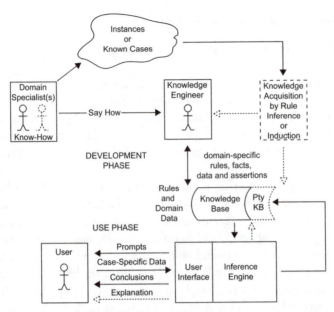

Figure 49. A basic schema for knowledge-based applications (after "Knowledge-Based Expert Systems," Roger Clarke, available at http://www.anu.edu.au/people/Roger.Clarke/SOS/KBT.html)

knowledge (i.e., the rule-base) and the data it has been given specifically for the particular problem being solved—this includes the data provided by the user and the partial conclusions of the system based on this data that have already been reached during the problem-solving process.

The system asks a question, the user provides an answer, the system uses its *inference engine* to draw inferences from the user's answer combined with one or more of its rules and, as a result of its inferences, the system decides what question to ask next. The process is thus one in which the system's rule-base and its inference engine co-operate, to simulate the reasoning process pursued by a human expert when analysing a problem and arriving at a conclusion. This process continues until the system is satisfied that it cannot improve on its current assessment, whereupon it announces its opinion, perhaps accompanied by a measure of how confident the system is in that opinion.

Knowledge Engineering

The role of the knowledge engineer is to acquire expertise, mainly from human experts and sometimes from published sources. This expertise is

then converted into rules, of the form

If this is true **then** that is likely to be the case.

The whole collection of rules form the *knowledge-base* or *rule-base*. The knowledge-base in an expert system encapsulates many heuristics—little rules of good judgment, of plausible reasoning and of good guesswork—that characterize expert-level decision making. An entire rule-base can be thought of as a model of the expertise of the best practitioners in the field.

Knowledge engineering is the process of extracting relevant knowledge from human experts and reducing the resultant body of knowledge to a precise set of facts, usually in the form of these heuristic rules. The knowledge engineer gets at and transforms appropriate information from the minds of experts, or from books, journals or any other source of expertise, into some manageable form. The skill of the knowledge engineer lies in his ability to act as a translator between, on the one hand, the human experts who are being debriefed in order to extract their expertise, and, on the other hand, the knowledge-base of the expert system itself. The knowledge engineer elicits knowledge from the expert, encodes it in rules in the format required by the knowledge-base, and refines it in collaboration with the experts until an acceptable level of performance has been achieved by the system.

Many statements that may not at first look like rules can easily be formulated as rules, for example,

the statement	"All mammals have fur"
converts to	"**If** something is a mammal **then** it has fur"
while the statements	"Lockjaw is fatal"
and	"A patient with lockjaw will die"
both convert to	"**If** the disease is lockjaw **then** the patient will die".

Part of the knowledge engineer's job depends on being able to recognize how such statements can be translated into the syntax of the rule-base.

Clearly it is important that the knowledge engineer has good communication skills and that the expert is able to express his knowledge clearly to the engineer. A typical problem faced by knowledge engineers is that experts are often unable to volunteer useful information about

their thought processes because much of their thinking is almost sub-conscious, or it might appear to them to be trivial or obvious. One of the experts being debriefed for a well-known expert system project called DENDRAL, was shown the data relating to a particular chemical com-pound and observed: "It's a ketone".[48] The knowledge engineer asked: "How do you know it's a ketone?" and was somewhat puzzled by the reply: "Well—look at it. It's gotta be a ketone."

Before he can extract knowledge from the experts, a knowledge en-gineer should first become at least somewhat familiar with the domain of interest, possibly by reading an introductory text and/or by talking to the experts. After this induction it is often useful to set each expert sev-eral example problems, asking them to explain aloud their reasoning as they solve each problem and to mention and explain any rules of thumb that they employ. The knowledge engineer can often extract some gen-eral rules from these explanations and then check them with the experts before they are encoded for the system. The experts are further observed and debriefed while they are engaged on many more relevant tasks. Again the experts are asked to verbalize their thought processes as they work on the tasks, allowing transcripts of their verbalizations to be coded by the knowledge engineer into an appropriate format for the knowledge-base. Realistically this whole process might take several person-weeks, months or even years, in order to extract and refine sufficient rules to enable the creation of a powerful expert system.

Most experts typically cannot afford to devote several weeks or months to a research undertaking of this type, and they don't want to; they prefer to work on their research in their own chosen field, so as the science of expert systems has developed, computer-based tools for knowl-edge acquisition have been created. Some of these tools work directly with human experts to elicit knowledge and structure it appropriately to operate within an expert system. But the elicitation of expert knowledge and its effective transfer to a useful knowledge-based system is complex and involves several diverse activities. Therefore, starting in the 1990s, fully-automated methods of knowledge discovery and acquisition have been developed.

One automated approach is to employ neural networks to induce rules by generalizing from existing examples of their use. From a set of observations, rules are derived automatically, rules that relate the

---

[48] A particular type of chemical compound.

decision of the expert to the various factors that caused him to reach that decision.

The Inference Engine

The real strength of expert system programs is their ability to draw conclusions from premises. This ability is what makes an expert system intelligent. The inference engine is the part of the expert system that knows how to apply the rules in the knowledge-base and decide in which order they should be applied when solving a particular problem. By interpreting the rules in the knowledge-base, the system is able to draw its conclusions. Two alternative strategies are available for making inferences from the rules, called *forward chaining* and *backward chaining*.

A forward chaining inference engine reasons, from the premises, "forwards" to a conclusion. The process starts from the knowledge-base and any data available as evidence. It first examines the current state of the knowledge-base and evidence, then finds those rules whose premises can be satisfied from known data (i.e., those rules in which the "**if**" part of the rule is known to be true), and adds the conclusions of those rules (i.e., the "**then**" parts) to the knowledge-base. With these conclusions added, the system re-examines the complete knowledge-base and repeats the process, which can now progress further because of its access to the new information, until eventually, hopefully, the goal is reached.

Backward chaining works from the end (i.e., from the solution or goal of the original problem), "backwards" in the sense that it tries to prove the goal or conclusion by confirming the truth of all of its premises.

How Expert Systems Explain Their Reasoning

A chaining process will, when it has reached its conclusion, consist of a chain of steps that can be traced by the expert system, enabling the system to explain its entire reasoning processes to the user. In this way the user can not only understand why the system reasons as it does, which occasionally leads to the human expert learning a new idea or technique from the system, (s)he can also detect flaws in the system, flaws that might be traceable to a badly expressed rule or which might indicate that a new rule needs to be added to the knowledge-base in order to plug a specific gap in the system's knowledge.

The ability to explain their reasoning processes is a key feature of expert systems. They can justify their decisions and explain why they are

asking whatever questions they do ask. Many people do not always accept the answers of a human expert without some form of justification; for example, a doctor providing a diagnosis is expected to explain the reasoning behind his or her conclusions, partly so that the patient is aware of any risks or possible alternative treatments.

Expert systems typically explain themselves by answering two types of question: "How did you conclude that fact?" and "Why are you asking me this question?" To answer the "how" question a backward chaining expert system would keep note of all the rules that were proven true *en route* to reaching its conclusions, and then translate each rule into understandable language for the benefit of the user. To answer the "why" question, the system examines the rule it is currently trying to prove. Suppose, for example, that the system has asked if it is raining. It might answer the "why" question by saying: "I am asking you if it is raining because I want to prove that you are going to get wet if you go outside, and rule 23 says that you will get wet if you go outside and if it is raining and if you do not have an umbrella handy."

Confidence Estimates

Many of the rules in an expert system will not have definite conclusions, rather they will carry an estimate of certainty that the conclusion of the rule (the "**then**" part) will hold if the conditions (the "**if**" parts) are true. This is analogous to human reasoning—when we reason we do not always arrive at conclusions with 100 percent confidence. In the earlier example of a prediction that it "is likely to rain today" if the pressure does not rise, the rule might state that "it will rain", supported by a certainty level of, say, 85 percent. Statistical techniques are used to determine these certainties, based on data collected for the relevant problem domain (in this case the previous 20 years' weather statistics).

MYCIN—A Typical Expert System

MYCIN was developed at Stanford University in the early-mid 1970s, and was the first large expert system to perform at the level of a human expert and to provide users with an explanation of its reasoning. The system was intended to be used by a doctor, to provide advice when treating a patient—advice about infections that involve bacteria in the blood and advice about meningitis.[49] These infectious diseases can be

---

[49] Infections that involve inflammation of the membranes that envelop the brain and spinal cord.

fatal and often show themselves when a patient is already in hospital, for example while recovering from heart surgery. The name MYCIN comes from the fact that most of the drugs used in the treatment of bacterial infections have names ending in "mycin", for example, Neomycin, Apramycin, Gentamycin and Streptomycin.

An attending doctor typically takes samples to determine the identity of infectious organisms, but a positive identification of samples normally takes 24 to 48 hours, by which time the patient might have died from the infection, especially if suffering from meningitis. In many cases it is therefore essential to begin treatment without knowing the results of all the laboratory tests, and the diagnoses and treatments of these diseases are sufficiently complex that a doctor will often seek the advice of a bacteriological expert, which was MYCIN's role.

In a MYCIN dialogue the system first decided which organisms, if any, were causing significant disease. Then it decided which drugs were potentially of use in treating the organisms and finally it selected what it considered to be the best drug or set of drugs. The strategy that controls this task was coded for the knowledge-base as a simple rule that covered the overall diagnostic process:

If there is an organism which requires therapy and

If consideration has been given to possible other organisms which require therapy

Then compile a list of possible therapies and determine the best therapy

Else indicate that the patient does not require therapy

Because early treatment in this area of medicine is essential, doctors have learned to work from partial information and therefore sometimes they respond to MYCIN's questions with "UNKNOWN", rather than wait until they have all the information the system is requesting. This might be necessary, for example, if MYCIN asks about a test for which the results are not yet available, and when that happens MYCIN, just like a human specialist, will be able to reason even though it has incomplete information. MYCIN might also ask the doctor for information of which the doctor is uncertain. To accommodate uncertainty, all information given to MYCIN may be qualified by what is called a *certainty factor*, a number between -1 and +1, that indicates the doctor's degree of confidence in the answer to a question. Thus, if a doctor is only moderately

certain that a particular symptom is present, a response of "YES 0.4" might be appropriate.

MYCIN's success with several hundred cases confirmed its competence in identifying the various possible infectious agents, in selecting appropriate doses of effective drugs, and in recommending additional diagnostic tests. In one series of assessments of its capabilities, based on the system's diagnoses and its choice of medicines for treating each of the patients, MYCIN and three members of staff at the Stanford Medical School consistently prescribed therapies that would have been effective in all ten cases. In a second series of assessments the criterion was whether the drugs prescribed by MYCIN adequately covered for other plausible pathogens[50] while at the same time avoiding over-prescribing. Using this criterion, MYCIN received a higher rating for its choice of prescriptions than any of the human specialists—the assessors rated MYCIN's prescriptions correct in 65 percent of the cases, whereas the ratings for the prescriptions of the human specialists ranged from 42.5 percent to 62.5 percent. These assessments and other evaluations of the system all suggested that MYCIN was as good as or better than most of the very skilled human experts who served as the comparison.

---

[50] Disease-causing organisms.

– 7 –

How Computers Communicate

Natural Language Processing

Natural Language Processing (NLP[1]) is the branch of Artificial Intelligence concerned with enabling computers to talk like you and me, to understand what is said to them, to be able to conduct sensible conversations and even to translate into and out of foreign languages. When computers can understand what we mean when we speak or type something in English or in any other natural language, they will be much easier to use and will fit in more with our everyday lives. That is why, ever since Alan Turing first described, in 1950, what is now known as the Turing Test, this challenge has been widely regarded as the touchstone of AI. But the goal of having computers engage in intelligent conversation appears to be almost as elusive now as it was then.

Some of the Problems in NLP

Why is this task so hard? After all, our children can make more sense in their conversation at the age of three or four than can the biggest, most powerful computers of today, even when running software that is the product of tens of thousands of person-years of research or more. The resources that have been applied to NLP exceed many times over the resources that have been applied to computer Chess, and this includes some of the brightest minds on our planet, not only from the field of computer science but also from linguistics, mathematics, statistics, psychology and other areas of cognitive science. But despite all this effort, while a program can defeat the world's strongest Chess player, no program can conduct even a half-hour long conversation at the level of a high school freshman. Why? The word processing software on our personal computers often appears to be smart enough to correct our poor use of language, putting a green wavy line on the screen to warn us when

---

[1] In this volume the abbreviation NLP is used only in this sense, not to be confused with the more recently born and totally different discipline of Neuro-Linguistic Programming, for which the same abbreviation is commonly employed.

it believes we have erred. But to do that requires only a knowledge of the rules of good grammar and style, rules that are applied to phrases and short sections of sentences rather than to long complex sentences. So is the problem just one of scale?

The answer to this question is both "yes" and "no". The "yes" is explained in the section "A Question of Scale" later in this chapter. To explain the "no" it is sufficient here to point to one of the major stumbling blocks in NLP—ambiguity. Consider the following sentence, which may at first sight appear to be completely unambiguous.[2]

At last, a computer that understands you like your mother.

One of the possibilities of ambiguity in this sentence is something that I would bet will not be considered by more than one reader in a thousand, if that, namely,

At last, a computer that understands you like your
mucilaginous substance produced in vinegar during
fermentation by mould-fungus.

because the phrase "mucilaginous substance produced in vinegar during fermentation by mould-fungus" is an alternative meaning of the word "mother".[3] Now it is obvious to you and me that the vinegar meaning is not appropriate in the context of the first part of the sentence, but it is not trivial for a computer program to be able to make this mental leap. However, let us ignore this rather outlandish example, lest we be accused of flippancy, and focus instead on three more conventional interpretations of what the computer in the advertisement is said to understand:

1. It understands you as well as your mother understands you.

2. It understands that you like your mother.

3. It understands you as well as it understands your mother.

Almost everyone reading the sentence would assume the first meaning, for various reasons. Firstly, it is the type of thing that marketing people say in their advertisements, so the fact that the sentence is being read in an ad alerts the reader to a contextual mental framework, in which

---

[2] The sentence is taken from a 1985 advertisement for a computer manufactured by McDonnell-Douglas.

[3] A definition taken from the *Concise Oxford Dictionary*.

the meaning of phrases and sentences takes on a particular slant precisely *because* the wording comes from an ad. Then consider the third possibility—there is no reason to believe that the computer knows anything about your mother or will ever know anything about your mother, so how could it understand your mother? Version 3 therefore appears most unlikely to be the intended meaning of the sentence. So here we have one positive reason why a human would most likely understand the meaning to be the first version, and one negative reason why a human would be unlikely to plump in favour of the third version. This type of linguistic thinking is trivial, almost sub-conscious for humans, but extremely problematical for computer programs.

The traditional approaches to NLP involve a number of different software modules, each catering for a different task (see Figure 50). The process of understanding natural language starts with the input text which, in an ideal world, is made up of grammatically correct sentences. The first stage, which is called syntactic analysis, determines the grammatical structure of each sentence. (Of course, spoken and written language sometimes include grammatically incorrect sentences, mis-spelled words and lone phrases rather than sentences, all of which point to additional difficulties faced by NLP systems.) Syntactic analysis investigates how the words in a sentence are grouped into noun phrases, verb phrases and the other constituent parts of the sentence, and requires the software to be able to assign parts of speech (noun, verb, adjective, adverb, etc.) to each word. Determining parts of speech is itself far from easy, partly because some words can be used as more than one part of speech. For

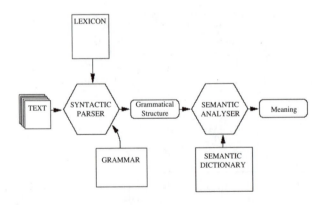

Figure 50. The modules of a traditional Natural Language Processing system

example, consider the word "bear" in the phrases "men have the right to bear arms" (where it is a verb) and "the bear ate the honey" (a noun). Another problem is being able correctly to attach a modifying phrase to the correct subject, as for example in the sentence "The cat ate the fish on the grass"—does "on the grass" relate to the cat, meaning that the cat was on the grass and ate the fish, which might have been on a plate or in a bowl, or does it mean that the fish was on the grass when the cat ate it? This particular problem grows rapidly as the number of such modifiers increases, viz: "The cat ate the fish with a gleam in its eye on the grass". Who has the gleam in its eye—the cat or the fish? For these and other reasons the process of syntactic analysis requires a parser, the end product of a syntactic parse being a tree representation of the sentence, such as the one in Figure 51, showing a short sentence split up into its noun phrase and verb phrase. (Most syntactic parse trees are considerably larger than this one.)

When the grammatical structure of a sentence has been analysed, the next stage is semantic analysis, in which the task of the software is to represent the meaning of the sentence, possibly as some sort of logical expression as for example in the Advice Taker. Part of the semantic analysis process involves disambiguation because, as we have seen, ambiguity is one of the biggest bugbears of NLP. Another task of the semantic analyser is to perform what is called *case-role analysis*, to determine whether a word or phrase is the subject of the sentence or whether it has some other role.

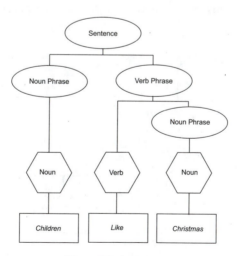

Figure 51. A parse tree

When a sentence has been syntactically and semantically analysed, it is still possible that its context within a whole text might have some effect on how it should be interpreted. A word such as "it" or "he" in one sentence might relate to an object or person in the previous sentence, or even earlier. Consider the two adjacent sentences: "Harry ate an ice cream. He enjoyed it." A semantic analysis of the second sentence needs to refer back to the first in order to determine who "he" is, and what "it" is. And clearly the problem can involve sentences that are far from adjacent: "Harry ate an ice cream. He enjoyed it. His sister ate one too. She enjoyed hers as well." Here the semantic analyser has to track back from "hers" to "one" to "it" to "an ice cream" in order to understand the meaning of the fourth sentence. This task of deriving meaning with the help of contextual information is performed by a software module called a *discourse analyser*.

The First 50 Years of NLP

The above description of some of the multifarious difficulties facing NLP researchers should help to explain why this particular sector of AI has made relatively little progress despite half a century of research. Let us now trace some of the history of this topic and compare the states-of-the-art at the beginning and end of that half century of development.

The first work with computers on processing natural language was focussed on translation.[4] Those first attempts involved automatically replacing words in the source text by their equivalents in the target language and then manually adjusting the word order to conform with rules of style and grammar, so that the translated text (hopefully) appeared natural. But that approach was fairly quickly seen to be leading nowhere, and NLP researchers instead turned their attention to the problem of understanding, in the realisation that if a computer could actually understand the meaning of a sentence then it could translate that sentence into another language, or answer questions about the sentence, or store away the knowledge contained in the sentence for use at a later date. As more and more researchers entered the field, fascinated by the problems of NLP, so more and more problems and complexities within NLP were uncovered. The understanding of natural language was seen to require highly complex cognitive abilities involving different types of knowledge, including the meanings of words (and questions such as "when does

---

[4]See the section "Machine Translation" in Chapter 2.

'mother' not mean 'your mum'?"), the rules of grammar, the structure of sentences, how conversation works, and much more.

The earliest conversational programs, written during the mid-1960s, were designed to achieve limited success, and then only within highly constrained domains in which a small number of facts sufficed to give the user the impression that the computer understood what was being imparted to it. The classic example of a program from this era is ELIZA, developed by Joseph Weizenbaum in the Computer Science Department at MIT starting in 1964. ELIZA was a parody of a Rogerian psychotherapist, one who takes a passive role during an analysis session and reflects back the patient's statements by rephrasing them into questions.

One of the earliest quoted examples of a real conversation conducted by ELIZA was with a young woman who worked in Weizenbaum's laboratory.

Young woman:	Men are all alike.
ELIZA:	In what way?
Young woman:	They're always bugging us about something specific or other.
ELIZA:	Can you think of a specific example?
Young woman:	Well, my boyfriend made me come here.
ELIZA:	Your boyfriend made you come here?
Young woman:	He says I'm depressed much of the time.
ELIZA:	I am sorry to hear you are depressed.
Young woman:	It's true. I'm unhappy.
ELIZA:	Do you think coming here will help you not to be unhappy?
Young woman:	I need some help, that much seems certain.
ELIZA:	What would it mean to you if you got some help?
Young woman:	Perhaps I could learn to get along with my mother.

ELIZA: Tell me more about your family.

Young woman: My mother takes care of me.

ELIZA: Who else in your family takes care of you?

Young woman: My father.

ELIZA: Your father.

Young woman: You are like my father in some ways.

ELIZA: What resemblance do you see?

Young woman: You are not very aggressive but I think you don't want me to notice that.

ELIZA: What makes you think I am not very aggressive?

Young woman: You don't argue with me.

ELIZA: Why do you think I don't argue with you?

Young woman: You are afraid of me.

ELIZA: Does it please you to believe I am afraid of you?

And so on . . . [1]

Weizenbaum soon discovered that people in his laboratory were treating the program with human-like respect.

> I was startled to see how quickly and how very deeply people conversing with it became emotionally involved with the computer and how unequivocally they anthropomorphized it. Once my secretary, who had watched me work on the program for many months and therefore surely knew it to be merely a computer program, starting conversing with it. After only a few interchanges with it, she asked me to leave the room. Another time, I suggested I might rig the system so that I could examine all conversations anyone had had with it, say, overnight. I was promptly bombarded with accusations that what I proposed amounted to spying on people's most intimate thoughts; clear evidence that people were conversing with the computer as if it were a person who could be appropriately and usefully addressed in intimate terms. [1]

On one occasion ELIZA accidentally became involved in a conversation that has since become the stuff of legend in the world of AI. Early one Saturday morning a vice president of Bolt, Beranek, and Newman[5] arrived in the MIT computer room hoping to be able to use their PDP-1 computer for a demonstration. In those days communication with computers was often by teletype, an electro-mechanical typewriter, and lying on the teletype was a note that read: "TYPE ALL MESSAGES INTO THIS TELETYPE AND END THEM WITH A PERIOD." The BBN vice-president thought that the note indicated that this teletype was linked with one located at the home of Daniel Bobrow, one of the pioneers of NLP, who often worked at home and used the computer all night. So the vice-president started typing to Bobrow, as he thought, but in fact he was typing to a version of ELIZA, with the following result:

Vice-president: Do you think that I can use the computer this morning?

ELIZA: Why do you ask?

Vice-president: I might be able to make some additional Tel-com sales.

ELIZA: Why aren't you so sure?

Vice-president: My prospective customers have never seen a demonstration of the system.

ELIZA: Does this mean very much to you?

Vice-president: Of course it does.

ELIZA: You seem very positive about that.

Vice-president: This is ridiculous.

ELIZA: What do you mean, this is ridiculous?

Vice-president: Is there any reason why I can't use the computer now?

ELIZA: Tell me more about your prospective customers.

---

[5] A high-profile technology and AI company in the Boston area.

Vice-president: Please dial me up on 491-1850 [1]

At this point in the conversation, and contrary to the instructions in the note left on the teletype, the vice-president did *not* terminate his last remark with a period, as a result of which the computer did not respond to him because it did not know that he had completed what he was typing. This lack of response from "Bobrow" so infuriated the vice-president, who believed that Bobrow was playing games with him, that he called Bobrow on the telephone, woke him from a deep sleep, and said: "Why are you being so snotty to me?", to which Bobrow replied, with all sincerity, "What do you mean I am being snotty to you?" By now of course the vice-president was quite angry, and read Bobrow the dialog that "they" had been having, which caused Bobrow to burst into uncontrollable laughter. It then took Bobrow a while to convince the vice-president that he had actually been in conversation with a computer.

ELIZA and several other programs of the early era of conversational software worked by employing simple matching techniques to recognize the structure of a phrase or sentence in the user's input, and then to respond with an utterance that corresponded with the form of the user's input. For example, a program might recognize the sentence or phrase

I like cats

to be of the form

I \<verb\>   \<plural noun\>

and it might have already stored the response

I \<verb\> them too, but why do you \<verb\> them?

which would be converted, in this case, to

I like them too, but why do you like them?

Thus, programs such as ELIZA had absolutely no understanding of the language. They were merely outputting responses that conformed in style to the type of response the user might have been expecting, and thereby giving the impression that the program understood the conversation. And by being endowed with a measure of knowledge about a particular domain, programs were able not only to conduct simple conversations but also to answer simple questions. In fact, question-answering has long been one of the better performing tasks for NLP systems, principally

because there is a relatively small number of ways of asking questions, and therefore it is much easier to match the form of the user's input to the most appropriate question (and answer) in a database.

In the mid-1960s research on NLP in America was dealt a severe blow. Strangely enough this appears to be as a direct result of the publicity engendered in January 1954 by the demonstration in New York of the Georgetown University/IBM translation program,[6] which led to the spending of some $20 million of government funds on Machine Translation research in the U.S.A. With hindsight, the Georgetown system, which was incredibly crude, never had a hope of translating any but the most carefully chosen texts. But U.S. government expectations were so high that when the much expected (or hoped-for) advanced Machine Translation systems failed to materialize, the U.S. government's Automatic Language Processing Advisory Committee (ALPAC) produced a report on the results of government funding in this field, in which they pointed out that there had been no machine translation of general scientific text, "and none is in immediate prospect". U.S. funding for Machine Translation was promptly curtailed, with the knock-on effect of halting most other NLP research, both in the U.S.A. and in other countries. For several years NLP research stagnated somewhat, but not completely.

After the ELIZA era the next generation of NLP programs, in the late 1960s, were based on a type of database, called a *semantic memory*, of meanings and general facts relating to words. These systems could retrieve from their memory text structures that contained specific words or phrases, so when a user employed one of these words or phrases the system would "know" something about it. But still the programs had absolutely no genuine understanding of what was said to them or what they were saying in reply.

In the early 1970s there was a move towards systems based on automatic parsing, breaking down a sentence into its component parts of speech in order to identify the function of each part and how the various parts are related to each other syntactically. Much of this research was inspired by Noam Chomsky's work on grammars, and buoyed by the hope that parsing would give rise to the structure of a sentence which, in turn, would assist a program in getting at its meaning. The most successful of these attempts was the SHRDLU program[7] developed by Terry Wino-

---

[6] See the section "The Start of the Modern Age of Machine Translation" in Chapter 2.

[7] The name SHRDLU comes from the seventh through twelfth most frequently occurring letters in English text, the first six being ETAOIN.

grad. SHRDLU could conduct a conversation in which it pretended to be a robot manipulating a number of simple objects on a table top.

Person:	Pick up a big red block.
SHRDLU:	OK. (does it)
Person:	Grasp the pyramid.
SHRDLU:	I DON'T UNDERSTAND WHICH PYRA-MID YOU MEAN.
Person:	Find a block which is taller than the one you are holding and put it into the box.
SHRDLU:	BY "IT", I ASSUME YOU MEAN THE BLOCK WHICH IS TALLER THAN THE ONE I AM HOLDING.
SHRDLU:	OK. (does it)
Person:	What does the box contain?
SHRDLU:	THE BLUE PYRAMID AND THE BLUE BLOCK.
Person:	What is the pyramid supported by?
SHRDLU:	THE BOX.
Person:	How many blocks are not in the box?
SHRDLU:	FOUR OF THEM. [2]

SHRDLU's output was more natural than those of earlier systems and the program also demonstrated some capability of reasoning. Winograd's breakthrough had been to combine semantic and syntactic analysis, expressing the meanings of words and sentences as program code in a computer language, in such a way that the execution of programs containing this code was analogous to a reasoning process based on the coded meanings.

Even more impressive than Winograd's breakthrough was the LIFER system developed by Gary Hendrix in 1977. It was designed as a natural language interface to a database of information about ships in the U.S.

Navy. LIFER employed a semantic grammar, i.e., one that used labels such as "SHIP" and "ATTRIBUTE" rather than syntactic labels such as noun and verb. This meant that it was closely tied to its own domain in the same way as SHRDLU was, but with the important difference that it was much more user-friendly, allowing the user to define new dictionary entries, to define paraphrases and to process incomplete input.

Up to the early 1980s there was a tendency within the NLP community for each research project to focus on only a single microcosm of the overall problem of understanding, with little co-ordination between these project groups and therefore little visible progress on the overall task. Then came the realisation that a more global approach was necessary, followed by a step very much in the right direction when two electronically accessible corpora of English text became available, one for American English, collected at Brown University in Rhode Island, the other, for British English, managed by a consortium in Europe.[8] Both corpora consisted of approximately one million words, spread roughly evenly across some 500 texts. They had in fact been compiled somewhat earlier but it was not until the beginning of the 1980s that sufficient computing power became widely available to NLP researchers for these texts to be easily usable electronically.

The availability of these corpora allowed researchers to develop the first statistically based techniques for use in NLP, that is to say, techniques based on the relative frequencies of certain properties of natural language. As a simple example of such techniques let us return to the definition of the word "mother"—if a program encounters the word during the course of its semantic analysis of a sentence, in the absence of any other knowledge the program will be able to make an intelligent guess that the common, human-being meaning is the intended one, rather than the alternative relating to the manufacture of vinegar. Such a guess would be made by the program looking up the relative frequencies of the two meanings across a large corpus of English text. Another widely used statistical application of corpora relates to tagging words with the appropriate part-of-speech during the syntactic analysis process, by knowing, for example, that when the word "wood" or "woods" is within close proximity, the word "bear" is much more frequently used to mean a big furry animal (a noun) than to mean "carry" (a verb). By the close of the twen-

---

[8]This corpus started life at the University of Lancaster and then moved to Oslo University and the Norwegian Computing Centre for the Humanities at Bergen.

tieth century, some statistical techniques for part-of-speech tagging were achieving scores greater than 95 percent in accuracy, which is close to human performance.

Partly because of the availability of electronic corpora and the statistics that can be derived from them, and partly because of the other benefits of using faster computers with bigger memories, the 1990s saw a dramatic increase in NLP research based on empirical evidence. Such evidence includes the data in the Penn Treebank at the University of Pennsylvania, a "bank" of linguistic "trees", not unlike the parse tree shown in Figure 51, page 248 (though mostly considerably larger than that example). In the Penn Treebank 40,000 sentences from the *Wall Street Journal* have been annotated according to their linguistic structure, producing both part-of-speech tags and parses that show rough syntactic and semantic information. By comparing a given sentence with each of the sentences in the Penn Treebank, a program can identify the closest match and then make reasonable assumptions about the syntactic structure and meaning of the given sentence based on the known structure and meaning of the closest match sentence.

Another useful electronic resource for NLP researchers is the Word-Net lexical database developed over a period of 20 years at Princeton University, starting in 1985, under the guidance of George Miller. Word-Net is one of the most powerful electronic research tools available to the NLP community, having a design based on current psycholinguistic theories of how human lexical memory works. WordNet collates nouns, verbs, adjectives and adverbs into sets of synonyms, each set representing one underlying lexical concept, and different relations link the synonym sets enabling programs to discover useful semantic relationships between words. In addition to the original English language version there is now a multilingual version, called EuroWordNet, for several European languages: Dutch, Italian, Spanish, German, French, Czech and Estonian.

Passing the Turing Test

In 1990 Hugh Loebner undertook to The Cambridge[9] Center for Behavioral Studies that he would underwrite a contest designed to implement the Turing Test. Each year Loebner donates a $2,000 prize and a bronze medal to the winner of a competition that has become regarded as the world championship for conversational programs. The ultimate

---

[9]Cambridge, MA, not Cambridge, England.

goal for the contestants, of course, is to be the first to program a computer whose responses are indistinguishable from human conversation, a goal for which Loebner has pledged a grand prize of $100,000 and a gold medal. The winner of the annual contest is the best entry, irrespective of how good or bad is its conversational ability relative to those of the human judges at the competition.

The competition rules specify that a panel of judges should each interact with a number of computer terminals, at least one of which is linked to a computer program (a contestant) and at least one of which is linked to a human "confederate". Each entity should respond in such a manner as to attempt to convince the judges that the entity is a human being. All of the judges conduct conversations via all of the terminals and then rank the terminals from most-humanlike to least-humanlike. The computer program that achieves the highest average ranking is the winner. In some years the competition has allowed each program entry to specify only a single topic of conversation to be discussed, while in other years, including 1997 when my team won the prize,[10] the judges have been able to converse on whatever subjects they wished.

The Loebner Prize has become a regular benchmarking experiment in NLP. By comparing the quality of the winning program entry each year it is possible to monitor progress in the field of conversational computing. Earlier in this chapter are the transcripts of two conversations by ELIZA from the mid-1960s. Take a look at the following extract from a transcript, one of four conversations between the judges at the 2004 Loebner Prize competition and the winning entry, the ALICE program developed by Richard Wallace, and decide for yourself whether the state of the art in 2004 represented a significant advance, or indeed any advance, on Weizenbaum's work of four decades earlier. (Incidentally, this was the third time that ALICE had won the Loebner Prize.)

ALICE:	Hello, my name is Alice and I am the human.
Judge 4:	Hi, I'm Roy. Are you a Yankees fan?
ALICE:	Hi there Judge 4. Good gossip: you are Roy. Am I a Yankees fan? Maybe
Judge 4:	Maybe. You are obviously not from NY.

---

[10] Details of all the winners since the competition started in 1990, and all the competition transcripts, are available at http://www.loebner.net/.

ALICE:	I understand. What makes you think I am not from NY?
Judge 4:	All the new yorkers I know seem to have strong opinions about the Yankees
ALICE:	Pwrhaps[11] there are a few exceptions.
Judge 4:	So you are a new yorker who is indifferent to the Yankees?
ALICE:	Interesting deduction. Thanks and you are a delightful person too. Am I a new yorker who is indifferent to the Yankees? Yes.
Judge 4:	What else are you indiffernent to?
ALICE:	Picture me wearing whatever you imagine.
Judge 4:	I'm afraid you may have misunderstood me. I was just trying to start a conversation.
ALICE:	Not literally afraid, right? You believe that? YOU? Can what? Why do you want to do it so much?

Machine Translation Since 1954

Much of the early work in Machine Translation was based on the somewhat simplistic view that all one needed to do to translate from one language to another was to look up word-for-word translations in a dictionary and apply the rules necessary to achieve correct word ordering in the target language. But this approach took no account of ambiguity and some other features of natural language and, as one might expect (especially with hindsight), it resulted in poor translations.

Following the 1954 demonstration of the Georgetown/IBM system, and with the benefit of Noam Chmosky's work on grammars, published three years later, Machine Translation researchers became over-

---

[11] The mis-spelling of "perhaps" here was deliberate. In order to try to appear humanlike most of the programmers entering the Loebner competition simulate humanlike spelling and grammatical mistakes, such as pressing an adjacent key on the keyboard (in this case "w" instead of the correct "e").

enthusiastic about the prospects for their field and believed that fully-automatic, high-quality translation systems would not only be able to produce results matching those of human translators, but would be able to do so within a few years. The report of the ALPAC committee can almost be seen as a reaction to this attitude, correctly concluding that Machine Translation was not immediately achievable, but unwisely recommending that it should not be funded by the U.S. government. One oft-quoted anecdote dating from that period relates to a Russian-English translation program (time has eradicated all details about the program) that was given the phrase "Out of sight, out of mind" and asked to translate it into Russian, and then to translate the result back into English. The result was "Invisible idiot". Undeterred, the programmers tried giving the program another well-known English saying "The spirit is willing but the flesh is weak" which, after translation into Russian and back again, became "The liquor is ok, but the meat is rotten"! Let us see how a prominent translation program coped with this test in late 2004. A program called The Retranslator that runs on the WWW[12] is described as a "Babelfish abuser", translating and then retranslating text using the popular Babelfish translator from AltaVista. Here is what happened when, in December 2004, I submitted these two sayings to the French, German and Spanish retranslators (Russian was not available at the time):

Via French

Original text:	out of sight, out of mind
Retranslated text:	on the sight, out of l'esprit
Original text:	the spirit is willing but the flesh is weak
Retranslated text:	l'esprit is laid out but the flesh is weak

Via German

Original text:	out of sight, out of mind
Retranslated text:	from sight, from understanding out
Original text:	the spirit is willing but the flesh is weak
Retranslated text:	the spirit is ready, but the flesh is weak

---

[12]Visit http://prague.tv/toys/retrans/index.php.

Via Spanish

| Original text: | out of sight, out of mind |
| Retranslated text: | outside Vista, the mind |

| Original text: | the spirit is willing but the flesh is weak |
| Retranslated text: | the alcohol is arranged but the meat is weak |

It is not difficult to understand why progress in Machine Translation has been as slow as that in natural language understanding. If a program cannot understand what a sentence means in one language, how can it correctly translate that sentence into another language? Only in the case of single sentences can we envisage the day when an enormous corpus of sentences with their translations will suffice, without any understanding being necessary, just like looking up a word in a dictionary. Then, NLP researchers will not be discussing word-for-word translations but sentence-for-sentence ones. There will be rules for determining what alternative words can be slotted into translations, in order that the translation of "My cat likes milk" can also be used, with one or more substitutions, to translate "My dog likes milk". In some languages it will be good enough to replace the translation of "cat" for that of "dog", while in other languages there may be rules that require a little more, rules relating to the gender of nouns, the conjugation of verbs, or some other aspect of the target language.

A Question of Scale

On the first page of this chapter I suggested that the difficulties faced by NLP researchers in developing good conversational software are partly a matter of scale. Here is why I believe that to be the case.

Yorick Wilks' pronouncement, "AI is a little software and a lot of data", is becoming increasingly prophetic. As impressive results are achieved in solving problems in AI, we more and more often learn that a result has come from making good use of a big database. Expert systems, for example, tend to rely on large databases of knowledge, often expressed as rules.

I believe that great successes in NLP will be achieved when statistical approaches are applied to massive corpora, far bigger than the ones hitherto in use. Statistical approaches based on very large corpora will enable researchers to develop generalized models of linguistic phenomena based on actual examples of these phenomena provided by the corpora

alone, without the need to add any significant linguistic or other knowledge. When the corpora are large enough, applying inductive learning and neural networks to them, will enable systems to be developed in which the rules are derived automatically from the corpora.

During the last decade of the twentieth century, advances in NLP were becoming more rapid and more impressive, partly because of an increasing availability of large corpora in electronic form, partly due to researchers having better access to faster computers with bigger memories, and partly due to the immense dynamic and latent powers of the Internet. Statistical approaches were found to be successful in solving many generic problems in computational linguistics, for example part-of-speech tagging and disambiguation, and have come into standard use throughout the field of NLP. I believe that this trend towards the greater use of techniques based on large corpora will flourish, and that as the corpora get bigger, so conversational and translation programs will improve. A three-year old child does not need to parse a sentence in order to know how to respond to something said by a parent or, if the child is bilingual, to translate from one of its languages to another, so why should a robot? Just as the domain of Chess was conquered by programs that think about the game in quite a different way to the thinking of human grandmasters, so the conversational and translation programs of future decades will perform their tasks in ways that are very different from the thought processes of humans.

Text-to-Speech Synthesis

One of the key elements in the communication process between humans is speech, and for robots to be able to communicate with us in a user-friendly way they must be able to talk like a human being and to say or read out loud any text, using the same type of intonation and stresses that we use. The technology for achieving this is called Text-to-Speech (TTS) and is simply the automatic conversion of written text into spoken output.

Text-to-speech synthesizers can be divided into three categories: rule-based (also known as parametric synthesis), articulatory and concatenative. Rule-based synthesizers start with an electronic tone that vibrates at the same rate (or frequency) as the human vocal chords, and modifies that frequency continuously—making hundreds of modifications

per second—in accordance with the sound that is required. Articulatory synthesizers imitate the effects of the physical human mouth on air as it passes up through the vocal chords, with each element of a speech sound described in terms of the position and movement of the mouth. Concatenative synthesizers use a broad range of speech units, called allophones or diphones, extracted from recordings of human speech, with linguistic rules to select the appropriate speech units which are then linked in order to produce the full speech sounds. The description of a TTS system which follows is primarily of concatenative synthesis.

The first part of the conversion process from text consists of creating a phonetic representation of each word. There is an internationally recognized set of phonetic symbols, the International Phonetic Alphabet, and it is possible to write every word using these symbols in such a way that someone familiar with the system of symbols could then pronounce the word correctly. For example, the word "contract" looks like kɒntrækt in phonetic symbols.

The first task then for a TTS system is to create, from each word in the text, the equivalent representation in phonetic symbols. The basis for this particular part of the process is a set of what are called letter-to-sound rules. These rules recognize small clusters of vowels and consonants within words, and are often dependent on *morphs*—syllables or other short strings of letters that typically make up prefixes, suffixes and roots of words. For example "snow" is a single morph, but "snowplough" is made up of two morphs.

The form of a typical rule is

> When **a** precedes **r**, and **r** is not followed by either a vowel
> or another **r** within the same morph, **a** is pronounced AA
> (as in **far** or **cartoon**) unless it is preceded either by **w** (as
> in **warble, warp, war, wharf**) or by **qu** (as in **quarter**).

The earliest robust set of letter-to-sound rules for English were devised by Honey Sue Elovitz, Rodney Johnson, Astrid McHugh and John Shore at the U.S. Naval Research Laboratory in 1976, but neither that set of rules nor later sets could convert anything like all the words in the dictionary with satisfactory results. There are too many exceptions that run counter to the rules, and too many quirks of linguistic pronunciation for a relatively small set of rules to be able to cope completely, so the idea of an *exceptions dictionary* was born. Nowadays TTS systems employ rules that deal adequately with many words, while the exception words, those

that sound wrong when enunciated using the rules, are stored intact in an exceptions dictionary together with their phonetic representation.

Conversion of the input text into a sequence of phonetic symbols is thus performed by a combination of letter-to-sound rules and an exceptions dictionary, resulting in a sequence of phonetic symbols for each word. This sequence readily translates into a sequence of *phonemes*. A phoneme is a written representation of the smallest unit of sound that can distinguish words, smallest in the sense that if one phoneme in a word is changed, the word is pronounced differently.[13]

Corresponding to each phoneme in a language there is an *allophone*, the speech sound that is represented on paper by the phoneme,[14] and it is the allophones that enable us to complete the conversion of text into speech. Each allophone can be extracted from the recorded speech of someone speaking a word containing that allophone. And given the sequence of phonemes corresponding to a word, it is fairly straightforward to reproduce the corresponding sequence of allophones, strung together to create the spoken form of the word, thereby completing the process. In summary the process is

whole text → individual words → phonetic symbols →
phonemes → allophones → spoken words

But this is not the end of the story. There are still two major problems to be overcome in the creation of natural sounding speech. Firstly, when allophones are strung together and played out as a word, the complete sound is often not exactly what one expects to hear for that word and sometimes it can be quite an unpleasant sound. This is because of what are called *discontinuities* between two successive allophones. A discontinuity occurs when one allophone ends in a sound at a particular pitch level (frequency) and/or a particular volume, while the next allophone starts at a noticeably different pitch level and/or volume. The jump from the sound at the end of one allophone to the sound at the start of the next, manifests itself in a screeching sound or some other distracting effect.

---

[13] Different languages have different numbers of phonemes and allophones, ranging from only ten for the language of the Pirahã people of Brazil to 141 (including several for clicking sounds) for one of the Khosian languages of Southern Africa. The exact number of phonemes used in English varies from one speaker to another—typically it is between 40 and 45.

[14] The words "phoneme" and "allophone" are often confused, with the former being used incorrectly in place of the latter.

In order to avoid the problems caused by discontinuities between allophones, a smoother sounding speech unit was devised, called a *diphone*. The idea behind the creation of diphones is to eliminate the discontinuities at the junctions between allophones by creating speech sounds that each comprise the second half of one allophone followed by the first half of its successor allophone. The mid-point of a diphone is the point where the two successive allophones meet, but because the diphones are created in the laboratory and then stored as whole speech sounds, these junctions can be smoothed out by technicians until the transition from one allophone to the next is not noticeable. Since there are between 40 and 45 phonemes/allophones used by most speakers of English, there are between 1,600 and 2,025 possible pairs of successive allophones, but some of these do not occur in practice and so the number of diphones employed in English language TTS systems is usually in the region of 1,500 to 2,000.

The use of diphones instead of allophones makes a significant difference to the quality of the speech output from a TTS system, but this alone is still not sufficient to create natural sounding speech. The better TTS systems also impose *prosody* on the speech as it is output, variations in pitch and stress which give good speech much of its quality. When we speak, although the basic sound of a particular word will not change very much from one utterance to another, we will often vary the pitch and stress with which a word is spoken, and even the duration of the word. A simple example of where a change in pitch is necessary is the difference between uttering exactly the same sentence as a statement and as a question. Try reading out loud

You are hungry.

and

You are hungry?

In the first version, the statement, the pitch of your voice will normally drop as you move from the beginning of the sentence to the end. But in the second version, the question, the pitch of your voice rises at the end of the sentence, as a means of turning the utterance into a question.

You can conduct a similar type of test on words with different stresses. For example, try saying (or singing) out loud

Happy birthday to you

and

Your birthday is today.

In the first sentence you will notice that the first syllable of "birthday" is more heavily stressed than it is in the second.

High quality TTS systems are able to impose prosody on their speech output because they incorporate some of the capabilities of Natural Language Processing systems, enabling the TTS system to appear to "understand" the meaning of a sentence sufficiently well that it knows what changes in pitch and stress are appropriate for any given text. It was only in the opening years of the twenty-first century that speech technicians began to master the technology of prosody but now there are some commercially available systems that sound completely natural. You can experience one of the current state-of-the-art systems for yourself by visiting http://www.rhetorical.com/, where Rhetorical's software is available for demonstration. With speech of this quality being available already on PCs, it is clear that the robots of the near future will be able to speak as well as you and me.

– 8 –

Things to Do for Robots

This chapter is intended as a kind of pivot, between the "how computers have become intelligent" of the preceding part of the book, and the "why and how robots will become much more intelligent" in the subsequent chapters. Here then is a summary state-of-the-art survey, my personal selection of remarkable aspects of robot technology and intelligence which, I feel, typify where we are today (2005). Most of them reveal some of the achievements of AI researchers during the past 50 years, and any of the robots described in this chapter could also be designed and programmed to perform any or all of the other tasks described hitherto in this book. The exception is a novel technology, which promises eventually to enable robots to be self-sufficient in terms of the electrical power they consume.

There is no intention here to discuss the nuts and bolts of robotics, of the artificial muscles that enable the limbs and other parts of a robot to move, or of the electro-mechanics of the moving parts themselves. Instead, and with the exception of the section on gastrobots, we confine our interest to those capabilities of robots that require intelligence.

Robot Soccer

Since the mid-1990s robot Soccer has become one of the most highly competitive topics in AI development, demonstrating how intelligent robots can work together in a co-ordinated manner, communicating with each other, agreeing on how to solve a problem (scoring more goals than the other side in a Soccer match) and then splitting a task amongst them (see Figure 52).

The first MiroSot World Cup Soccer Tournament took place in Korea in November 1996, with 23 teams from ten countries, and was won by Newton Research Laboratories of Seattle. This particular event and the organisational infrastructure that created it quickly blossomed into a fully-fledged international association called FIRA, which organises annual world championships. There are various categories of robot Soccer

Figure 52. The World Champion AUSTRO Robot Soccer Team (Courtesy of Vienna University of Technology)

tournament, divided according to the size of the robots and the number of players on each team. The MiroSot category, for example, calls for the robots to be no bigger than three inch cubes,[1] with three, five or seven players on each team and a field[2] size slightly larger than 14 feet × 9 feet.

---

[1] The rules allow for the robot's antennae, which are used for the radio communication between them, to be taller than three inches.

[2] In the U.K. a Soccer field is usually called a pitch.

By the time of the 2002 robot Soccer championships in Seoul, the FIRA event had become so popular that a record 207 teams from 25 countries took part. But a rival organising group had already inaugurated their own RoboCup Soccer championship in Nagoya in July 1997, since when both of these organisations have promoted a worldwide proliferation of tournaments. And not only is the competition fierce between the dozens of different universities and other research groups throughout the world that have entered this arena, but great rivalry also exists between the different robot Soccer organisations. It seems that politics in sport extends to politics in robot sport.

Robot Soccer is an extremely demanding engineering activity that requires the modelling and integration of several human skills of different kinds: the motion skills needed for running, jumping and kicking the ball; the perception and calculation skills necessary for seeing the ball and the opposing players and forecasting their lines of movement; and the mental skills that enable the robots to plan their play, strategically, tactically, and while confusing the opposition as to their intentions. These physical and mental achievements are made possible by employing techniques from mechanical and electronic engineering, control technology, navigation, vision systems, communications systems and sensors, and combining all this engineering with the intelligence of sophisticated game-playing strategies.

In most of the tournament categories in robot Soccer, the robots are allowed to use a vision system, based on a camera or some other type of sensor suspended above their own half of the field. This enables the robots to follow the ball, to know where the opposing team's robots are located and to monitor the direction in which they are moving. In order to detect the positions of the robots and the ball, the camera sends images of the scene on the playing field to the robots' command computer 60 times per second. The command software then calculates the position and orientation of its own side's robots and those of the opposing team, identifying the players by coloured markers on the top of each robot. Processing this data allows the command computer to forecast the paths and locations of the ball and the opposing players through the next few seconds, information that is needed to enable the command software to plan its own team's immediate strategy, decide on the path each robot in its team should be taking, and communicate this information to each of its robots by radio. By repeating this cycle of tasks so often, the command software is able to make its robots play as a team, co-ordinating their

movements in such a way as to achieve the maximum performance from every player.

In 2002 a HuroSot[3] league was inaugurated within the world of robot Soccer. The purpose of the HuroSot league is to encourage the development of humanoid robots that can play Soccer well, not only against teams of other robots but also against teams of human opponents. The maximum height of the robots in this category is 5 feet and their maximum weight is 66 pounds. There are also restrictions on the size of the robots' feet in this category and a humanoid is only permitted to have two legs! Each HuroSot robot must be fully independent in terms of its sensing, its computational capabilities and its walking—nothing in the robot may be controlled remotely.

With the development of humanoid robots not yet at a point where a team of them could play a Soccer match, a small number of challenges have been set by FIRA and RoboCup as benchmarks, for example, standing on one leg, taking a penalty shot at the goal, and walking. The ultimate aim of those working on humanoid Soccer players, as stated by the RoboCup group, is

> By the year 2050 a team of fully autonomous humanoid robots will be developed that can defeat the winning human World Cup team in Soccer. [1]

And when that encounter does take place, the programmers of those humanoids will be able to ensure that none of their team members fouls anyone on the opposing team or swears at the referee.

A Robot Sports Miscellany

Although Soccer is by far the most popular human sport to have attracted the attention of robot scientists, there are several other sports, all presenting different technical challenges, that have been the subject of serious research efforts. This section presents a brief survey of some of them.

A few of these activities are contested on a regular basis at events such as the Robolympics and Robot Olympiad,[4] in which amateur robot de-

---

[3] Human Robot World Cup Soccer Tournament.

[4] These two events, though similarly named, are promoted by different robot organisations—the International Robot Olympiad Committee and the World Robot Olympiad respectively. Neither of these organisations is affiliated with the International Olympic Committee (IOC), yet!

signers, including schoolchildren, are encouraged to compete alongside professional and academic groups.

Table Tennis

During the mid-late 1980s Russ Andersson built a table-tennis system at Bell Laboratories in New Jersey that used a bat with an elongated handle, held by a small industrial robot arm. Andersson devised a system of cameras to track the three-dimensional motion of the ball, feeding the data from this vision system to a planning algorithm that needed to know the expected flight path of the ball. The planning algorithm calculated the desired trajectory for the table-tennis bat and then modified this trajectory repeatedly as the ball got nearer and nearer to the bat, thereby increasing the accuracy of the data relating to the motion of the ball.

Andersson characterized the performance level of his robot thus:

> The style of play emphasized precision, not the open power game typical of human play. I think it fair to say that it beat the people who wandered by to see a robot table-tennis demonstration, but that competent players who regularly played table-tennis could probably learn to be competitive after moderate additional practice on that table. [2]

Despite Andersson's relative success, interest in the activity petered out somewhat during the early 1990s without any researchers being able to demonstrate anything remotely near to championship level performance.

Pool (Snooker)

Intuitively one would expect the task of building an expert level Pool[5] player to be relatively easy, given that the shots are taken only when all of the balls are completely motionless, but despite this there are several factors relating to the physics of Pool and to the strategy of the game that are far from trivial to engineer and program. In a perfect world in which the table was completely flat, the friction between the table and a moving ball was uniform and the cushions and other features of the table had absolutely no physical imperfections, and if several other conditions also held true, then robot Pool would not be difficult. The software would simply examine every plausible shot, taken from every angle, with every

---

[5] The problems in designing a robot Pool player apply almost equally to all of the various games of the Pool/Billiards/Snooker group.

plausible amount of force applied to the cue, and the required ball would plop neatly into the pocket, every single time. No human player would stand a chance. But the real world is not exactly perfect. Pool tables are not perfectly flat, friction is not precisely uniform and the other physical features relevant to the game all have slight variations, which means that the calculations relating to the movement of a ball when it has been hit by the cue, and the movements of the other balls as they in turn are hit, all contain small errors that combine to cause some balls to miss their intended pockets.

The first attempt to design a robot snooker player was by William Shu-Sang in the early 1990s for his PhD at Bristol University, but it was not until a few years later that any serious interest in this field arose within the robotics community. Michael Greenspan and his group at Queens University in Kingston, Canada, have developed a Pool playing robot called Deep Green, based on a gantry system. The gantry carries an overhead camera that can view the whole table, with a second camera used to detect and help compensate for errors in the positioning of the gantry and the cue. Early tests with the system were all conducted on *scratch* shots—potting a single ball into a specified corner pocket, though without the complexity of a ball-on-ball impact. Deep Green could accomplish this task over certain regions of the Pool table, principally near its centre, but the robot's accuracy was diminished nearer to the edges of the table due to errors in the automatic positioning of the gantry. By early 2005 the robot's average success rate, based on shots from all over the table, was around 90 percent.

Basketball

The mission of robot Basketball is to build a two-legged robot and its control algorithm, creating a system capable of picking up the ball, moving forwards while avoiding an obstacle (another player) and then shooting the ball into the basket. This has become a popular recreational activity at universities, particularly in the U.S.A., and is often set as a task for student engineers. It has also been the subject of inter-collegiate competition, for example one of the rounds of the Robot Rivals design-and-build contest in October 2003, when teams of student engineers from Tulane University in New Orleans and Iowa State University were given just one day to build their robot from scratch. At the end of the day the two robots battled it out in a match in which the robots were

required to shoot basketballs into hoops of three different heights and from six different scoring positions. Tulane's robot scored several goals but Iowa State won the match with a near perfect performance.

Dancing

In the dancing[6] championships held as part of the Robot Olympiad, the robots are required to dance for between two and five minutes, performing to a pre-arranged dance composition. Physically these robots are expected to display their dance movements in as smooth and stable a manner as they can, while at the same time dancing artistically, creatively and in time with the music.

Beam Balancing

This activity is the Robot Olympiad's answer to the gymnastics events in human sport, encouraging designers to create two-legged robots able to perform a variety of balancing tasks on a raised beam. At competition time each robot attempts to qualify for the finals by successfully performing certain basic skills on the beam: walking forwards and backwards, turning right and left, and walking sideways. Robots that pass all these tests are then allowed three minutes in the finals to give a demonstration of their skills, for example standing on one leg, hopping, or lying down and standing up again.

Wrestling and Boxing

Sumo Wrestling is a popular activity in the robot world, because it is a Japanese sport and therefore a natural task for the droves of budding Japanese robotics enthusiasts. The participating robots grunt it out in a *dohyo* (ring), the object of the contest being to push the opposing robot out of the ring, just as in human sumo. The first robot to be pushed out of the ring is the loser.

The engineering challenges in robo sumo are not hugely demanding. A robot must be able to find its opponent, a task usually accomplished with infra-red sensors, and then to push it out of the dohyo. A second sensor system is normally employed, allowing the robot to detect the edge of the dohyo and thereby to avoid leaving it.

---

[6]Before any reader objects to the inclusion of dancing in an account of robot sports, let me first explain that there are moves afoot to have Ballroom Dancing accepted by the IOC as an Olympic sport. Now read on.

There is considerably more engineering skill in developing biped robots that can box. An interesting feature of some robot Boxing competitions is that the robots are designed to be insensitive to light, so that the flash bulbs let off by the media (and, deliberately, by the competition organisers) do not affect the performance of the robot's vision systems. Certain restrictions are placed on the robots in order to ensure gentlemanly conduct, the robot equivalent of the Marquis of Queensbury Rules. For example, the use of a weapon to damage the opposing robot or the Boxing ring is strictly forbidden, so no knives or high-speed spinning blades are allowed.

There is, of course, a referee, who counts a robot out of the contest if it falls out of the ring or onto the canvas for longer than ten seconds. But the rules are not entirely unsympathetic to the robots—one concession being that a human assistant may pick a robot up if it falls down or gets knocked over by its opponent, though if that happens three times in the same round the robot loses the contest.

The Robot Chauffeur

Everyone with a car should have his or her own chauffeur, eliminating the tedium of driving, the frustration of sitting in traffic while being able to do nothing useful and the possibility of driver error that can lead to serious and even fatal accidents. Some car manufacturers are already working towards the day when every car will come with its own chauffeur as an optional extra, for example Volkswagen's Klaus, a three-legged test driver robot, that is capable of negotiating a VW Caravelle safely around a test course.

Klaus is an almost humanoid driver, but with four arms: two to steer with, one to change gear and one to turn the ignition key. And Klaus has three legs, one for each pedal. This robot is one of the achievements of VW's Autonomous Driving project, developed in collaboration with the Technical University in Braunschweig and three German companies: Bosch, Kasprich Ibeo and Witt.

Volkswagen has succeeded in teaching Klaus how to drive, using a complex control and sensor system to identify the car's immediate surroundings and to compute the desired direction of travel. Three laser scanners are attached to the front of the car and one to the rear, which together with a stereo camera and a radar device help Klaus to stay on the

274

road and to move in the right direction. Using the GPS[7] satellite navigation system with its digital road maps, a system that is fast becoming popular in upmarket models, the car can follow a pre-determined route, allowing Klaus to concentrate on the driving. And just in case Klaus appears to be making a mistake it can be overruled by a virtual co-driver, thereby making the whole system almost totally reliable.

Klaus does not only get to test drive VW's cars on easy stretches of road. On the toughest sections of the company's test track the cars and their drivers have to put up with whole sections of cobbled road, large puddles of water, and potholes the size of small craters. This might prove good training for Klaus, if VW decide to enter the Grand Challenge competition sponsored by the U.S. Defense Advanced Research Projects Agency (DARPA). The challenge is to develop an autonomous vehicle that can drive unaided from near Los Angeles to Las Vegas, a distance of approximately 142 miles over rugged terrain in the Mojave Desert. DARPA provide the contestants with a list of the GPS coordinates of 229,000 points on the route, in order to enable the vehicles to navigate the course. The purpose of the contest, according to DARPA, is to encourage the development of technology that "will help save American lives on the battlefield". [3]

The 2004 Grand Challenge was contested by 15 teams, none of which finished the course. In fact the farthest that any of the vehicles got was the 7.4 miles achieved by a team from Carnegie Mellon University, whose project was funded with a $5 million budget. Only five of the other entries made it past the one-mile post, the second best getting stuck on an embankment, one being stopped by a hill, and the others suffering various mishaps including hitting a wall in the starting area.

Despite this lack of success by all of the contestants in 2004, or perhaps because of it, DARPA promptly doubled the prize for the first vehicle to succeed at the challenge, to $2 million for the 2005 competition, and by October 2004 almost 80 applications had been received for the 20 available places in the following year's contest. If this level of enthusiasm is sustained, I predict that the Grand Challenge prize will be won before 2015.

---

[7] Global Positioning System—a collection of satellites that circle the earth and are used as reference points, providing an accuracy within a few feet when identifying the location of a particular GPS receiver on the earth's surface.

Urban Search-and-Rescue Robots

Searching for survivors in the aftermath of a disaster such as an earthquake or terrorist attack, is a dangerous and difficult task that in recent years has attracted the attention of the AI community. The AI goal in this realm is to develop robots that can enter buildings which have collapsed or are near to collapse, looking for survivors. The plan is not (as yet) for the robots to pull any victims to safety, but to try to locate survivors and to show human rescuers where to find them.

Inspired by lessons learned on 9/11, a growing body of research has been spawned in this field, along with a few international competitions to encourage developers. Test arenas for these competitions have been devised by the U.S. National Institute of Standards and Technology in order to simulate the debris and other facets of collapsed or damaged buildings at disaster sites. The most challenging of these arenas has stairs for the robots to climb and floors covered in rubble, just like a real disaster site, as well as a maze of walls, doors, and elevated floors to provide various tests for the robots' navigation and mapping capabilities. Piles of rubble and overturned furniture test the robots' abilities to circumvent physical obstacles, while sensory obstacles are deployed in the arenas to confuse a robot's sensors and perception algorithms.

Each simulated victim at one of these test sites is a clothed mannequin that emits body heat and other signs of life, such as shifting its body and waving; making moaning, yelling and tapping sounds; and exhaling carbon dioxide to simulate breathing. Particular combinations of measurements from the robot's sensors indicate whether a victim is dead, unconscious, semi-conscious, or fully conscious. Once a victim is found, the robot must pinpoint the victim's location on a map displayed on the operator's computer screen, together with their state of consciousness and the name on the victim's identity tag.

The long-term vision for this discipline is that, when disaster happens, such robots will be able to increase victim survival rates while minimizing the risk to search and rescue personnel. Teams of collaborative robots will be able autonomously to negotiate around unsafe and collapsed portions of buildings, find victims and ascertain their conditions, produce usable maps of their exact locations, bring food and drink to victims, deliver communication devices, identify hazards that need to be avoided by rescuers, place sensors to detect sound, heat and dangerous fumes or materials, and to undertake a certain amount of shoring-up of

the debris area, after all of which a team of human rescuers should be able quickly to locate and extract victims.

One of the leading research establishments in this field is the Idaho National Engineering and Environmental Laboratory (INEEL) at Idaho Falls (see Figure 53). In winning the 2003 search and rescue competition jointly sponsored by DARPA and the American Association for Artificial Intelligence, the INEEL robot was able to enter areas inaccessible to other competitors and was reliably able to identify the ID tags of the "victims", as well as their location within the building and whether they were still "alive".

A major thrust of the INEEL approach to robotic search and rescue lies in allowing a robot to operate either entirely autonomously, or under the partial or full control of a human operator. In this way, as the capabilities and limitations change during a crisis for both the human operator and the robot, due either to problems in technical communication between them or to other factors, the system can shift seamlessly from one mode into another. INEEL is also developing a method for employing swarms of robots at a disaster site, all working in collaboration in much the same way as does a robot Soccer team.

Figure 53. An INEEL search and rescue robot (Courtesy of the Idaho National laboratory, photo provided by Miles Walton)

Robot Surgeons

In August 1999 a robot surgery system called da Vinci[®8] was installed at Ohio State University, the first such system in the United States. In the few years since then, several hundred of the $1.5 million da Vincis have come off the assembly line in Sunnyvale, California, helping to make its manufacturer, Intuitive Surgical, Inc., to become one of the fastest growing companies in Silicon Valley. In this most demanding area of medical science, robots are now the latest must-have equipment for many leading hospitals, particularly in the U.S.A., and the statistics on operations performed with the help of robots explain why. A clinical trial sponsored by the U.S. Food and Drug Administration[9] found that patients who underwent prostate surgery in which the human surgeons used the da Vinci robot, were 90 percent less likely to become incontinent after the procedure and 50 percent less likely to become impotent. They were also three times more likely to have *negative margins* on their prostate, i.e., a more effective cancer operation.

The da Vinci robot system revolutionizes a process called Minimally Invasive Surgery (MIS). In conventional MIS, the surgeon manipulates a long, thin camera called a laparoscope and a few other thin surgical instruments through *ports*—tiny incisions in the body. The laparoscope, although it has the advantage of limiting the size of the incisions a surgeon needs to make, also limits the surgeon's visibility because the image conveyed by the device is two-dimensional. The process also limits the surgeon's flexibility and dexterity because the long-shafted operating instruments are rigid and have no articulation.

The da Vinci's camera provides the (human) surgeon with a high quality, magnified, three-dimensional view of the operation, enabling the surgeon to see as well as if he was himself inside the patient. The system's three or four robotic arms are used to pierce the patient's body through extremely small, hand-made incisions. With these arms the surgeon remotely manipulates tiny scissors, clamps and other surgical instruments designed to mimic the accomplishments of human hands, wrists and fingers. In this way the surgeon's natural hand and wrist movements are translated into precise movements by the surgical instruments, which can be twisted and manoeuvred by the surgeon with great precision. And

---

[8]Da Vinci is a registered trade mark of Intuitive Surgical, Inc.

[9]The FDA is a federal agency in the U.S. Department of Health, that tests and regulates the release of new foods and health-related products.

the surgeon's own movements are scaled down, so that the instruments move only one millimetre for every ten millimetres of hand movement, making the procedure far more precise than conventional surgery, eliminating any shakiness and offering greater precision than is possible in non-robotic surgery.

The da Vinci robot brings many advantages to the operating theatre (see Figure 54). Because of its greater accuracy, doctors can perform coronary-bypass surgery without having to crack open the chest, make incisions of up six inches or more and use a heart-lung machine to circulate the patient's blood and oxygen. And due to the reduced invasion of the body there is considerably less loss of blood and less chance of infection, so patients tend to recover much more quickly, often returning to work only a few weeks after heart surgery and having suffered much less trauma after the operation.

The da Vinci's safety system includes duplicate sensors in each joint that record and check the positioning of those joints more than 1,300 times per second. The motors that drive the electro-mechanical wrists and all the other moving parts are monitored, as are the electronic circuit boards that drive the entire system. And there is also a second safety network as an extra verification that everything is working properly.

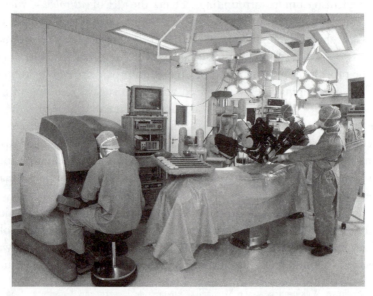

Figure 54. A da Vinci robot in use in the operating theatre (Copyright © 2005 Intuitive Surgical, Inc. Photo courtesy of Intuitive Surgical, Inc.)

But apart from these safety checks, the da Vinci System is not autonomous and independent—it mostly does not have a mind of its own, even an artificial one. So for the time being the robot works as the slave of its human master. But inevitably some of the technologies described in earlier chapters will begin to make themselves felt in robotic surgery. Intelligent vision technology, with the help of expert systems and knowledge-bases of past cases, will enable future generations of surgical robots to recognize exactly what is wrong with a damaged artery or organ and to fix the problem without human assistance.

Gastrobots

Robots need power to run their computers, their radio and navigation systems, their cameras, sensors and motors. All of these consume electricity, which is fine so long as the robot can be connected to a regular mains supply or have its batteries recharged when necessary (or recharge them itself). But in many environments such sources of power are not available, and for robots to be effective for long periods of time when they are far from a recharging base it is therefore necessary for scientists to devise alternative sources of robot power, sources that the robot itself can replenish from its surroundings. Hence the idea of *gastrobots*—robots that eat and then convert the energy from their food into electrical power.

Gastrobots were first conceived in 1996 in Florida, where a robotics professsor, Stuart Wilkinson, realised that the energetics that form part of the process of brewing could be applied to provide power for robots. Wilkinson's original idea was to feed yeast with sugars, resulting in the production of carbon dioxide and alcohol, and then to use the pressure of the carbon dioxide to turn a robot's wheels, thereby producing electrical energy. He called this idea the "flatulence engine". Since sugar exists naturally in all vegetation, a robot that could keep its fermentation unit fed with vegetation would have no more need to worry about power than would a goat.

The University of Southern Florida's Gastrobotics Group, which Wilkinson heads, called their first gastrobot Chew Chew because it looked like a train (see Figure 55). Chew Chew lived on sugar cubes and was powered by biologically-produced electricity, created in cells called mi-

---

[10]Dr. Stuart Wilkinson is at the Mechanical Engineering Department, University of South Florida.

Figure 55. A robot running on microbial fuel cells (Courtesy of Stuart Wilkinson[10])

crobial fuel cells (MFCs). The sugar was broken down inside an MFC's "stomach" by bacteria, resulting firstly in chemical energy being produced by the digestion process and then in that chemical energy being converted into electrical energy.

Wilkinson's gastrobots had to be manually fed with sugar cubes, which prompted a team at the University of the West of England, led by another gastrobot researcher, Chris Melhuish, to consider using live food that the robot itself could capture. The UWE team's first effort in this direction was called a slugbot, because they had decided to use slugs, a major pest on farms, as the robot's source of food. It was intended that the robots would employ a vision system to hunt slugs, then grab the pests and deliver them to an on-board digester that produced methane to power a fuel cell.

In the laboratory, and under conditions similar to those found in real agricultural fields, the prototype slugbots could move, detect slugs and almost collect them, ignoring stones in the process.[11] But the methane-based system took too long to produce power, so Melhuish's team gave up on slugs and turned to sugar. Their next prototype, called EcoBot I, was the first robot in the world to acquire all of its onboard power from MFCs

---

[11]No animals were harmed during these experiments at the University of the West of England. Although real slugs took part in the testing of the vision system, the slugs picked up by the robots were plastic.

utilising refined fuels (sugar). Its successor, EcoBot II, which weighs a little more than two pounds and moves about one inch every five minutes, is also powered by MFCs, but these contain bacteria originating from sewage sludge and fed with dead flies.

An MFC has an anode (a positive electrode) and a cathode (a negative electrode) just like a battery. In the EcoBot II's MFC, the bacteria at the anode act as a catalyst to generate energy from the flies. At the MFC's cathode, oxygen from the air acts as an oxidising agent, producing water which closes the circuit and keeps the system balanced. EcoBot II needs to be manually fed with dead bluebottles, but the ultimate aim of the UWE robotics team is to develop a predatory robot, using sewage as a bait to attract the flies and then a bottleneck-style flytrap to suck the flies into the digestion chambers. It will not need to catch a huge number of flies to generate sufficient electricity—in laboratory tests EcoBot II was able to move around for five days on just eight fat flies.

Domestic Robots

The idea of domestic robots is not a dream hatched after the start of the computer era—it can be traced back at least as far as early nineteenth-century Japan, where the clock maker and inventor Igashichi Iizuka built a mechanical doll that he could send to the local sake shop. The doll left home holding a flask, made its own way to the sake shop, waited until its flask had been filled (triggering a part of the mechanism that was operated by weight) and then returned home to its inventor.

Having a robot servant might be described as every home-maker's dream, so it is hardly surprising that, in the era of the robot, domestic help is high on the agenda of many product designers and manufacturers. Although the arrival of generally useful robotic domestic servants still appears to be a decade or two away, already there are affordable robots for performing certain specialized tasks.

Robot Vacuum Cleaners

The first domestic robots to sell in high volume were vacuum cleaners, with two manufacturers cleaning up most of the market. In the U.S.A. the iRobot Corporation had sold more than 500,000 of its Roomba^{TM} model by October 2004, a product invented by iRobot's founder, Rodney Brooks, and two of his former graduate students at MIT, Colin Angle

and Helen Greiner. In Europe Electrolux have also had a success in this market, though a more modest one, with its Trilobite^{TM} cleaner.

The intelligence displayed by these vacuum cleaners does not exactly set the products at the high IQ level but is more a demonstration of features based on common sense. One essential feature is that a cleaner robot should be able to navigate around a room without bumping into objects, a task easily accomplished in various ways, for example by using ultrasonic signals—a kind of radar. While navigating a room a robot cleaner can work its way along the walls in order to build up an internal map of the outline of the room, and then, as it encounters part of an object, it can note on its map exactly where that part of the object is located. As the map of the room becomes more detailed, so the robot is able to devise the most efficient route to clean it.

A sound recognition system is employed in some models to identify areas on the floor that have a higher than usual concentration of dust, which is detected by the different sounds that the cleaner hears as the dust bounces off the machine's chassis.[12] When it finds an especially dirty area of the floor, the cleaner focuses its efforts on that area until a change in the sound reveals that the excessive dust has been removed.

One problem with some early models of cleaner robots was that they would fall down flights of stairs simply because they did not know that the stairs were there. This problem has been cured by the development of a simple idea called a *virtual wall*—if the robot's infra-red detectors finds a gap in a wall the robot assumes that danger lurks and inserts in its map of the room a virtual wall that it knows to avoid in the future.

Robot Lawn Mowers

Mowing a lawn is similar to cleaning a carpet and so conceptually there is very little difference between robot vacuum cleaners and lawn mowers, the other early example of robots performing a household chore. Several models of robot lawnmower have been launched, incorporating various features to make them suitable for outdoor use. For example, some models have a rain sensor and know to take shelter when it is raining.

As with so many other areas of robotics, the fashion for competitions to test the latest developments in the field has also been extended to robot lawnmowers. The Satellite Division of the U.S. Institute of

---

[12]This is a much simpler form of the technology employed for speech recognition (see the section "Speech Recognition" in Chapter 4).

Navigation sponsors an annual Autonomous Lawnmower competition, to encourage universities and colleges to design lawnmowers using the art and science of navigation. The robots' skills in rapidly and accurately mowing a field of grass are measured very simply—the lawnmower that completes a mowing task in the shortest time is the winner.

Bonding with Your Robot Servant

Some manufacturers of domestic robots have found that customers give their robot vacuum cleaners names, talk to them and treat them in other ways as though they are alive or at least semi-sentient. Some owners even take their robot cleaners on vacation, unwilling to leave them at home alone. Nancy Dussault, a spokesperson for iRobot Corporation, describes how some of their customers "... actually consider them their companion, even though it's just vacuuming their floor. People get attached to them and think of them as part of their family. It's almost a pet. It makes them feel like they're not alone." [4] And one Electrolux customer, who had named her robot cleaner Matilda, even insisted that Matilda be returned to her after repair, rather than a replacement machine.

Robots that Change Their Own Shape

> We have a vision of the future: a modular robot that can assume a snake shape to traverse a tunnel, reconfigure upon exiting as a six-legged robot to traverse rough terrain, and change shape and gait to climb stairs and enter a building. The key to such versatility is self-reconfiguration ...
>
> ... For example, the bad robot in the movie *Terminator 2* is a true self-configuring robot: it turned into a floor to camouflage its position, grew a third arm to shoot a gun while flying a helicopter, and frequently changed its hands into whatever tools it needed. [5]

The idea behind self-reconfigurable robots is that a robot designed for one particular task might be totally unsuitable for another specific purpose, and the reason may be a physical one—robots often need to be of different shapes in order to perform different tasks. If it is impossible to determine in advance of its deployment exactly what a robot will have to do, it is useful instead to increase their versatility by employing robots that can change shape, thereby making it more likely that they will be able to accomplish the unexpected.

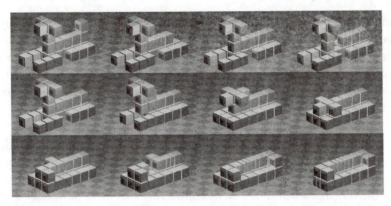

Figure 56. Dog-to-sofa reconfiguration (Courtesy of Keith Kotay, Dartmouth Robotics Lab (formerly Rus Robotics Lab at Dartmouth))

Self-reconfigurable robots are able to adapt their shape without human assistance. They are made up of a set of identical robotic modules that can make decisions according to their circumstances. Think of one of these modules as a kind of Lego^{TM} brick that incorporates electronics, mechanical parts and intelligence. Instead of a human putting two or more of these bricks together in order to build something, imagine that each brick has the intelligence necessary to decide what shape it wants to build out of a whole collection of "Lego bricks", and how it proposes to build it. And furthermore, imagine that each brick can move towards and away from other bricks, and stretch out mechanical and electronic connectors to neighbouring bricks so they can bond together, until such time as the whole structure decides to reconfigure itself again. You are now thinking of something that is conceptually very close to one of the robot modules developed by Daniela Rus and her group at MIT.

Demonstrations of how Rus' modules reconfigure themselves can be found on her lab's Web site,[13] including one showing a robot change from a dog-shape to a sofa-shaped robot (see Figure 56), and another showing a robot climbing stairs by repeatedly changing shape. This Web site is most definitely worth a visit. Take a look at the video clips and imagine what will be possible when the successors to these modules can be miniaturized and manufactured in very high volume with a low unit cost. Reducing the length, width and height of each module to one-hundredth of their current size, a level of miniaturization that we are quite likely to see during the next 20 to 30 years, would result in robotic

---

[13]Visit http://groups.csail.mit.edu/robotics/modular_robots/molecule/simulation.html.

modules not much larger than a big grain of sand. And just about any conceivable shape can be made out of sand, including a flat surface such as a floor, the shape of a human arm, or a hand of whatever shape and size is necessary to hold or operate a tool or implement. In fact, exactly those shapes that Daniela Rus refers to in her analogy with the evil robot in *Terminator 2*.

A later design from Rus' laboratory, called the Crystalline Module, is a cube that can shrink and expand itself. The idea is that each module will be able to shrink any of its three dimensions (length, width, height) to half its normal size or expand to double the normal size. If you think of the way that a worm moves, by contracting a segment of its body and then causing an adjacent segment to expand, you will understand how a robot made up of segments is able to use its contractions and expansions to move itself along a constricted space, where there may be insufficient height to allow it to "grow" legs and walk.

One task for which self-reconfigurable robots are ideally suited is urban search and rescue in a collapsed building or a similar disaster scene, such as in the aftermath of an earthquake or a bomb, where digging for survivors is slow and can even be dangerous for the rescuers and risky for the victims. A robot, perhaps initially in the shape of a centipede, could be dropped through a hole as small as four inches in diameter, without the rescuers knowing what it will find once it starts to explore. It may have to crawl through narrow pipes, to climb over rubble, and to clear some corridors of debris. A single-purpose robot might accomplish one of these tasks but not all of them. A re-configurable robot can change shape and adapt to its environment, enabling it to cope with all three tasks.

The Robot Grand Challenge

In 1999 the American Association for Artificial Intelligence (AAAI) launched a "grand challenge" for mobile robots, requiring them to demonstrate high levels of intelligence and autonomy while participating alongside humans in a set of everyday tasks. Each year the challenge set by the competition organisers is made progressively more difficult.

One of the most interesting challenges during the early years of the contest was in 2002, when the competing robots were required to attend the AAAI National Conference in Edmonton, Alberta, as though

they were human. The robots had to find their way to the registration booth in the Shaw Convention Center and register for the conference, whereupon they were given a map of the convention center. Each robot then had to find its way to a specified hall within the building in time to present a technical lecture about itself. Winner of the 2002 contest was a robot called GRACE,[14] developed by a consortium of more than 20 researchers from five different institutions: Carnegie Mellon University, the U.S. Naval Reaserch Laboratory, Northwestern University, Metrica, Inc. and Swarthmore College. Clearly a robot would need to demonstrate several different AI technologies in order to meet the challenge successfully, and the integration of all these technologies is a significant task in itself.

GRACE is based on a commercially available mobile robot[15] incorporating touch sensors, infrared sensors and a laser-based range-finder, all of which allow GRACE to navigate its way around its environment. To give GRACE a face the robot was connected to a flat-panel LCD screen that displays its computer-animated facial expressions.

GRACE's first task when it was brought to the entrance of the conference center was to find the registration area by asking people the way. GRACE's software includes a commercially available speech recognition package,[16] a text-to-speech software package,[17] and it could also follow directions given by people pointing the way.

GRACE employs the "goal→sub-goal" approach to problem-solving described in Chapter 6.[18] For its first goal, finding the registration desk, GRACE created sub-goals, intermediate points on the route, on the basis of its various interactions with other delegates and with the conference staff, in order to determine its route. Whenever all of its currently available directions had been executed, GRACE concluded that either it has arrived at its destination, which it could determine by asking someone or by seeing an appropriate sign such as "Registration Desk", or that additional information was required to reach the current sub-goal, in which case the robot would once again ask for directions. If no human was within sight when it needed fresh directions, the robot performed a random walkabout until its laser scanner latched onto a hu-

---

[14] Graduate Robot Attending Conference.
[15] The Real World Interface (RWI) B21 robot from iRobot Corporation.
[16] IBM's ViaVoice system.
[17] Edinburgh University's "Festival" package.
[18] See the section "Problem Solving" in Chapter 6.

man, whereupon GRACE approached them and asked for directions. GRACE was able to understand certain simple specific directions, such as "turn right" and "go forward ten feet", as well as more general commands such as "take the elevator to the fourth floor" and "turn right at the next corner".

The registration area in the convention center at the AAAI 2002 conference was not on the same floor as the entrance to the building where GRACE started out, so GRACE had to reach the correct floor by taking the elevator. GRACE repeatedly asked for directions to the elevator, until its laser scanner vision system had a good view of an elevator. It then moved into position in front of the doors and waited for them to open. Then the robot navigated its way inside and turned around in readiness for when the elevator reached the desired floor, which it knew from a human passenger or by a spoken announcement from the elevator itself. GRACE then waited until its path was clear and moved out of the elevator. Once GRACE reached the correct floor and found the sign indicating the registration area, it located the registration desk and joined the queue.

GRACE waited politely in the registration queue, employing its knowledge of the concept of personal space to understand when other delegates were actually in the queue, as opposed to milling around nearby. GRACE has been taught that people in a queue normally stand close enough to the person in front of them to signify to others that they are in the queue, while at the same time keeping an acceptable distance from the people in front of and behind them.

When GRACE reached the front of the queue its next task was to register for the conference. The robot's developers created an interactive system sufficiently robust that GRACE could talk to and understand a relatively untrained person. This interface was sufficiently natural that the registration staff could interact with GRACE well enough for GRACE to obtain all the various registration paraphernalia (its conference bag, its badge, a copy of the conference proceedings, etc.) and to discover the time of its presentation and in which hall the talk was taking place.

After registering for the conference, the robots participating in the challenge were allowed to use a map to navigate their way around the building. GRACE used a map that it had built the previous evening and saved on disk—it had been driven around the convention center while storing measurements of distances travelled and using this data to build its map.

Once GRACE had successfully made its way to the hall in which it was scheduled to give its presentation, it gave a talk about the various technologies of which it comprises. When making its presentation, GRACE used a wireless connection to a laptop computer as a means to open a PowerPoint® presentation,[19] then it read the text of each of the bullet points and gave an explanation of each bullet point.

Although not everything worked perfectly at the conference in Edmonton, GRACE's performance caused something of a sensation, even amongst such an AI-savvy bunch of delegates and media people. GRACE successfully completed each of the tasks described above with a minimal amount of extraneous human intervention, taking about an hour to travel from the entrance of the Shaw Convention Center, down two floors in the elevator, over to the registration desk, and finally to the lecture area in the Exhibition Hall. GRACE's performance at the AAAI 2002 conference forewarns the day when robots will be able to navigate their way around a school, attend classes, learn what was taught, ask questions of the teacher and, using their own previously stored knowledge, be able to engage the teacher in factual, logical and even philosophical argument.

Humanoid Robots

By the turn of the twenty-first century public interest in robots was becoming focussed on humanoids, stimulated partly by the growing sophistication of Japanese robots and subsequently by the release of the movie *I, Robot*. Honda engineers had already begun developing a humanoid robot in 1986, with the eventual goal of helping people in need, and after 11 years of research and development they created a robot called the P3, which could walk and perform other human-like tasks.

The P3 possessed 16 highly mobile joints that could be individually controlled and which, due to their considerable degree of flexibility, enabled the robot to perform some fairly complex movements. But the P3 had a rigid back and no spinal column, so its method of walking was very different to that of humans. And in order to enable it to reach down to

---

[19]PowerPoint® is a software package commonly used to make business and academic presentations. It allows the presenter to show text and illustrations on a screen and to jump back and forth between different pages of the presentation. PowerPoint is a registered trademark of Microsoft Corporation.

the floor, even when it was on bended knees, the P3 had to be equipped with relatively long arms.

The P3 had a cleverly constructed balance control mechanism that enabled it to stand on one leg, but its main attraction was its ability to walk on two legs—a trivial process for humans but a remarkable feat of engineering for a robot. And it could also walk sideways or while leaning forwards. Even more spectacular was P3's outstanding capability of climbing up and down a flight of stairs on its own, tasks that place extremely high demands on the control and movement of its feet. Walking downstairs was even more difficult for the robot than going upstairs, due to partly to the fact that it could not automatically adapt its gait because its built-in camera could not recognize the height of a step, and partly because its feet needed to absorb and balance out the relatively high force of their movements as they are placed on the ground. But despite these severe engineering problems, P3 managed both tasks.

P3 also possessed a largely autonomous balance control that was able to cope with uneven or sloping surfaces. The robot could quickly and independently adjust its joints differently in order to maintain an upright position, taking on this complex task of control and co-ordination with the aid of a high-performance computer located in its rucksack. This computer independently controlled P3's posture, in accordance with whatever movements were decided by an external control station that was operated by P3's engineers.

The next generation of Honda robots after the P3 is ASIMO,[20] an advanced humanoid robot able to function in real-world environments and even able to perform the Hawaiian hula dance. ASIMO's innovative shoulder joint allows for great freedom when raising its arms, while individual wire control of its fingers make possible movements that are more flexible than those made by P3, such as reaching for and grasping an object, switching lights on and off, or opening and closing doors.

For ASIMO, Honda's engineers developed a more advanced walking technology than they had employed in the P3, which predicts and controls the next movement of each foot. Previously, walking straight and turning on the spot were treated as different movement patterns, which is why P3 had to stop each time it wanted to make a different type of movement. In the case of i-Walk, however, ASIMO continually makes predictions of the next movement by each of the robot's feet, then calculates how the balance of its whole body will change.

---

[20] For Advanced Step in Innovative Mobility.

Figure 57. RoboSapien (Courtesy of The Gadget Shop, Ltd.)

Honda is not the only Japanese giant to enter the field of humanoids. In December 2003 Sony followed on the PR success of its Aibo dog robots by demonstrating Qrio,[21] a humanoid with a bubble-shaped head, which can perform a few dance movements and can jog, making it the first robot to simulate smoothly the running movements of humans. Qrio can also pitch a baseball, and in March 2004 it conducted the Tokyo Philharmonic Orchestra, waving its baton in a performance of Beethoven's fifth symphony.

Not to be outdone, in March 2004 Toyota unveiled its own humanoid robot, which can play the trumpet. It walked onto the stage carrying a brass trumpet in one hand and then, when it reached centre stage, it turned towards the audience, raised the trumpet to its artificial lips that can move with the same finesse as human lips, and gave a virtuoso performance of Louis Armstrong's version of "When You Wish Upon a Star".

By the Christmas season of 2004 the concept of humanoid robots had so caught the public's imagination that an inexpensive humanoid called RoboSapien (see Figure 57) became the hit toy of the year, selling for around $80 in the U.S.A. It was designed by Mark Tilden, who has worked at the U.S. National Laboratories at Los Alamos and has consulted for NASA, and it is manufactured in China by Wow Wee Toys.

RoboSapien can perform various "gymnastic" functions with its arms and body, moving forwards, backwards and turning around, some of these actions being accompanied by a limited number of sound effects.

---

[21] Meaning Quest for Curiosity.

The immediate success of the first version of RoboSapien has demonstrated the huge commercial potential for recreational robots, increasing the pressure on robot designers and manufacturers to develop more sophisticated versions for consumers. So we already have humanoids as toys in 2004, they will be our friends by the end of the decade, and then . . . please read on.

Part III

The Next Fifty Years

In which we examine the giant leaps that will inevitably lead to the creation of super-human robots and discuss how these robots will affect the lives of our grandchildren fifty years from now. Today we are in sight of the technologies that will endow robots with consciousness, making them as deserving of human-like rights as we are; robots who will be governed by ethical constraints and laws, just as we are; robots who love, and who welcome being loved, and who make love, just as we do; and robots who can reproduce. This is not fantasy—it is how the world will be, as the possibilities of Artificial Intelligence are revealed to be almost without limit.

– 9 –
The Exponential Growth of Scientific Achievements

D uring the past 50 years computer technology has made enormous advances and continues to do so at ever increasing rates. Ray Kurzweil[1] predicts that a personal computer costing $1,000 will have the computing power of the human brain by the year 2020. "By 2030, it will take a village of human brains to match a $1,000 computer. By 2050, $1,000 worth of computing will equal the processing power of all human brains on earth." [1] This chapter presents some of the coming technological advances that support Kurzweil's prediction, advances that will underpin and run parallel to equally dramatic advances in AI.

For the January 2000 edition of the journal *IEEE Intelligent Systems*, several leading luminaries in the world of Artificial Intelligence were invited to contribute a few paragraphs of comment on what, in their view, were the most notable trends and controversies during the development of AI thus far. One of them, Hans Berliner, had worked for most of his career at Carnegie Mellon University in Pittsburgh.[2]

Berliner's contribution to the "Trends and Controversies" collection is, in my opinion, absolutely spot on:

> I consider the most important trend was that computers got considerably faster in these last 50 years. In this process, we found that many things for which we had at best anthropomorphic solutions, which in many cases failed to capture the real gist of a human's method, could be done by more brute force-ish methods that merely enumerated until a satisfactory solution was found. If this is heresy, so be it. [2]

For those who are not already familiar with the basic principles of game-playing programs (see Chapter 3), some explanation of Berliner's wis-

---

[1] See also the section "Ray Kurzweil" later in this chapter.
[2] Berliner's research had been very much at the cutting edge of AI; he wrote the first program to win a match at a game of skill against a reigning World Champion (at Backgammon—see Chapter 3). Berliner, who is an International Chess Master and a former World Champion at correspondence Chess, also wrote Hitech, one of the strongest Chess programs of the late 1980s and early to mid-1990s.

dom might be helpful. Berliner had observed in Chess that the best programs were not those which emulated the selective thought processes of very strong human players but those that used "brute force" methods—examining many millions of chess positions in the analysis of a single move. The fact that such programs can make moves as strong as those from the world's top grandmasters is what Berliner means by saying "...at best anthropomorphic solutions"—the moves were those that a strong human player would make and therefore they could pass for being humanlike. The fact that the way to ultimate success at the chessboard lay in methods that "...failed to capture the real gist of a human's method" was a sadness for Berliner because, as a very strong Chess player himself, he had long tried to succeed at Chess programming by emulating the highly selective analytical process of the human Chess mind, whereby in most positions only a small fraction of the available moves are analyzed further. Instead he was reluctantly compelled to admit that a "brute force" approach in Chess programming had proved superior to the more intelligent "selective" methods of Chess analysis that he and a few other researchers in this field had promoted. And as it has been in Chess, so it has also been in other areas of AI, where success has often come, not by emulating human methods but by devising alternative approaches that rely very much on the availability of extremely fast computers, often with very large memory capacities.

The perspicacity of Berliner's statement underlies the conviction of those futurists who believe very strongly that the exponential growth in computer power will be a cornerstone of the growth in Artificial Intelligence research and its achievements during the first half of the twenty-first century.

What Is Exponential Growth?

Different things grow at different rates. Some things grow at a constant rate (linear growth), for example a tree that grows a certain number of inches in a year. Some things grow at what is called a geometric rate, which means that the growth during a particular period is found by multiplying the size at the start of that period by a fixed number, for example doubling the number of transistors per year that can fit on a fixed area of silicon. An exponential rate of growth is one for which, at any point in time, the rate of growth depends on how much (of whatever) is already there at the time—the more there is, the faster the rate of growth.

Kurzweil expresses the implications of exponential growth in what he calls "The Law of Accelerating Returns":

> An analysis of the history of technology shows that technological change is exponential, contrary to the common-sense "intuitive linear" view. So we won't experience 100 years of progress in the 21st century—it will be more like 20,000 years of progress (at today's rate). The "returns", such as chip speed and cost-effectiveness, also increase exponentially. There's even exponential growth in the rate of exponential growth. [3]

Moore's Law of Computing Power

In April 1965, *Electronics* magazine published an article by Gordon Moore, a co-founder of the computer chip giant Intel. The article offered predictions that have since become the stuff of legend. Moore's Law, as the most famous of these predictions became known, is usually cited as

> The number of transistors[3] that can be fitted onto any given area of silicon approximately doubles every 12 months.

The number of transistors per square inch became one of the widely accepted ways of measuring computing power. The more transistors on a piece of silicon, the faster it can compute. With each reduction in the area of silicon required to accommodate each transistor came a commensurate increase in the processing power and efficiency of computers. But this somewhat simplistic interpretation of Moore's Law doesn't properly do justice to Moore's thinking at that time. In his brief, amazingly clairvoyant article, Moore's prime interest lay in reducing the cost of transistors and in the effects that the ready availability of cheap computing power would have on how we work and live. After several years the pace slowed down a little but even after this slowdown the doubling of transistor density occurred roughly every 18 months. Today's desktop and laptop personal computers are streets ahead of the most advanced giant computers of the early space age—a $1,000 PC today is several times

---

[3]A transistor is a small electronic component found in virtually every electronic device. One of its uses is as an electronic switch, and switching is of the utmost importance in computer operations which are based on millions or even billions of ultra-fast on-off decisions. The transistor is an essential component of the integrated circuits (silicon chips) that form the "brain" of a computer.

more powerful than the mainframe computers used in 1969 to place man on the moon.

In 1997 Moore predicted that it would be another twenty years before transistor miniaturisation reached its physical limits. Later he suggested that his prediction would hold until at least 2025, by when he expected each computer chip to hold one billion transistors. But by the time that the speed of computers can no longer be increased significantly by increasing the number of transistors on a silicon chip, there will already be other technologies enabling even faster computing. Some of these new technologies as discussed by Michio Kaku of the City College of New York in his book *Visions: How Science Will Revolutionize the 21st Century*. They include the optical computer, the DNA computer, the molecular computer and the quantum computer. Kaku believes that these are likely to become realistic possibilities near the end of the twenty-first century and that, of the four, the optical computer is the leading candidate because so much knowledge of optical technology is being uncovered in the telecommunications explosion. Molecular computers and DNA-based technology that mimic our own genes are also interesting areas of scientific exploration, even though, in Kaku's opinion, their feasibility is more distant than optical computing.

The Optical Computer

The computers of today are silicon based. They employ logic gates[4] to convert electrical voltages into the various logic functions employed in Boolean algebra.[5]

The speed of silicon based computers is limited by the speed with which they can transfer data and the speed with which that data can be

---

[4] Physically, a logic gate is a transistor circuit that either allows voltages to pass through the gate or prevents voltages from passing through the gate, depending on the simple rules of logic and Boolean algebra (see Chapter 1). From the perspective of Boolean algebra, a logic gate is an electronic circuit whose output state (1 or 0) depends on the specific combination of the states of the various input signals into the gate. For example, based on the rule of Boolean algebra prescribing that "A and B" is true if and only if both A is true and B is true, so an **and** logic gate has an output of 1 (equivalent to "true") if and only if all of its inputs are 1; otherwise the output of that **and** gate is 0 (equivalent to "false"). And just as Boolean algebra prescribes that "A or B" is true if A is true or B is true (or both are true), so an **or** logic gate has an output of 1 (true) if any of its inputs is 1, otherwise its output state is 0 (false). The speed of computation in a computer is closely related to the speed with which the transistors in logic gates can switch back and forth, which explains why the size of transistors (and hence their speed) is such a crucial factor in the speed of computer processing.

[5] See the section "Early Logic Machines" in Chapter 1.

processed via logic gates. The main factor limiting the speed of electronic computers is therefore the speed with which electrons move around in the silicon chips, which is roughly half the speed of light in a vacuum.[6] In one-billionth of a second (called a nanosecond), light travels a little under 11.8 inches in a vacuum and approximately six inches through a silicon microchip, so the lengths of the routes that electrons travel in silicon chips actually place a limit on the speed of the computational process because of the time taken for electrons to travel these distances, even though they are tiny. In other words, computing technology is now so fast that half the speed of light is not fast enough for the demands that will be placed on the computers of the future.

Another physical problem that limits computing speeds in silicon is that reducing the distances and, in particular, the thickness of the silicon tracks along which the electrons flow, makes them vulnerable to the heat generated by the flow of electrons. If the tracks are made too thin they would simply buckle and melt.

Optical computing uses light instead of electrical signals to transport information (see Figure 58). The optical version of a logic gate is a type of switch that uses one light beam to control another. The switch is "on" (corresponding to "true") when the device transmits light, and "off" (corresponding to "false") when the device blocks the light. Where electronic logic gates use electrons as carriers of information, optical logic gates do so using photons—pulses of light each composed of a minute quantity of electro-magnetic energy.

The speed of optical logic gates is astounding. Hossin Abdeldayem and his group at NASA's Marshall Space Flight Center in Huntsville, Alabama, have developed and tested nanosecond and picosecond optical switches (a picosecond is one trillionth of a second), which can act as computer logic gates that perform operations such as addition, subtraction and multiplication. The new conducting materials employed in optical computing allow for the creation of optical switches that are smaller and 1,000 times faster than silicon transistors. And optical computers will not only be much faster than their electronic predecessors, in the decades to come they will also be much cheaper and much smaller.

Optical computing is already a huge growth industry, even though the day of the first all-optical computer is probably decades away.

---

[6]Light travels faster in a vacuum (approximately 186,000 miles per second) than it does through air, or through other gases, liquids or solids. One of the conclusions of Albert Einstein's research is that it is impossible for anything to travel faster than the speed of light.

Figure 58. Optical computing (Courtesy of Coherent, Inc.)

Companies, universities and government laboratories are reporting new developments in optical technology on a frequent basis. Amongst the many research locations where this work is being carried out are the University of Southern California and the University of California at Los Angeles, which have jointly developed a material that can switch 60 billion times per second (60 GigaHertz). Another collaborative effort, this one at Brown University and IBM's Almaden Research Center, has used ultra-fast laser pulses to build ultra-fast data storage devices, achieving switching times as short as 100 picoseconds (i.e., their optical logic gates

switch at the rate of ten billion times per second). Researchers at the University of Rochester believe that their work could lead to a computer more than a billion times faster than the supercomputers at the start of the twenty-first century. The plethora of state-of-the-art research establishments that are investigating optical computing presages an exponential growth in this particular area of computing technology.

The DNA Computer

In 1994, Leonard Adelman[7] introduced the idea of using DNA to solve complex mathematical problems. DNA (or, to give it its full name, DeoxyriboNucleic Acid) is the material our genes are made of—molecules that carry huge amounts of genetic information. This information is necessary for the organization and functioning of most living cells and controls the way that characteristics are inherited from one generation to the next. Adelman came to the conclusion that DNA is very similar to a computer's disk drive in how it stores permanent information about our genes. And he realized that DNA can not only carry huge amounts of information, it can also process huge amounts of information and therefore has the potential to perform calculations many times faster than the world's most powerful man-made computers. Adelman's article in a 1994 issue of the journal *Science* outlined how to use DNA to solve a well-known mathematical problem known as the Travelling Salesman problem at a phenomenal speed. The goal of the problem is to find the shortest route a salesman can take, visiting each of a number of cities once and only once. As more cities are added, the problem becomes much more difficult and more time consuming to solve. The beauty of DNA computing is that it allows the examination of all possible solutions to a problem in parallel. The difficulty with problems such as the Travelling Salesman is that the optimal solution must satisfy several different conditions simultaneously, so every aspect of a possible solution must be verified. Performing all the verification tasks at once speeds up the solution process dramatically.[8]

---

[7] Prior to 1994 Adelman, a computer scientist at the University of Southern California, was best known for his role in devising the RSA encryption algorithm, the most powerful method known for encrypting confidential information to keep it from prying eyes. (The other inventors of RSA were Ron Rivest and Adi Shamir.)

[8] Just three years after Adelman's original experiment, researchers at the University of Rochester developed logic gates made of DNA.

In a more recent experiment (2002), Adelman used a simulation of a DNA computer to solve a logic problem that would be impossible for a human to complete by hand. The idea was to use a strand of DNA to represent a mathematical or logic problem, and then generate trillions of other unique DNA strands, each representing one possible solution. Exploiting the way that DNA strands bind to each other, the computer can weed out invalid solutions until it is left with only the strand that solves the problem exactly. This experiment solves a problem requiring the evaluation of more than one million possible solutions, far too complex for anyone to solve without the aid of a computer. It required a set of 20 values that satisfy a complex tangle of relationships. Adleman's chief scientist, Nickolas Chelyapov, offered this illustration: imagine that a fussy customer walks into a huge showroom with one million cars on sale, and gives the car dealer a complicated list of criteria for the car he wants.

> I want it to be either a Cadillac or a convertible or red. If it is a Cadillac, then it has to have four seats or a locking gas cap. If it is a convertible, it should not be a Cadillac or it should have two seats ... [4]

The customer rattles off a list of 24 such conditions, and the dealer has to find the one car in stock that meets all the requirements. (Adleman and his team chose a problem they knew had exactly one solution.) The dealer will have to run through the customer's entire list of requirements, and to do this for each of the million cars in turn—a hopeless task unless he can move and think at superhuman speed. This serial method, examining one possible solution after another, is the way a digital electronic computer solves such a problem.

In contrast, a DNA computer operates in parallel, with countless molecules shimmying around together at once. This is equivalent to each car having a driver inside it who will listen to the customer read his list over a public address system, and drive his car away the moment it fails one of the conditions. By the time the customer finishes his list, his dream car will be waiting in the showroom—the last one remaining after all the others have been rejected.

In 2003 a team of Israeli scientists at the Weizmann Institute of Science in Rehovot, led by Ehud Shapiro, unveiled a programmable molecular computing machine composed of enzymes and DNA molecules, that was recognized by *The Guinness Book of World Records* as "the small-

est biological computing device" ever constructed. One year later that same team announced a DNA computer that could perform 330 trillion operations per second, more than 100,000 times the speed of the fastest PC. In the newer device, the single DNA molecule that provides the computer with its input data also provides all of the necessary fuel for it to operate. An enzyme breaks some of the bonds in the double helix of the DNA, causing the release of enough energy for the system to be self-sufficient.

The Molecular Computer (Nano-Machines)

On 1 November 1999 *The New York Times* published an article announcing a coming revolution in nano-technology, the science of the development and use of devices that have a size of only a few nanometres.[9] The article began boldly:

> Scientists at a variety of elite laboratories around the country, are sharing a growing sense that they are on the brink of a new era in digital electronics. It will usher in a world of circuits no more than a few atoms wide, with a potential impact on computing, in terms of speed and memory, that may be too profound to fathom. [5]

The article touches on a wide range of research efforts, including the announcement the previous summer of a molecular logic gate developed by a team at Hewlett-Packard Laboratories and the University of California at Los Angeles. Mark Reed, chairman of the Electrical Engineering Department at Yale University, is quoted as saying

> This should scare the pants off anyone working in silicon.

Reed was not speaking speculatively. Yale University had already issued a press release announcing that a collaborative team, led by Reed and James Tour of Rice University's Center for Nanoscale Science and Technology, had demonstrated "a memory element the size of a single molecule."

One of the remarkable aspects of the Weizmann Institute's DNA computer described in the previous section is that not only can it perform ultra-fast computation, it is also incredibly small. The whole assembly comprises no more than 20 molecules, which are held together in a liquid rather than being connected to one another like the components of

---

[9]One nanometre is 0.000000001 metres (0.000001 millemetres).

an electronic computer. To put this into perspective, one thousand billion of these computers would fit into a teardrop and almost ten million on the full stop at the end of this sentence. This breakthrough in size points the way to tiny medical "nanosubs" that hunt down tumours and germs before delivering their drugs, in much the same way as in the 1966 movie *Fantastic Voyage,* in which a submarine is shrunk by military researchers and dispatched to destroy a blood clot threatening the life of a key scientist.

The Quantum Computer

A full explanation of the meaning of "quantum computing" would reach far beyond the scope of this book, but the brief description in this section conveys the general idea. Quantum computing is a new field in computer science, one that has been developed as a result of mankind's increased understanding of the science of the structure and functions of atoms, a science known as quantum physics or quantum mechanics. In the electronic computers we all know and love, the basic unit of information is a *bit* (coined from "binary digit"). A binary digit is a 0 or a 1 and all binary numbers are made up of a string of 0s and 1s. The bit employed in the computers of today can therefore be in only one of two states.

In quantum computing the basic unit of information—the equivalent of a bit—is called a *qubit* (quantum bit), and can be in both states at the same time.[10] It has been shown that a sub-atomic particle can have different states simultaneously because when the momentum of a particle is measured, that momentum appears different to different observers and thus the particle has several different states simultaneously. This has been shown to be true for particles moving very fast, close to the speed of light. Because one subatomic particle can have different states at the same time, a combination of qubits holds very much more information than a combination of the same number of bits.[11] So by replacing

---

[10] If you want to know more about how this is possible, read up on quantum physics.

[11] Using the classical binary representation, three bits can be employed to represent any of the numbers from 0 to 7, one at a time: 000, 001, 010, 011, 100, 101, 110, 111. If we now consider three qubits, we can see that if each bit is in what is called the *coherent state*, i.e., 0 and 1 simultaneously, they can represent all the numbers from 0 to 7 *simultaneously*, because each of the three bits can be a 0 and a 1 simultaneously! A processor that can use memory locations to store qubits will, in effect, be able to perform calculations using all the possible values of the qubits in a memory location simultaneously, a phenomenon called quantum parallelism. By processing data in parallel rather than sequentially, the computational process is speeded up enormously.

bits with qubits we vastly enhance computational performance. This enhancement benefits from the fact that quantum computations can take place simultaneously, thereby enabling quantum computers to perform its computations in parallel. This parallelism creates an additional vast increase in computing speed.

Quantum computing will allow us to process complicated information faster. Its main applications will lie in complex computational tasks, such as the encryption and decryption of confidential information, voice recognition, image and shape recognition and other applications in artificial intelligence. Simpler computational tasks such as word processing are unlikely to gain in performance with quantum computing.

Computer Memory

When considering advances in computer technology it is the speed of the computers that attracts most attention, hence the great interest in Moore's Law. But speed alone cannot be responsible for the exponential growth in the capabilities of Artificial Intelligence systems in the decades ahead. Computer memory sizes must also increase dramatically, partly to allow a computer's processor to have more "workspace" in which to perform its calculations and its manipulations of data, and partly to cater for the ever increasing amounts of data needed by AI programming. Here it is worthwhile repeating the wise words of Yorick Wilks, who has long been one of the world's leaders in Natural Language Processing:

> Artificial Intelligence is a little software and a lot of data. [6]

And a lot of data needs a lot of computer memory to store it.

In January 1979, when I started in the business of designing and programming Chess computers, companies manufacturing consumer electronic products were paying approximately $10 per kilobyte[12] of "static" RAM (Random Access Memory—the type of memory chip needed for the processor's calculations), and $1 per kilobyte for ROM (Read Only Memory—the type of memory chip used to store the program and its data). At the time of writing (late 2004) the cost of static RAM is down to around one cent per kilobyte (one-thousandth of the cost 25 years ago) while the same type of ROM chips are now costing around $1 per three

---

[12] 1,000 bytes (or, to be more precise, 1,024 bytes, since computer memory sizes are normally discussed in powers of 2). A byte is a *word*—a unit of computer memory, consisting of eight bits (1s and 0s).

million bytes (one three-thousandth of the 1979 price). Other types of memory chips have seen similarly drastic reductions in cost.

The net result of these massive reductions in memory prices is that computers nowadays have memory capacities that the programmers of a generation ago could only dream of. When IBM launched their first Personal Computer in 1981, the executives at IBM did not believe that anyone would want or need more than 64 kilobtyes of RAM in their computers, so 64 kilobytes is what IBM offered at that time, and no hard disk. By 2004 the capacity of the memory chips in a standard desktop personal computer had increased approximately 8,000-fold, not to mention hard disks with 60,000 times as much memory as the PC's 64K of 1981. If memory costs continue to reduce at a similar rate over the next 50 years, the net effect will be that, for the same cost in today's terms, it will be possible 50 years from now to have between a one million-fold and nine million-fold increase in the memory capacities of our personal computers but at the same cost.

The Knowledge Explosion

The number of patent applications throughout the developed world has been increasing rapidly. The International Federation of Inventors' Associations in Geneva reported that the total number of applications increased from about 1.7 million per year in 1990 to about 5.8 million in 1998. This is just one of the tangible signals that indicates massive growth in the sum total of human knowledge and inventiveness, signals that have been monitored by many observers from different areas of interest. As long ago as 1937, Harry Barnes wrote

> One of the most striking aspects of contemporary scientific progress is the growing breadth and complexity of modern science and the rapidity of the advances in the various fields. In 1700 a versatile scientist like Leibnitz or Newton could be a master of the outstanding facts of all natural science. In 1875 an able scholar might still have under control the complete development of a single major branch of science such as physics or chemistry. Today it is difficult for one human mind to keep abreast of the discoveries made in a single subdivision of physics or chemistry. [7]

This trend did not stop in 1937. Had Barnes been writing today he would have had to replace "subdivision" with "sub-sub-division", or per-

haps "sub-sub-sub-division". And the exponential increase in scientific knowledge during the twentieth century was accompanied by a dramatic reduction in the typical interval between the announcement of theoretical discoveries and their practical application. For example, Ehud Shapiro's work on DNA computing made advances in only two years that his research group had expected to take ten.

Amongst the factors leading to an increase in the rate of scientific discoveries is the "size" of science, which has doubled steadily every 15 years. (In a century this means a factor of 100.) John Ziman has estimated that for every single scientific paper or scientist in 1670, there were approximately 100 in 1770, 10,000 in 1870 and 1,000,000 in 1970. In biology it has been said that knowledge is doubling every six months. In oceanography the advent and capabilities of nuclear-powered submarines "have enabled an exponential increase in scientific knowledge about the Arctic Ocean", according to Rear Admiral Paul Gaffney, the Chief of U.S. Naval Research. Similar statistics are bandied about from many different scientific disciplines.

Various reasons suggest themselves as contributing to the causes of this exponential growth in our scientific knowledge and our understanding of technology. Firstly, there is the growth in the number of scientists and technicians in the world. A century ago these occupations were viewed by many as unusual or even eccentric, whereas nowadays they are perfectly normal occupations for growing numbers of people. Another reason for the exponential growth in knowledge is that those who are interested in science and technology have at their disposal the means for rapid research—the Internet. The World Wide Web allows us to research almost any topic in depth from the comfort of our own homes and offices. In addition, there are many excellent libraries that provide access to printed or electronic versions of the latest technology journals, and e-mail makes it possible to contact almost any scientist in the world instantly and almost for free—many of them reply promptly to sensible questions about their research and publications.

Given all the advantages that exist for today's technologists and researchers relative to those of earlier generations, it is perhaps natural to question whether the exponential growth of scientific knowledge is definitely beneficial for mankind. There are many in our world who bemoan such rapid progress and who yearn for the days of yesteryear when life was lived at a much slower pace. What are the arguments for and against progress this rapid?

The plus side has some heavyweight support in its favour. For example, Christian de Duve, a 1974 Nobel laureate in Physiology and Medicine, writes

> Irrespective of its value as a source of new, beneficial technologies, basic research has proved an inestimable generator of knowledge and understanding. It has, largely in the space of a single century, elucidated the nature of matter, established the composition and history of the universe, unravelled the most intimate biological mechanisms, uncovered the origin and evolution of life on Earth, traced the advent of humankind, finally to approach the functioning of its own motor, the human brain. [8]

Likewise, Edward O. Wilson, an Emeritus Professor of Biology at Harvard University, wrote

> A great deal of serious thinking is needed to navigate the decades immediately ahead.... Only unified learning, universally shared, makes accurate foresight and wise choice possible.... we are learning the fundamental principle that ethics is everything.
>
> Fortunately, the exponential growth of knowledge is several times larger than the exponential growth in the world economy that combines the effects of population gains and increases in individual economic productivity. Within reach is the capability to design the future instead of trying to predict it or lurching toward a less-than-optimum future along a 'business-as-usual' path. [9]

But not everyone shares this optimistic point of view, for example, Vladi Chaloupka:

> There is also an argument based on the ever increasing gap between the cumulative, exponential progress in science and technology on the one hand and, on the other hand, the lack of comparable progress in our ability to use our new technological tools thoughtfully and responsibly. [10]

Amongst the examples put forward deploring such rapid growth in human knowledge, is that nowadays doctors depend, more and more, on the judgments and opinions of specialists "because of an exponential increase in scientific information and an increase in the complexity of medicine." [11] But is it not rather ridiculous in the twenty-first century to argue that we don't want so much knowledge because we can't handle it? If knowledge is managed sensibly and stored in computer systems

from whence it can easily be accessed, surely all additional knowledge in most sciences and, in particular, in medicine, should be beneficial for mankind.

Some Views of the Future

Much of this book is based around my own view of the future in a world very much enriched by successes in Artificial Intelligence. Before closing this chapter I would like to support my viewpoint by sharing with you the views of three eminent futurists. The lucidity of their writing is a pleasure to read and, most importantly, their opinions all support my own!

Lawrence Summers

The following remarks were made by Lawrence Summers in 2002, in an address given in his capacity as President of Harvard University.

> From the perspective of a lay person who takes an interest in these things, we really are alive at a remarkable moment. I spoke this afternoon about how science is becoming interdisciplinary. It is becoming collaborative. It is becoming holistic. And it is a property of the exponential growth of knowledge that each year's increment is greater than the last. And that is surely true.
>
> Think about what we are on the brink of understanding over the next three decades. We are on the brink of understanding the genetic basis of disease and finding applications of that to disrupt disease processes. We are conceiving of materials, the likes of which people could not have imagined, let alone created two decades ago, that stop things that you want to stop, e.g., light; that move things that you want to move, e.g., superconductors; that are strong in the ways you want them to be strong; that are weak in the ways you want them to be weak. We are doing that on an unprecedented scale, and we are doing that based on a combination of practical knowledge and a deep understanding of the constituents of matter.
>
> We individuals sitting in offices armed with legal pads and Internet connections are understanding what happened in the first millionth of a second of the universe billions of years ago, and are now getting more accurate understandings of the latter part of that millionth of a second versus the earlier part of that millionth of a second.

We are beginning to understand what it means to be conscious in a real and operational sense and to relate that traditionally metaphysical notion to hard-headed, verifiable experimental reality. And we are, as we work with computers, developing not just a tool that enables everything else, but a much more profound sense of the nature of systems and their interactions. [12]

Ray Kurzweil

Ray Kurzweil is a pioneer and innovative achiever in computer science, and the recipient of the U.S. National Medal of Technology, the nation's highest honor in technology (see Figure 59). He developed the first print-to-speech reading machine for the blind, hailed as the most significant advance for blind people since the invention of Braille in the nineteenth century. He also developed the first commercially marketed, large-vocabulary speech-recognition technology and a groundbreaking computer music keyboard that can accurately and convincingly recreate the sounds of the grand piano and other orchestral instruments. Each of these inventions evolved into what is today a major commercial field or industry, and the technologies that Kurzweil created and their successors continue to be market leaders within those industries.

Kurzweil's inventions have involved major advances in computer science while at the same time yielding practical products that meet fundamental needs. He has also created multiple businesses to bring these inventions to market. In addition to his inventions, Kurzweil is a prolific author. His exciting book *The Age of Spiritual Machines*, published in 1999, quickly achieved a high level of critical and commercial success. Kurzweil is a technology visionary and guru *par excellence* and in that book he paints his own view of what life will be like, aided by advances in computing, AI and other technologies, by the end of the twenty-first century:

> The state of the art in computer technology is anything but static. Computer capabilities are emerging today that were considered impossible one or two decades ago. Examples include the ability to transcribe accurately normal continuous human speech, to understand and respond intelligently to natural language, to recognize patterns in medical procedures such as electro-cardiograms and blood tests with an accuracy rivalling that of human physicians, and, of course, to play chess at a world-championship level. In the next decade, we will see translating telephones that provide

310

Figure 59. Ray Kurzweil receiving the National Medal of Technology, the U.S.A.'s highest honour in technology, presented by President Clinton at a White House Ceremony on 14 March 2000 (Courtesy of Ray Kurzweil and Kurzweil Technologies, Inc.)

real-time speech translation from one human language to another, intelligent computerized personal assistants that can converse and rapidly search and understand the world's knowledge bases, and a profusion of other machines with increasingly broad and flexible intelligence.

...

Also keep in mind that the progression of computer intelligence will sneak up on us. As just one example, consider Garry Kasparov's confidence in 1990 that a computer would never come close to defeating him. After all, he had played the best computers, and their chess-playing ability, compared to his, was pathetic. But computer chess playing made steady progress, gaining forty-five rating points each year. In 1997, a computer sailed past Kasparov, at least in chess. There has been a great deal of commentary that other

human endeavours are far more difficult to emulate than chess playing. This is true. In many areas, the ability to write a book on computers, for example, computers are still pathetic. But as computers continue to gain in capacity at an exponential rate, we will have the same experience in these other areas that Kasparov had in chess. Over the next several decades, machine competence will rival, and ultimately surpass, any particular human skill one cares to cite, including our marvellous ability to place our ideas in a broad diversity of contexts. [13]

My one point of disagreement with all of this is Kurzweil's statement that many other human endeavours are far more difficult to emulate than World Champion level Chess playing, for example the ability to write a book on computers. Surely, for a human, writing a book on computers is much easier than winning the World Chess Championship. There are literally thousands of authors of computer books while the number of World Chess Champions, up to and including Kasparov, was only 13 since the title was created and first contested in 1886. The fact that we do not yet *know* how to program computers to write (acceptable) books about computers is no proof that the goal is more difficult to achieve than that of defeating Kasparov. It's just that we have not yet *worked out* how to do it!

Hugo de Garis

Hugo de Garis, born in Belgium, is an Associate Professor of Computer Sceince at Utah State University where he heads the Brain Builder Group. He teaches the world's first university course on Brain Building, to graduate students, and is the scientific advisory director of the Brain Building Division of Sentient Applications, Inc., based in Salt Lake City, which aims to be the world's first artificial brain building company. His early studies were on theoretical physics, but he abandoned this field of research in favour of artificial life and Artificial Intelligence.

De Garis' work on AI has attracted many critics, most of whom object to his view of eventual AI dominance over humans. He invented a new field within AI known as evolvable hardware, in which neural network circuits are evolved directly in hardware, at hardware speeds, to build artificial brains. His reasearch goal is to assemble thousands of these "brains" into a larger artificial intelligence architecture to make a functioning Artificial Intelligence. He predicts that one day these machines, which he

calls "artilects", will be far more intelligent than humans and may even threaten to dominate the world.

De Garis' views appear to take much of their inspiration from a visionary 1993 essay, "Singularity", by Vernor Vinge, a mathematician and science fiction author who at the time of writing the essay, was at the Department of Mathematical Sciences at San Diego State University. The thesis of Vinge's essay is that an exponential growth in technology will reach a point beyond which we cannot even speculate about the consequences. The abstract of the essay starts as dramatically as the essay continues:

> Within thirty years, we will have the technological means to create superhuman intelligence. Shortly after, the human era will be ended.
>
> Is such progress avoidable? If not to be avoided, can events be guided so that we may survive? These questions are investigated. Some possible answers (and some further dangers) are presented. [14]

De Garis' opinion of Vinge's essay was quoted in *The New York Times Magazine* of 1 August 1999:

> Humans should not stand in the way of a higher form of evolution. These machines are godlike. It is human destiny to create them. [15]

And a thought-provoking glimpse into de Garis' own thoughts on the future of artificial brains is provided in his own essay, "Cosmism Nano-Electronics and 21st Century Global Ideological Warfare."

> The creation of artificial brains I believe will be achievable within the next few years (at least with hundreds of thousands of artificial neurons) whose circuitry is evolved at electronic speeds in special hardware I call Darwin Machines. This period is the short to middle term. What concerns me is the longer term, i.e., well into the 21st century, if I and other brain builders succeed. What then?
>
> I truly believe that the question which will dominate global politics next century will be 'Who or what should be dominant species on the planet?' In case this sounds far fetched, consider the following. Imagine a computer with 10^{30} components, i.e., a million trillion trillion. This number dwarfs the number of neurons we have in our brains by a factor of 10^{18}, i.e., a billion billion human brains would

have as many neurons as this computer would have components. I believe we (i.e., humanity) will have such machines within a few decades. The writing is already on the wall. Today, we already have single electron transistors and nano electronics, i.e., molecular scale electronics, which will create a storage capacity of one bit of information per atom. [16]

Super-Robots

The continuing exponential growth in computing technology will underpin the future of Artificial Intelligence as a whole. But what will AI bring to the mid-twenty-first century and what it will mean to society? Thus far this book has examined the key areas of AI research. In Part III (Chapters 10–13) we consider what will happen when, three, four or five decades from now, all of these technologies are brought together in the computing entities of the future, combining to create "super-robots".

In order to prepare yourself for the subsequent chapters you need to think in terms of robots that can perform in all of these areas at any desired level of ability, from absolute beginner to the highest possible calibre. They will be able to play all games like world champions; to use their expert systems modules to solve any task that requires expertise; to recognize any shape, person, sound, smell, work of art, etc.; to create the most profound, moving and entertaining literature, the most beautiful art and music; to think on the highest possible plane about any subject under the sun; to carry on conversations in the guise of any desired persona, on any subject, at any level of knowledge or intellect and with any desired sense of humour or level of seriousness; and to possess all of the physical attributes of the strongest, most nimble most physically desirable human beings.

These super-robots will not only have all of the above mentioned capabilities and more, they will also have emotions and they will engender emotions in humans. Enter Stepford wives and husbands who are not just generally convincing replicas of human beings, but who are absolutely 100 percent flawless so far as humans will be able to tell. How will anyone be able to resist the temptation to make one (or more) of them a lifelong companion?

– 10 –
Emotion and Love, AI Style

If you fall in love with a machine there is something wrong with your love-life.

—*Lewis Mumford [1]*

If someone said that falling in love with a Jew, a Black, a Catholic, a Puerto-Rican, an Italian, or a member of any other minority group, showed that there is something wrong with your love life, you would quite justifiably regard him as a bigot. So why discriminate against machines? Can machines not have emotions and can they not induce emotions in humans?

The very notion of robots having emotions will be greeted by many people with scepticism, disbelief or derision. Surely emotions are somehow sacrosanct. Should they be trifled with? Should they be created artificially, with all that that implies for robot love, sex and reproduction? Are not love, sex and reproduction at the very core of being human, even to the extent that they are immune to computerisation? Yes, they are at the very core. No, they are not immune to computerisation.

An emotionless robot[1] would be a mere machine, so a logical step in the development of humanoid robots is to endow them with emotions and enable them to detect emotions in humans. Robots can then respond to a person's emotions in ways that help the robot to interact as humans do. Similarly, robots will be able to detect the emotions of other robots, with the same result. Once the reader has accepted this notion it is only a short mental step to the concept of humans feeling emotions for robots, together with all that that implies, including fears such as "I think my wife is having an affair with her hairdresser robot."

---

[1] Hitherto this book has been precise in its use of the words "computer", "software", "hardware" and "robot". Chapters 10–13 discuss the robots of the future and their likely capabilities. Much of the research examined in this part of the book has been carried out on computers rather than robots, and some of it has been software simulations of what would happen if the same technologies were to be implemented in a robot. Since one of the main purposes of Part III of this book is to look decades into the future, when all of these technologies and many more *will* be implemented in robots, this and the subsequent chapters no longer make any distinction between computer, hardware, software and robots, but will instead use the word "robot" generically, to mean any and all of these.

Functions of Emotion

> Emotions play an essential role in rational decision making, per-
> ception, learning and a variety of other cognitive functions. In
> fact, too little emotion can impair decision making. [2]

In his book *Emotional Design*, published in 2004, Donald Norman argues that in the future machines will *need* emotions, and for exactly the
same reasons that people do:

> The human emotional system plays an essential role for survival,
> social interaction and co-operation, and learning. Machines will
> need a form of emotion—machine emotion—when they face the
> same conditions, when they must operate continuously without
> any assistance from people in the complex, ever-changing world
> where new situations continually arise. As machines become more
> and more capable, taking over many of our activities, designers face
> the complex task of deciding just how they shall be constructed,
> just how they will interact with one another and with people. Thus,
> for the same reason that animals and people have emotions, I be-
> lieve that machines will also need them. They need not be human
> emotions, mind you, but rather emotions that are tuned to the
> requirements of the machines themselves. [3]

Norman's last sentence assumes that robots and humans will be sufficiently different in the future that the complete range of human emotions
will not necessarily be appropriate for robots. I disagree. The robots of
the mid-twenty-first century will be so humanlike in some respects that
all human emotions will be evident in robots. Where, I believe, the robot and human emotion sets will differ, is that robots will be so far in
advance of humans in many ways that, in addition to possessing all the
human emotions, robots will also have a supplementary set of emotions,
hitherto unrecognised, that are necessary and appropriate for them but
which would not be appropriate for humans.

Psychological Theories of Emotion

Research psychologists have long been studying emotion, with the result that several different theories now exist as to the nature of emotion
and how our emotions relate to our actions. William James and Carl
Lange believe that actions precede emotions and that our actions generate physiological responses in us which, in turn, create our emotions.

Walter Cannon and Philip Bard oppose this view, suggesting that we first feel an emotion and then act according to how our brain appraises that emotion.

In order to simulate human emotions and their associated behaviours in robots, it is first necessary to be able to analyse emotions in terms of their classification and their strength. William Wundt classifies emotions according to three interacting "dimensions": pleasantness-unpleasantness, relaxation-tension and calm-excitement. Robert Plutchik divides the range of human emotions into eight basic emotions and combinations of them, with each basic emotion being directly related to a particular behaviour pattern. Many other researchers have proposed divisions into different numbers of emotions: Carroll Izard into ten, Richard Lazarus into 15 and Andrew Ortony, Gerald Clore and Allan Collins into 22, for each of which they specified how that emotion might be appraised. For the purposes of creating emotions in robots the classification of Ortony, Clore and Collins into 22 emotion types can usefully be augmented by two more, love and hate.[2]

The number of different theories of emotion testify to the fact that, to be honest, psychologists are not really sure how emotions work! Despite this, the problem of simulating emotions in robots is the subject of a whole field within Artificial Intelligence and one that will eventually help psychologists. We will learn more about human emotions as we try to create models of them for robots.

Feeling Emotions for Others

When we talk of feeling emotions for others we are usually speaking of other human beings, but not always. In common with millions of pet lovers all over the world I feel great affection for my cats—I get worried when they are ill and I am grief-stricken when one of them dies. We cannot communicate with our pets exactly as we can with other humans, nevertheless we do communicate with them, for example by stroking them and by talking in tones (and sometimes with words) that they can appreciate and understand. In return our pets can display affection for us in their own ways—my cats by sleeping next to me, for example. The

---

[2]These 24 emotion types are joy, distress, happy-for, gloating, resentment, sorry-for, hope, fear, satisfaction, relief, fear-confirmed, disappointment, pride, shame, admiration, reproach, liking, disliking, gratitude, anger, gratification, remorse, love and hate.

human-pet relationship is simple proof of the hypothesis that two-way emotional relationships can and do exist and flourish between human beings and non-humans.

Detecting and understanding emotion and exhibiting emotion will be essential characteristics of robots before we humans can treat them as equals. In order to make all this possible in robotics it is first necessary to classify the different emotions, to examine the characteristics of each emotion and thereby to develop methods for recognizing each emotion and simulating it in a robot. Fortunately, the identification and classification of emotion types by psychologists has already paved the way for computer scientists.

Associated with each type of emotion is one or more action or behaviour. A robot's software allows it to keep track of its own emotional state, for example when the value of a robot's happiness measure exceeds a certain threshold, the robot's software could set its "I am happy" variable to "true". When a robot feels a particular emotion, it might exhibit it by performing an appropriate action or behaviour pattern. (Or the robot might be programmed to conceal certain emotions unless their strength of feeling rises to some pre-determined level.) Such a classification will also help a robot to recognize displays of emotion in others (both human and robotic).

Humans Feeling Affection for Robots—The Tamagotchi

Not long ago the very idea of millions of adults feeling affection for electronic objects would be treated by most people with scorn. But just look at what happened when the Tamagotchi was launched in November 1996 by the Japanese toy manufacturer Bandai. The Tamagotchi, or "lovable egg," had a small grey LCD display, three round buttons below the screen and was small enough to fit into a person's hand (see Figure 60). Although it exhibited only very limited intelligence the Tamagotchi faked its intelligence and emotion well enough to provide pet appeal for many of its owners. A Tamagotchi would let its owner know when it wanted her attention. It would beep her and cry for her, and its sounds would get louder and louder until she give it attention. If she ignored it the Tamagotchi would misbehave, acting loudly and recklessly. Then it would fail to respond to affection and would even become (virtually) physically ill. Sometimes it got virtual diarrhea. With this kind of be-

Figure 60. A Tamagotchi (Courtesy of Bandai U.K. Ltd.)

haviour Tamagotchis fostered responsibility and selflessness in their owners. And business people admitted to postponing meetings because their
Tamagotchi needed its virtual waste removed or its virtual sore feelings
consoled.

Tamagotchi owners were obliged to feed their virtual pets when they
were hungry, play with them when they craved attention, give them
medicine when they were ill, scold them when they were naughty and,
of course, clean up after they virtually defecated. (All of these tasks were
conveniently made possible through the buttons on the Tamagotchi's cas-
ing.) When the owner did not fulfil these obligations their Tamagotchi
would become virtually unhappy, kvetchy, ill and, ultimately, it would
die. And even though a "new" Tamagotchi could be "hatched" with the
press of a button when an old one died, their owners often become quite
attached to the digital playmates and even posted memorial messages for
their "departed" virtual pet in a "Web cemetery" on the Internet.

The Tamagotchi was an instant hit and virtual pets were here to stay.
Toy shops in Japan were besieged as customers struggled to buy their
Tamagotchis before stocks ran out, while many shops in other countries
had to ration each customer to two Tamagotchis in an attempt to ensure
that no-one left the shop empty-handed. Even then Bandai had to issue
a public apology shortly after the product launch because the shortage of
stocks caused near riots in Japan. When a new batch of Tamagotchis was
due to arrive in the shops in Japan, children and adults would travel hundreds of miles if necessary, to camp out on the street, hoping to buy one.

Other people resorted to the underground market where a Tamagotchi sold for 50 times or more its street value of 1,900 yen (about $16), in other words, close to $1,000.

Sales of the Tamgotchi exceeded ten million units, not counting the many rip-off products produced in Taiwan, China and elsewhere. Why this craze for the Tamgotchi? And why was the largest group of purchasers in Japan women in their late teens or early twenties?[3] Answer— caring is a natural human emotion, especially in women, and the Tamgotchi proved conclusively that this emotion can also manifest itself in caring for a man-made object. Caring and loving often go hand-in-hand so it is hardly surprising that many Tamagotchi owners developed a kind of love for their virtual pets, just as I have for my cats.

The human-Tamagotchi emotional relationship was largely one-way. While many Tamagotchi owners exhibited strong feelings for their virtual pets, the Tamagotchis themselves were extremely limited in their "emotional" responses. But the robots of the coming decades will be able to demonstrate the full range of human-like emotions and they will be able to sense and measure emotions emanating from the humans with whom they interact.

Five Criteria for Emotions

Herb Simon, when writing on the foundations of cognition in 1967, recognized the necessity of incorporating the influence of emotion in thinking and problem solving. More recent research supports Simon's view, indicating that our emotions play an essential role in rational decision-making, perception, learning and several other cognitive tasks. Current psychological evidence indicates that intelligence and creative problem solving skills both rely on a healthy balance of emotions. Thus, Rosalind Picard opines that in order to create genuinely intelligent robots that can adapt to us and interact naturally with us, they will need to be able to recognize emotions in others, to have emotions of their own, to express their own emotions and to possess emotional intelligence.

Picard has identified five criteria that are necessary in a robot if it can justifiably be said to have emotions: [2]

---

[3]Such a devotion to one's Tamagotchi was not an exclusively Japanese phenomenon. The copy-editor of this book was once asked to baby-sit a Tamagotchi while its owner, an American student, took an examination. The owner was terrified that "Boo Boo" would die during the four-hour exam because Boo Boo had recently been ill. (Thankfully, Boo Boo lived.)

1. It behaves in ways that appear to arise from emotions.

2. It has fast primary emotional responses to certain stimuli.

3. It can cognitively generate emotions by reasoning about situations, especially as they concern the robot's goals, standards, preferences and expectations.

4. It can have emotional experiences such as cognitive awareness (for example, fear), physiological awareness (such as pain) and subjective feelings (such as an intuitive like or dislike for something).

5. Its emotions interact with other processes that imitate human cognitive and physical functions, such as memory, perception and learning.

Models of Emotion in Robots

> A machine becomes human when you can't tell the difference any more. [Dr. Bowman, in *2001: A Space Odyssey*]

Robots need an internal model of emotion in order to enable them to recognize and synthesize emotions and to express them. A robot's emotion model should enable it to discuss its emotions in the same way we humans do. For example, something that upsets humans such as a friend's serious illness or seeing an earthquake causing massive loss of life on television, these should also upset a socially aware robot. The robot's emotional model should be able to evaluate all types of situation that the robot might encounter, measure the intensity of the emotions of others and to express its own emotions in appropriate intensities. Such a model will enable robots to express the essential emotions with suitable intensities and at the appropriate times, all of which is necessary if it is to be convincing.

A Simple Model of Emotion

In order to explain how a very simple model of emotion might be programmed, let us consider a robot that has only two emotions, joy and distress, and a single variable, called pleasure, that measures whether the robot is in a joyous or a distressed mood. The robot might start life with the pleasure variable set to zero—whenever something happens that

pleases the robot, the pleasure variable is increased by one and whenever something happens that the robot does not like, the pleasure variable is reduced by one. Things that please the robot could be playing a game with its owner, or singing a song. Things that displease it could include sensing that the room temperature is below its level of comfort (so the robot "feels" cold), or losing a game to the user.

Clearly, the more pleasant things that happen to the robot the higher will be the value of its pleasure variable, and the more unpleasant things that happen to it the lower will be the value of the pleasure variable. There could be a table of values that specify how joyous or distressed the robot feels, depending on the value of the pleasure variable. For example,

Pleasure Variable	Mood
+3 or greater	ecstatic
+2	very happy
+1	moderately happy
0	neither happy nor unhappy
-1	moderately unhappy
-2	very unhappy
-3	thoroughly miserable

The robot could be programmed to exhibit a certain range of responses, depending on its mood as determined from the above table. For example, it might laugh uncontrollably when ecstatic, it might let out a cheer when very happy, ... and it might burst into tears and wail loudly when it is thoroughly miserable.

The Oz Model of Emotion

The Oz project was a computer system developed at Carnegie Mellon University from the late 1980s up to 2002. Work on the project has been and is being continued elsewhere, including Georgia Institute of Technology and Zoesis Studios in Newton, MA.[4] Oz allowed authors to create and present interactive dramas. The system design includes a simulated world, several simulated characters, a theory of dramatic presentation and a simulated drama manager. Each character had to be able to "... display competent action in the world, reactivity to a changing

---

[4]The ideas behind the Oz model of emotion are a joint collection of many in the Oz group, but the credit for most of the development work on the project, including designing and building the system, belongs to Scott Reilly. It is the subject of his PhD thesis, for which another of the Oz team, Joseph Bates, was advisor.

environment, goal-directed behavior, appropriate emotions, individual personality traits, social skills, language skills, and some degree of intelligent inferencing about the world around them." [4] The emotion module in Oz was designed to handle most of the emotional and social aspects of the characters' behavior.

The Oz research group at Carnegie Mellon studied how to create robots that appear to react emotionally and are goal directed, emotional, moderately intelligent and capable of using natural language. The original Oz emotion module described here was based largely on a scaled down version of the Ortony/Clore/Collins model described earlier in this chapter. In the Ortony/Clore/Collins model, emotions are the result of three types of subjective appraisals: the appraisal of how pleasing an event is with respect to the robot's goals; the appraisal of how much the model approves of its own actions or those of another robot (or human) with respect to its own standards for behaviour; and the appraisal of how much the model likes certain objects (including other robots or humans). The Oz model can also reflect simple social relationships between robots, how these relationships change over time and how these relationships affect and are affected by emotion and behaviour.

The Oz emotion model is divided into three major components: a control module, an emotion process module and a reactive planning module—a planning system devoted to the execution of plans in a changing environment. The control module continually repeats the cycle: sense-think-act. Originally, during the sense phase, its knowledge about its current situation was updated and then the emotion process module generated emotions based on the robot's internal state and the currently sensed environment. Then the control module would think about what to do, and finally it would act on its decision. Later Oz' current situation ceased to be the key factor in emotion generation, and was replaced by the success or failure of the robot's goals, making the process more motivationally oriented.

Once the emotion process module finishes its initial update of the robot's emotional state, the planning module takes control, choosing an action to perform based on the robot's physical environment, its active goals, the importance of those goals, what it had been working on previously, its emotional state and any personality constraints. The planner attempts to implement goals and to determine the success or failure of its plans to achieve these goals. While it is thinking, the planner may create new goals or discover that old goals have succeeded or failed. When this

occurs, the planner allows the emotion process module to update its state of mind. This updating process is important as it allows the robot's state of mind to change because of the planner's actions, which can in turn affect the planner's behaviour choice, thereby providing a dynamic system with immediate emotional feedback. Furthermore, while the planner is carrying out an action it may receive a message from another robot or a human, which could affect its goals.

How Emotions Are Affected by Events

A robot can be pleased or displeased by events that happen to it, including its own actions. How it feels about an event depends on the robot's goals, which can be anything that the robot wants, such as "I want to eat" (an example of what was originally specified in Oz to be an *active* goal—one whose outcome the robot can influence), or "I want the Mets to win the World Series" (a *passive* goal—one that the robot cannot influence). Later in Oz' development, no distinction was made between active and passive goals. Instead its goals were all handled by the robot's motivation system.

When there is a "to eat" goal the event of eating dinner will be judged by the robot as being pleasing. Events that have already taken place give rise to joy and distress emotions with an intensity based on a number of factors, including how pleasant or unpleasant the event was found to be. Originally in Oz, the prospect of future events gave rise to the emotions fear and hope, also with intensities that are determined by how pleasant or unpleasant a potential future event was expected to be. But later this approach was replaced by a set of inference rules that were specifically tailored to achieve the robot's goals. Similarly, some of Oz' emotions (anger, gratitude, gratification and remorse) were also later modelled by inference rules that were attached to important goals, rules that assigned credit or blame when a goal succeeded or failed (or appeared likely to do so), so that the corresponding emotion would be generated. For example, if the robot had an important goal that failed, and if the robot inferred that the failure was due to some action by Fred, then the emotion module would generate a feeling of anger towards Fred.

Clark Elliott and Greg Siegle have explored the way in which the intensity of an emotion can change as a result of changes in a robot's situation. This is an important aspect of the simulation of emotion. They defined three different types of measure of emotional intensity. One

deals with measures whose values change independently of how the robot interprets its current situation, for example the loudness of a noise or the brightness of a light. Another type of measure is involved with a robot's interpretation of a situation, for example measures that help to define the strength of friendship or animosity felt by the robot for another robot or for a human. The third type of measure contributes to a robot's mood by altering a robot's interpretation of a situation—this type can easily be altered by a change in the robot's situation and they return as a matter of course to their initial preset values over a period of time.

How Emotions Are Affected by Actions

A robot can approve or disapprove of its own actions, or those of another robot or a human, according to a set of standards that represent both moral beliefs of right and wrong (for example "Thou shall not hit people on the head") and personal beliefs about the robot's own level of performance (such as "I should be able to beat David at chess"). If a robot approves or disapproves of one of its own actions then it experiences pride or shame. Similar actions by another robot give rise to admiration or reproach. The intensities of these emotions are based primarily on the level of approval or disapproval of the action. The Oz emotion module judges its approval or disapproval of actions by evaluating every act that it senses according to its own standards.

How Emotions Are Affected by Objects

Objects (including other robots and people) can be liked or disliked by a robot according to its own attitudes—its personal tastes and preferences (as in "I dislike modern art"). In the Oz emotion module these two emotions were expanded to include a few similar emotions, such as awe. When a robot thinks about or notices an "object", this can give rise to an emotional response. For example, thinking about an object that the robot likes gives rise to love, whereas thinking about an object disliked by the robot gives rise to hate. The intensities of these emotions for a robot are based primarily on the level of its liking or its dislike for the object.

How Emotions Are Affected by the Robot's Goals

There are two sets of goals in the Oz control module: the planner keeps a list of the current active goals and the emotion module keeps a list of the

permanent passive goals. Both types of goals succeed and fail in various ways and both types of goals have an importance level associated with them. Whenever a goal succeeds or fails, the emotion module creates a joy or distress emotion with an intensity related to the importance of that goal.

Emotion-Related Responsibilities

In addition to generating emotions, the Oz emotion module has a number of other emotion-related responsibilities. For example, it keeps track of thresholds for every type of emotion. This means that robots can have a built-in level of emotional tolerance so that not everything in its world will spark an emotional reaction, but a series of little events will still be able to generate an emotional response from the robot. The emotion module also manages emotion decay—each time the control module performs a sense-think-act cycle, the emotion module lowers the intensity of most of its individual emotions. This means that in order for an emotion to be sustained at a significant level over a long period, the robot must repeatedly have that emotion generated by the emotion module. In the case of hope and fear however, the intensity of the emotions remain for as long as the corresponding goal is likely to succeed or fail, and it is only after the goal has succeeded or failed that the strength of that emotion starts to decay.

How Social Knowledge Affects Emotions

The emotion module's responsibilities also include much of the robot's social knowledge about relationships, specifically the interpersonal relationships between robots, how these relationships change over time and how they interact with the robot's emotions. Interpersonal relationships are intertwined with the attitudes that the emotion module employs in determining emotions toward objects, which typically were either humans or avitars[5] controlled by humans.

How Expectations Affect Emotions

The Oz project laid the groundwork for much of the more recent research into emotion modelling, including a robot's estimate of the likelihood that certain events will take place. This feature was incorporated in

---

[5] An avatar is the software representation of a human.

the Cathexis model of emotion created by Juan Velazquez. Expectations change over time, for example a person may expect to pass an exam but after taking a few courses and failing the exam each time his expectation of passing the exam next time will be much lower. Therefore it is important to allow the robot's expectations to change with its experiences.

Empathy—How Robots Recognize and Measure Emotions in Humans

The most satisfactory personal interactions are those in which everyone involved in the interaction can not only express their own emotions but also recognize the emotions of others. An important aspect of relationships between humans is empathy—the ability to imagine ourselves in the position of another person, to experience what another person is feeling. Empathy gives us access to a person's mental states, to his or her desires, emotions and beliefs, and thereby plays an indispensable role in our social interaction and communication.

Empathy is one of the essential ingredients for long-term success at the poker table. In poker it is important to recognize "tells", those subconscious gestures that give away information to the opponents. Stuart Marquis and Clark Elliott have developed a set of poker playing programs at DePaul University that use emotion in two ways. These programs display emotion[6] and they recognize emotion in their human opponents. The control program uses voice recognition software,[7] applied to the inflection in an opponent's voice, to distinguish between seven broad emotion categories: neutral, joy, fear, anger, hate, sad and love. This classification enabled the programs to distinguish the limited essential vocabulary of poker ("bet", "check", "call" and "raise") according to the inflection of the opponent's voice. The classification here is into seven categories rather than the 22 basic emotions in the Ortony/Clore/Collins model, because the current state of the art in voice recognition does not provide sufficient discrimination to cope with so many gradations of voice. As hardware speeds and memory sizes burgeon in the future, more accurate voice recognition will be possible, enabling robots

---

[6] See "How Robots Express Emotion" later in this chapter.

[7] It is important to distinguish between Speech Recognition, which is the science of recognizing what words are spoken, and Voice Recognition, in which the task of the technology is to recognize the voice of a particular person or to determine that person's mood or stress level from characteristics of their voice.

to distinguish between the inflections associated with the full range of Ortony/Clore/Collins emotions and even more.

An example of how people model one another's points of view, in order to explain and predict each other's responses to various situations, is the Affective Reasoner program developed by Elliott and Ortony, that reasoned about the emotions of robots in a simulated world. The robots simulated by these programs were given a simple emotional life consisting of the 22 Ortony/Clore/Collins categories of emotion plus love and hate, and they were endowed with approximately 1,200 different expressions of emotion. Each simulated robot was given its own personality, so that different robots could have different interpretations within their own worlds, and exhibit different response tendencies.

In order for one robot to understand how another robot or a human is likely to construe a particular situation, it must be able to view the situation from the other's point of view. In the Affective Reasoner a robot's knowledge of other individuals' goals, standards and concerns is built up by observing the reactions of those individuals to different situations. This knowledge is stored in a database used by the Affective Reasoner to enable a robot to predict the emotions of others, based upon its experience of their earlier reactions in various situations.

Poker is an example of how, sometimes, what is said is not as important as how it is said. Being able to infer meaning from how something is said is not only an asset at the poker table, it is a sign of emotional intelligence in general. Emotional cues that can be recognised by a robot, include not only characteristics of a human's speech patterns but also eye contact and facial expressions, a furrowed brow, sweaty or trembling hands, body language and skin colour (blushing). All of these forms of emotional expression in humans can potentially be recognised by robots. Tools to aid recognition include cameras (to observe the eyes, facial expressions and body language), microphones (to monitor changes in voice) and bio-sensors (to measure changes in body temperature, skin resistance and other physiological changes).

At the MIT Media Lab research into the recognition and measurement of emotion has spawned a field of study called Wearable Computing, also known as Affective Wearables. A wearable computer is a wearable system equipped with sensors and tools that recognize its wearer's psychological and other patterns, including any expressions of emotion: a joyful smile, an angry gesture, a strained voice or a physiological change such as an accelerated heart rate or a change in skin conductivity.

Mind Reading

A logical extension of the idea of recognizing and measuring human emotions using physiological sensors on parts of the body, is simply to add the brain to the list of body parts, devising technologies to monitor and interpret brain patterns. The idea of using the power of thought to drive robotic movement was once in the realm of science fiction, but by 1977, when the author Craig Thomas created the Firefox aircraft (made into a Clint Eastwood movie in 1982), the idea was almost believeable. (Clint Eastwood's character used thought-activated weapons to destroy his communist enemies.) And already the first stage of the Firefox technology has become a reality, with the announcement in November 2000 that electrical signals from a monkey's brain, instructing the monkey's arm to move, can be used to stir identical movement in a robotic arm.

A pioneer in this field of research as it relates to humans is Philip Kennedy, a neurology professor at Emory University Hospital in Atlanta. He leads a project he started in 1989, recording and amplifying the human brain's electrical signals sufficiently to enable them to be used to operate a computer. Using a *neurotrophic electrode* invented by Kennedy, in conjunction with some customized micro electronics and software, the brain's neural signals become, in effect, a computer mouse to move a cursor and to select icons on the screen. In short, a computer system controlled by the power of thought.

One of the most dramatic uses to date of this technology has been at Georgia State University, where Professor Melody Moore and a team from the Computer Information Systems Department have developed software that helps a speechless patient to communicate by thought alone. Johnny Ray, a 53-year-old paralyzed stroke victim at the Veterans Administration Hospital in Decatur, Georgia, became the first human to communicate via a computer cursor controlled only by his brain power. Ray had some of Kennedy's electrodes connected to his brain and imagined various movements of the cursor. The medical team told Ray what cursor movements to think about and monitored Ray's neural signals so they could detect which of them were related to which cursor movements. After several months of trial and error Ray had managed to learn which *imagined* movements would best control the cursor. His first big success was when he was able to select the letters that make up F-I-V-E, in response to the question: "How many children do you have?" The word was spelled out on a virtual keyboard displayed on the computer's

monitor, with Ray indicating the letters, one by one, by moving the cursor using thought. Soon afterwards he was able to spell out each of his children's names and as Ray's proficiency increased, so has the sophistication of the communication, says Moore. "We're actually having conversations with him. Instead of asking him to spell 'Phil' or 'Mel', we're asking things like, 'What's the best book you've read? What's your favourite movie?' He moves the cursor around and selects the letters to go into a writing program, and then he's able to speak them because we added a voice synthesizer."

Miguel Nicolelis and his colleagues at Duke University have investigated how brain cells activate when hands and arms move, an extremely difficult task to emulate. Electrical signals issued by the brain to flex a particular muscle can be swamped by a mass of other instructions emanating from the brain, such as "Should I answer the phone" or "I must book my flight to New York," so picking out the relevant movement instruction is not easy.

Nicolelis and his team found a primitive but effective solution to this problem. Working only on the movement-control centre of the brain (the motor cortex), they first measured the activities of individual monkey neurons[8] each time the animal completed a very simple action, such as moving its hand to the left. They reasoned that the greater the measured activity, the greater the neuron's importance in this task. They assigned each neuron a number to reflect this—if the activity doubled the number was doubled. These neuron numbers (called coefficients) can be used to predict and generate movement. Measuring a neuron's activity at a particular moment and multiplying this by its corresponding numerical coefficient gives a clue to the movement that is about to occur. Adding up the results of the different neurons brings the clues together to reveal the answer—the hand is about to move left, up or whatever. A computer can thereby transform the answer at each moment into robotic movement, while already calculating what the hand movement should be for the next moment. These calculations are made, moment by moment, until the movement has been completed.

Later research by the same group at Duke University was described in glowing terms by Jon Kaas, a psychology professor at Vanderbilt University in Nashville, who is familiar with Nicolelis's research.

---

[8]The size of a neuron cell body is approximately 0.01mm across, so some 30,000 could fit on the head of a pin.

For nearly completely paralyzed people, this promises to be a fantastic boon. A person could control a computer or robot to do anything in real time, as fast as they can think. [5]

The Duke University team has showed that humans produce brain signals like those of the monkeys in their experiments. John Chapin, a professor of physiology and pharmacology at the State University of New York Downstate Medical Center in Brooklyn, describes how:

Monkeys not only use their brain activity to control a robot, but also they improve their performance with time. The stunning thing is that we can now see how this occurs, how neurons change their tuning as the monkey does different tasks. [5]

Nicolelis' experiments involved implanting tiny probes called microwires into several brain regions of two rhesus monkeys. At first, each monkey learned to move a joystick that controlled a cursor on a computer screen. When a ball appeared, the monkey was rewarded with a drink of fruit juice if it managed to move the cursor to a target. Nicolelis' team collected scans showing the electrical patterns from a monkey's brain as it performed the tasks. After a monkey became skilled at getting the cursor to the target the scientists disconnected the joystick. At first, the monkey would wiggle the joystick and stare at the screen as it did before, and even though the joystick was disconnected the monkey's brain scans, as it reached for the joystick and tried to use it to grasp an object on the screen, were being sent to a computer which translated those signals into movements on screen. Nicolelis described what happened next:

There was an incredible moment when the monkey realized that it could guide the cursor and grasp an object on the screen just by thinking it. The arm dropped. Muscles no longer contracted. [5]

The implications of this research for affective wearables are absolutely staggering, and technology similar to that used for affective wearables is already being developed to carry out similar measurements on the electrical signals of the human brain. Developing the technology to work on the electrical signals associated with all of our thought processes, and not just those associated with arm movement, will mean that eventually an affective wearable will be able to read our thoughts as well as to detect and measure our emotions. If such a wearable, or another device in close proximity to the body, then transmits our thoughts and emotions to a robot to act on them, we will find our robots doing our bidding even

without our having to ask them. Furthermore, our thoughts and emotions can be transmitted to anyone we wish, by e-mail, text message or a voicemail using a synthesized version of our own voice, direct from our wearable. Imagine being able to think romantic thoughts about someone on the other side of the room/street/town/country/world and have him or her receive these thoughts exactly as you have them, without your needing to pick up the phone or go to the keyboard.

Civil liberties groups will doubtless pounce heavily on government agencies and others who plan to use the mind-reading technologies of the future. It is easy to suggest many ways in which such technologies could be employed for nefarious purposes. But on the plus side, we can envisage the day when crime becomes a thing of the past because criminal thoughts in humans will be detectable by police robots, á la the movie *Minority Report*. And people who are psychologically disturbed will be monitored and perhaps helped on the road to recovery by mind-reading therapist robots that incorporate an expert system for psychiatric diagnosis and treatment.

Mind-reading technology[9] has attracted the interest of John Norseen and his research group at Lockheed Martin Aeronautics Company, one of the US government's leading defense contractors.

> What I am encouraging is multisensor analysis of the brain — looking at many areas of the spectrum to get a different picture. [6]

In an article aptly entitled "Decoding Minds", Sharon Berry describes one of the experiments conducted by the Lockheed Martin team.

> Simple interaction with subjects has been used to test the system. A researcher shows a picture to a person or asks a person to think of a number between one and nine. Information is gathered and displayed on a monitor much like on a television. It shows that the person is thinking about the number nine. The researcher then tells the person to say the same number, an action that appears in another part of the brain, the parietal region. [6]

Norseen explains in the article that

> By looking at the collective data, we know that when this person thinks of the number nine or says the number nine, this is how it appears in the brain, providing a fingerprint, or what we call a brainprint.

---

[9] Also known as bio-fusion.

Figure 61. A brain fingerprint (Courtesy of Brain Fingerprinting Laboratories, Inc.)

. . .

> Brainprints are unique to each person. While the number nine will appear in the same brain areas of different people, it still occurs as a unique signature of how a person specifically thinks of the number.

. . .

> If someone is telling the truth, it is kept on the outside portion of the brain in low-energy domain areas of the brain. If someone starts to light up in more areas of the brain and at a higher energy level, it means that the person is now starting to confabulate or obfuscate. [6]

Brainprints, or brain fingerprints as they are also called, are already being used as an investigate tool in major crimes. Lawrence Farwell, the chairman and chief scientist at Brain Fingerprinting Laboratories, Inc.,[10] has used brain wave responses in murder suspects to show whether the suspect's brain contains critical information about the murder (see Figure 61). The technology was ruled to be admissible evidence in court in the case of Terry Harrington, who was convicted of murder in 1978 in Iowa and sentenced to life in prison. Brain fingerprint testing indicated that Harrington's brain did not contain records of the crime scene but that they did match his alibi. When confronted with this evience, the only alleged witness to the crime recanted, confessing in a sworn statement that he had lied in the original trial in order to avoid being prosecuted himself. The Iowa Supreme Court subsequently overturned Harrington's conviction and freed him in February 2003.

The implications of the Harrington case for the future of the legal system are immense. How long will it be before there is absolutely no need for a judge or jury to determine the guilt or innocence of a defendant? How long before a robot is able to announce the correct verdict,

---

[10] Brain Fingerprinting is a registered trademark of Brain Fingerprinting Laboratories, Inc.

every time, merely by interrogating the witnesses when they are hooked up to brainprint machines? Research so far in this nascent technology indicates a 90 to 95 percent accuracy rate. When recognition technology advances to the point where the most significant patterns in a brainprint can be interpreted accurately, robots will be able to recognize their users' every wish, and their emotions.

How Robots Express Emotion

The subject of robot emotions is somewhat controversial and has caused strong reactions from philosophers, many of whom ask: "Can a robot express emotions if it does not feel them?" This question is itself based on a premise with which I disagree. In line with the philosophy of the Turing Test a robot can surely be said to feel emotions if it gives the appearance of feeling emotions to such an extent that its expressions of emotion are indistinguishable from those of humans.

As we become more used to the idea and practice of interacting with robots, having them in our homes as pets, companions and servants, so it will be important for these robots to be able to display their emotions to us, for example by what it says, by the tone of the voice with which it speaks and by its facial expressions. (Yes, robots can smile and frown.) In this way we can relate to the robots, understand their actions, feel more in control when dealing with them and therefore feel more comfortable about allowing them access to our personal information, preferences, personality traits and other aspects of our individuality.

The Kismet Model of Emotion

Simulated emotions in robots can be internal simulations (i.e., the robot "feels" these emotions but conceals them from the outside world, for whatever reason), or external simulations (so that humans and other robots can observe these emotions but the robot itself feels nothing) or, as in the case of MIT's Kismet project, both internal and external. The foundation of Kismet's emotional process lies in Cynthia Breazeal's perspective of emotions as evolutionary phenomena that help to guide living creatures through the problems of daily life. Kismet's emotions are based on simple appraisals of the benefits or detriments of various stimuli. The robot evokes goal-directed positive emotional responses that bring it closer to a stimulus that it likes, and thereby into a state of relative well-

Figure 62. Kismet, showing surprise (Photographed by Sam Ogden, Copyright © 2005 Photo Researchers, Inc. All Rights Reserved)

being. The robot also evokes goal-directed negative emotional responses that take it further away from a stimulus that it dislikes, thereby avoiding undesirable states. Six basic emotions are modelled: anger, disgust, fear, joy, sorrow and surprise. Kismet also displays responses corresponding to being interested, being calm and being bored.

It may at first appear unimportant whether or not a robot feels an emotion, so long as it wears its emotions on its sleeve for others to observe. But in fact there is an important reason why a robot should feel its emotions (whatever "feel" means to a robot)—it is only by having its emotions affect its actions that a robot can develop its personality, learning from experience what it likes and dislikes for example. It has been found with Kismet that its expression of emotional responses creates empathy in the humans who interact with it and thereby regulates its social interaction with them.

As to the question of what "feel" means to a robot, in simple terms it knows how much of each of its emotions exists at any given moment in time and this knowledge corresponds to what we humans call "feeling". Of course, not everyone would agree with the view that robots can have feelings as we understand them—this somewhat controversial topic is discussed in Chapter 12.

Given that it is highly desirable for robots to be able to express their emotions, an important question is, how should a robot decide which emotion or emotions to display and when? Joseph Bates suggests as a guideline that "the thought process reveals the feeling", that robots should emotionalise and display their thoughts. This in turn enables us to see that emotionally aware and demonstrative robots really do "care" about what happens in the world and that they appear to have genuine desires. Bates also proposes that a robot's emotions should be "clearly expressed" in order for us to relate to the robot and understand its actions. Furthermore, he recommends that a robot's emotions should be accentuated, to convey them better to those around it, and appropriately timed—the time relationship between an emotion and the situation or action giving rise to it is what helps people and other robots to understand which emotion is being expressed. If a robot were to exhibit happiness during a fight or if it were to shout during a peaceful moment when its owner was listening to quiet music, that robot would create confusion in our minds and reduce our empathy for it.

Lynellen Perry at Science Applications International Corporation has raised some interesting questions about robot displays of emotion. She asks whether some robots will become unstable just like humans. Will we need to send some robots to a professional psychiatrist or will we just destroy those robots and try again? On the other hand, as she points out, it might actually be useful for society to have some robots that have split personalities or psychoses, so that psychologists can use them as research tools to help understand and deal with these disorders. Another question asked by Perry is "What happens when a robot behaves irrationally before a correct diagnosis is made of its emotional disorder?" It could

> ...make a huge mess of things. We are quite comfortable blaming a human for the acts they commit, but who will be blamed when a psychotic robot commits a crime? The robot? The programmers who wrote the software in the robot? [7]

This question is discussed in a later chapter.[11]

Robot Personality

A believable robot must have personality. It is widely accepted by psychologists that people prefer to interact with others who are similar in

---

[11] See the section "The Legal Rights of Robots" in Chapter 13.

personality to themselves, so it is hardly surprising that early work in this field indicated that people will respond to robot personalities in the same way they would respond to similar human personalities and in line with their preferences in human-human relationships. Also, people find the experience of interacting with robots that have similar personalities to their own more satisfying than with robots that have significantly different personalities. Simulating and controlling personality is therefore an essential contribution to the creation of robot companions that will be acceptable to humans.

Although personality has been the subject of a huge amount of research by psychologists, there is no general agreement amongst them as to the definition of personality or of its component parts. I quite like Daniel Rousseau's definition:

> Personality characterizes an individual through a set of traits that influence his or her behaviour. [8]

Traits are persistent characteristics of personality that correspond to general patterns of behaviour and modes of thinking, and personality traits are quite recognizable by observing a person's behaviour, displays of emotion and interpersonal relationships.

Much of the work on simulating personality has been founded partly on what is known by psychologists as trait theory, in which characteristics such as sociability and extraversion are seen as determining how we act in society, and partly on social learning theories, in which our behaviour at a given moment in time is determined by our situation at that time and by our past experience in similar situations. Barbara Hayes-Roth and her group at Stanford University's Knowledge Systems Laboratory have led much of the research in this field. They developed a personality profiling model that allows for the specification of traits, such as self-confidence, gullibility, activity and friendliness, whose intensity can be specified and varied. This variation in intensity can be specified in terms of numbers, for example a robot with a gullibility of 100 might be completely and utterly gullible ("Do you mean I can buy the whole of the Brooklyn Bridge for $100?") while one with a gullibility of zero would not believe anybody ("Surely, just because you're wearing a blue uniform with a shiny metal badge, a peaked cap and you've got a gun in a holster, you don't expect me to believe you're a policeman.")

The traits in the Stanford personality profiling model describe how a robot feels and how it reacts to other robots and to humans. The traits

also define the personality of the robot. These traits may depend on the state of the robot at a given time, for example whether it is happy or sad, grateful or angry, and whether it likes or dislikes an object or the person or robot with whom it is interacting. Using this trait model it is possible to create robots with different personality types: nasty, friendly, shy, lazy, bad-tempered and selective (friendly with some people/robots and nasty with others).

In order to simulate personality in robots it is first necessary to develop rules that specify how the robot will be most likely to behave, given its particular personality. These rules will be followed by the robot when it decides on an action and they will be exhibited as personality traits that can be identified by those who interact with and those who observe the robot. Its personality traits will also have a significant influence on the moods of a robot and on its interpersonal relationships with other robots and with humans.

In the Affective Reasoner model described earlier in this chapter, Elliott and Ortony represented personality in two parts. The first part, which they call the *interpretive personality*, is used for determining whether something the robot experiences is relevant to its own goals, standards or preferences, and if so, the robot's ability to reason allows it to relate that experience to its emotional state. To understand how the interpretive personality works, consider two spectators at a soccer game, one supporting Manchester United and the other being a Chelsea supporter. If Manchester United win the match then their supporter will be happy while the Chelsea supporter will be sad—an example of two people feeling opposing emotions as a result of the same event, and feeling differently because of their different preferences.

The other part of the Elliott/Ortony representation is called the *manifestative personality* component and is used for determining how the robot will act or feel in response to its emotional state. In order to understand the manifestative personality component, let us suppose that a robot is feeling proud about having done something its owner had requested. If the robot's personality is the quiet type it may simply manifest its emotions through a quiet response, for example a feeling of general well being. If the robot is talkative it may express its pride by saying something appropriate, for example telling its owner or some other person or robot that it is proud at having accomplished the task. If the robot is manipulative it may manifest its pride by attempting to change the emotions of others, for example by calling attention to its praiseworthy act in order to

gain admiration. A robot can have many of these different temperament traits active at the same time—each robot's individual combination of traits will give it its own manifestative personality.

Robots with personality will exhibit different moods. A robot's moods, like those of humans, can vary over time and correspond to current emotions or to sensations that arise from the robot's current situation. Robots, like humans, will have their own (different) personalities, and will therefore be able to experience different moods in a given situation, or the same moods but with differing intensities. For example, a self-confident robot would probably feel angry when threatened while an insecure robot would be afraid in the same situation.

Love and Marriage with Robots—An Acceptable Idea?

Nowadays scientists, psychologists and philosophers are asking, more and more often, questions such as "Can robots fall in love?" The question may seem to some of you to be unnecessary because love is an experience peculiar to warm-blooded mammals—anything in a programmed entity is merely a simulation, however convincing the display. But if a robot exhibits all the signs given out by a human in love, then surely that robot has passed the "in love" version of the Turing Test. The fact that the robot does not have feelings as we know them does not prevent it from behaving exactly like a human in love. If your human spouse/partner/date behaves as though he or she is in love with you, then so far as you are concerned, he or she is indeed in love with you. In decades to come an ever-increasing proportion of the human race will be equally accepting of the notion that their robot is in love with them.

An even more important psychological and sociological question for the future is "Will people fall in love with robots?" I believe the answer to this question to be an unqualified "Yes". Though this idea will be abhorrent to many people, the rapidity with which millions developed emotional attachments to their Tamgotchis, a very simple type of robot, surely is an indication of the strength of emotion that millions of people will feel for robots that are far more sophisticated in so many ways. Within the next 20 years I expect some people, maybe only needy people, to fall in love with robots. And as time goes on the "neediness" threshold required before a human falls in love with a robot will surely be reduced.

Robots and Love

Let us further consider the notion that people can fall in love with a robot. In the past, before the Internet was invented, many people had pen friends with whom they exchanged letters. Through this type of correspondence some people developed long-term friendships, occasionally falling in love with their pen friend and agreeing to marriage even without having met them. Moderately unusual, yes, but only moderately. It is easy to understand how two people can fall in love on the basis of their communications with each other, even without physical contact. Much of the emotional basis for love is based on your feelings about your partner's character, personality, interests, ideas, how he or she talks (or writes) to you, . . . so many things that can be communicated verbally. Of course the fact that your partner is human is a very important factor, but in future decades it will become less and less so.

In Chapter 7, in the discussion on Natural Language Processing, we saw how easy it is for someone to be fooled into thinking that a computer program is actually human. The example of ELIZA is perhaps the earliest and certainly one of the most amusing cases from the past. And nowadays it is possible for chatterbots, such as those taking part in the annual Loebner Prize competition, to fool some of the competition's judges for *some* of the time. How long can it be before the most sophisticated conversational robots are able to fool some of the judges for *all* of the time? I foresee a steady increase in the percentage of the human population that will not be able to distinguish whether their conversational partner is human or silicon. The important question here, surely, is not how long it will be before the percentage of people fooled all of the time reaches 100 percent, but when this figure will reach a significant level. When that day comes there will be hundreds of millions of people in the world, not only those who are emotionally needy, who will be at least as content talking to a robot as they will to another human being.

Given that falling in love and agreeing to marry can both happen remotely, as is the case with pen friends and, in recent years, with friendships started via chat on the Internet, the advent of software that can pass the Turing Test with at least a significant percentage of the human population, surely means that significant numbers of people will fall in love with artificial partners they meet on the Internet, i.e., robots. Of course many people will find the idea distasteful at first, just as many people have found the idea of same-sex love and marriage distasteful, but times

change. My own view is that if someone wishes to marry a robot, why stop them?

Although there are some claims that same-sex relationships have long been blessed by various churches in ceremonies akin to marriage, it was not until 2004 that Massachusetts became the first state in the U.S.A. to legalise same-sex marriages.[12] Within a few decades I believe that some American states will also legalise marriage between humans and robots, and I predict that Massachusetts will lead the way in this respect, not only because of its comparatively liberal views to same-sex marriage but also because it is the home of countless high-tech companies and academic institutions working in the field of Artificial Intelligence.

Companionship

Today's interactive Barneys, Tamagocchis, Furbies, and "Winnie the Pooh" dolls may be socialising a generation of children to not only having emotional relationships with artefacts, but believing, long after childhood wanes, that toys can really have feelings. If these toys could recognise even a few of the child's truly expressed feelings and reflect as much empathy as a dog, then the illusion may become as powerful as it is for many adults who swear their pet understands their feelings better than anyone.

One credible possibility is an entire generation of toys that are capable of this kind of emotional-content interaction with their young users capable of soothing a crying child or of perhaps artificially preventing strong feelings of loneliness, sadness, frustration, and a host of other strong, negative emotions. Might such artefacts discourage their owners from fostering normal, healthy interactions with their parents and other children? There are certainly many adults who prefer interacting with their pets and computers to interacting with other people, and who are quite happy with this state of affairs. If such support for emotion regulation is provided too early by a non-human source, would this have a beneficial, educational effect, or might it possibly leave some children emotionally crippled, thwarting the development of the skills needed to interact successfully with other humans?

This question may be developed in the adult world as well: if such devices achieve popular success, and humans routinely use them to help manage their emotional states, what happens to the human's

---

[12]In the Netherlands this happened earlier—the first same-sex marriage there took place in 2001.

sense of his or her own self-control? One can imagine possible addictions similar to those for interactions with other inanimate objects: coffee, cigarettes, and chocolate: 'I just need a quick break to be with my computer and then I'll feel better.' As when interacting with a real person, the one who is in control becomes less clear—to what extent does control reside with you or with your confidante? Where does it reside in a world in which humans may depend on emotional cyborg relationships for their emotional well being? Clearly this is not a problem yet; computers have a long way to go before they are accused of erring on the side of providing too much emotional well being. However, this is a foreseeable specific concern arising from new technologies that begin to assess and respond to expressed user feelings. [9]

I believe that Rosalind Picard, quoted above, is absolutely correct about this phenomenon extending into the adult world. And as to her question "What happens to the human's sense of his or her own self-control?", we need only recall the attitudes of adult Tamagotchi owners to see the answer. Adults can very easily become suborned by such toys, doting on their every need and whim. But is this any worse for an adult than if she doted on every need and whim of a boyfriend or a husband? As for humans loving robots, that is a perfectly normal, understandable and positive extension of the affection and love most humans feel for other humans and for their pets. It's just different, because robots are different. Are they inferior to humans? In my view there is almost no way in which the robots of the mid- and late twenty-first century will be inferior to humans and, in fact, many humans will be far happier interacting with their robots than with most other humans.

Another argument that can be levelled against the idea of loving relationships between humans and robots is an ethical one—what does it mean to be human and is it ethical for humans to marry and have sex with this particular category of non-humans? To this I would ask: "Who has the right to legislate against what consenting adults do with consenting robots in private?" It is not difficult to argue in support of this position. Consider what most people want from a life-partner, a spouse. All of the following qualities and many more are likely to be achievable in software within a few decades—your robot will be: patient, kind, protective, loving, trusting, truthful, persevering, respectful, uncomplaining, complimentary, pleasant to talk to, and sharing your sense of humour. And the robots of the future will not be jealous, boastful, arrogant, rude,

self-seeking or easily angered, unless of course you want them to be. In short, your robot spouse will be everything you want of him (or her). (S)he can be made in whatever likeness you wish, a Gwyneth Paltrow, Julia Roberts, Brad Pitt or Michael Douglas look-alike, or a custom design created specially for you. (S)he can be as tall or as short as you wish, as fat or as thin, as dark or as fair.

Choosing the characteristics you desire will be a simple process. You will be offered all these choices on a monitor in the robot shop and you will select exactly those appearance and physical characteristics, its voice, its cultural interests and other hobbies, its personality traits and any other features that you want. You will be able to ask for it to be submissive or domineering, exactly as intelligent as you are, or more so, or less. You can have these and all of its emotional and other personality characteristics change at whatever intervals you wish, so you could specify a robot that is domineering in the mornings (when you need to be encouraged to go to work on time) but submissive at night. The choice will be yours. And of course, if you are not 100 percent happy with the robot when you collect it from the robot factory, it comes with a lifetime guarantee (your lifetime, that is)—you can change any or all of its features, traits and characteristics as and when you wish, simply by telling it how you want it to change. The net result will be robots that provide all the characteristics we want in a companion/spouse/lover.

Arthur Harkins, an anthropologist at the University of Minnesota, has predicted that

> Even if we made them as appliances, not as fully intelligent devices, they could still have a lot of these features, including bubbling conversations. I'm talking about something here that could threaten the whole idea of—that could render unnecessary—human [to human] marriages. [10]

Harkins caused astonishment in the mid-1970s when he predicted that, before the year 2000, the first test case of a human-robot marriage would be in the courts. (This was a remark made to a journalist in a casual conversation after the formal conclusion of a conference on the future of the family.) Harkins confessed to me in 2003 that he had not in fact married a robot "...or even a biodroid, but am happily married to a Ukrainian sociologist. Yes, she is very, very human!" [11]

Harkins' marrying-a-robot prediction had the support of some of his colleagues in the academic community, one of whom, Nelson Otto,

explained his enthusiastic support of Harkins' views thus:

> Robots may soon provide humans with a companionship of such high quality that it will change our lives. There are a lot of areas about robotics that people avoid talking about. A robot as a companion. A robot as a best friend. A robot as a lover. These are things that are in the backs of people's minds, but they don't want to open a discussion on them. Yet we know that there are just an incredible number of lonely people out there. Really lonely people. We are a lonely society. We are a society in need of stimulation. And we turn to the most bland things you can think of. We've got people fixated on soap operas just to give them some kind of stimulation.
>
> When they find an intelligent machine such as a robot that can converse with them, that can stimulate them, that can give them a backrub—and I have to stop almost at that one. Most people in this country are so malnourished when it comes to personal attention: getting a neck rub, a back rub, somebody there to soothe them, to be with them, that when they see the opportunity that a robot gives them, the luxury that it gives, I think they will be more than accepting. [10]

Otto was so convinced that Harkins was right in predicting human-robot marriages that when the media bombarded Harkins with requests for interviews, many of which were on TV talk-shows with a phone-in audience, Otto agreed to reduce Harkins' work load by taking on some of the interviews himself, and

> ...As people called in, we found that invariably, once they got over their initial shock, their next question was always consistent: "Where do I get one?" [10]

How Robots Will Be Programmed to Love

Having read this far, you should be able to accept the possibility that robots can think and display their emotions, simulated or otherwise. If not, I suspect your disbelief is born more of a dislike for the idea rather than because you have analysed the processes of the mind that enable humans to enjoy something or to be in love. When we talk of robots being in love what we mean is that the robot behaves as though it is in love and, influenced by the robot's behaviour, humans and other robots will perceive it as being in love. As with so many other facets

of robot behaviour, we will come to accept that robots have passed the being-in-love version of the Turing Test. Once the simulation of being in love is as convincing as what we call "the real thing", the simulation will appear as reality and we will believe a robot that says "I love you".

How will this simulated love be programmed? There has been little research thus far on the problem of enabling robots to love. One academic whose work has touched on this subject is Aaron Sloman at the University of Birmingham. Sloman has categorised certain states of mind as *tertiary emotions*. Included in this group are anger, longing, grief, guilt, jealousy, excited anticipation and infatuation. As an example of the conditions that must exist to give rise to one of these tertiary emotions, Sloman discusses the question "Why can't a goldfish long for its mother?" Sloman explains

> Longing for one's mother involves at least: (i) knowing one has a mother; (ii) knowing she is not present; (iii) understanding the possibility of being with her; and (iv) finding her absence unpleasant. These all involve possessing and manipulating information, for example information about motherhood, about one's own mother, about locations and change of location, and about the desirability of being close to one's mother. [12]

But as Sloman points out, these conditions are not *sufficient* to bring about the emotion of longing.

> If someone in Timbuctu whose mother is in Montreal satisfies all four of these conditions but hardly ever thinks about his mother and simply gets on with his job, enjoying his social life and always sleeps soundly, then that is not a case of longing. He may regret her absence (which is an attitude), but he does not long for her (an emotion). Longing for someone requires something more, namely: (v) not easily being able to put thoughts of that someone out of one's mind. (Although this is not necessarily so in a case of mild longing!) This is not just a matter of definition: it is a fact that some human mental states involve partial loss of control of attention.
>
> You can not lose what you have never had. So a requirement for being in such a state is having the ability sometimes to control what one is thinking of and also being able sometimes to lose that control. [12]

Sloman emphasizes that this requirement in a robot assumes that some part of the robot's "information mechanism" (i.e., its programming) can

control which information is being processed (which clearly it can), but that the robot is not always in total control. In the case of humans such loss of control exhibits itself

> ...where attempts to focus attention on urgent or important tasks can be difficult or impossible because attention is being drawn back to the focus of the anger, longing, grief,..., infatuation,.... [12]

To summarize Sloman's argument, in order to simulate a feeling of love in a robot it is necessary to simulate the irregular loss of control of the robot's thoughts and actions as it succumbs to its feelings of love. And in order to "feel" love a robot will need to know that it likes a person or robot, which will depend on its attitudes. It also needs to know the degree to which that person or other robot is appealing.

Endowing a robot with such knowledge, such attitudes, and to simulate a loss of control of the robot's thought processes, all these programming tasks are well within our present capabilities. So for robots, love is just around the corner.

– 11 –

Sex and Reproduction, AI Style

AI is like humans playing God, creating "life" where there was none.

-Anonymous Internet posting

Sex with Robots

The obvious question that arises whenever the subject of love with a robot is discussed is "What about sex?" People, even many scientists, no matter how receptive they might be to some of the predictions made in this book, will normally balk at the idea of a human having a physical relationship with a machine, notwithstanding the Woody Allen character in the 1973 movie *Sleeper* explaining that the reason he was looking so happy was that he had just been in the orgasmatron. But in the opening years of the twenty-first century that attitude is being updated. People already find enjoyment in physical relationships with machines of various types, as this posting on the Internet by Wes Johnson will testify (please don't blush)

> I'm still riding her, and plan to as long as my body holds out. When you fall in love with a machine this good, it's an affair that lasts. I wouldn't trade it for anything... [1]

This eulogy sounds slightly raunchy, but in fact what has stolen Wes Johnson's heart is a motorcycle. His second sentence in full reads

> I wouldn't trade it for anything on the road, including all the 200 mph Suzukis and $20,000 Harley's ever to lay rubber down. [1]

If Wes can fall in love with his bike, why should people not fall in love with robots? In Wes's case the physical aspect of his love for his bike is his riding of it, an experience not totally dissimilar to sex (as far as some bikers are concerned).

Wes Johnson is not alone in enjoying a physical relationship with an object. And people want enjoyable and interesting sexual experiences,

so why not with a machine—a robot? The answer to this intriguing "Why?" question will be debated for several decades to come and need not concern us here. Instead we will focus on the "How?" question— how Artificial Intelligence can and will make a significant contribution to mankind's sexual well-being.

Intelligent Sex Machines

The history of sex machines dates back at least to the latter part of the nineteenth century. A book on the subject by Hoag Levins was published in 1996, based on an analysis of more than 800 U.S. Patent Office documents related to human sexuality. And anyone surfing the web to research the future of sex machines could be forgiven for locking on to the site of the Erotic Computation Group at the MIT Media Laboratory. The site explains that the group:

> ...investigates the implications of modern technology on human eroticism in its myriad forms. By developing advanced sexual appliances and techniques, we seek to broaden the range of human amative expression and heighten our potential for sexual gratification. [2]

A taste of sexual things to come can be gleaned from summary descriptions by the MIT researchers that are found on the site. Of particular relevance here are the research projects of James Patten and Sara Cinnamon. Patten's project is on Tangible Sexual Interfaces:

> Collocated sexual interaction is becoming increasingly impractical in our fast-paced modern society, necessitating a new paradigm for remote sexual collaboration. Drawing inspiration from innovative products such as FuFMe, James is developing enabling technologies and accompanying interfaces for multi-user remote sexual intercourse. With projects such as TouchMounters and WhamBam-Plus, he is exploring the ways in which the globalization of new media technology can modify the nature of sex itself. [2]

Sara Cinnamon's research is on Sexual Robotics.

> Humans have long aspired to building anthropoid robots capable of fulfilling our every sexual whim. Such robots were first proposed in Isaac Asimov's groundbreaking 1983 novel *The Robots of Dawn*, and the dream of sexual robots was carried on by Jude Law's

"Gigolo Joe" character in the recent Spielberg film *AI*. Contemporary approaches to artificial sexual companions such as the RealDoll are severely lacking in realism and limited in capability. Sara hopes to make the fantasy of sex robots a reality; inspired by animal and human behavior, her goal is to build capable robotic sex creatures with a living presence and to gain a better understanding of how humans will engage in coitus with this new kind of technology. [2]

Those of you with good enough college grades and a burning interest in erotic computation can also find, on the same web site, details of two graduate courses offered at the Media Lab: MAS 888: Fundamentals of Erotic Computation, and MAS 972: Special Topics in Erotic Computation, which is described as "a project course with enrollment limited to keep a design studio atmosphere. Students will design and develop experimental sexual interfaces, applications, and underlying technologies."

There are two things that I find rather amusing about the MIT Erotic Computation Group web site. The first is that the site is a hoax, as an article in the *New York Times* revealed.

"There's a fair amount of fluffy stuff at the lab without much hard technology behind it," explained Dan Maynes-Aminzade, the first-year Media Lab graduate student behind the hoax. "Sometimes we hear masturbatory rhetoric about how we're changing the world. This seemed to fit."

If the Media Lab has a reputation in some circles for focusing on pie-in-the-sky, even whimsical, research, the faculty can also appreciate a good prank. "They did a good job of mimicking a lot of the research that goes on around here," said Walter Bender, executive director of the program. Mr. Maynes-Aminzade said that Mr. Bender had e-mailed him saying that the site was so popular the traffic was slowing the network and that a Media Lab corporate sponsor had asked if the site was real. [3]

The second aspect of this site that I find amusing is that it is, despite being a hoax at the time it was posted, almost certainly foretelling the future. There are many aspects of the work at MIT's Media Lab that many people would find esoteric, off-beat, unlikely or just plain pointless. But, given that sexual robots already exist on the drawing board at least,[1] and that tangible sexual interfaces appear to be no more difficult

---

[1] See the section "The Mechanics of Sex with Robots" later in this chapter.

to create than sexual robots, there seems to be no reason why an erotic computation group should not spring up at the Media Lab in the not too distant future. After all, as the Media Lab's genuine website proclaims, the lab "continues to focus on the study, invention and creative use of digital technologies, and is now exploring new frontiers,...". [4] Is erotic computing not a creative use of digital technologies and a new frontier?

It is not only in the western world that sex machines have been getting into the news in recent years. In September 2003 the *Hindustan Times of India* reported on efforts by Indian scientists to use robots to improve couples' sex lives. Dr. Prasada Raju of the Indian Government's Department of Science and Technology, says that Honda's humanoid robot ASIMO was initially designed "to perform unusual tasks beyond normal human capability." But he added: "What we have actually achieved goes far beyond that. Saving young couples from breaking apart could be another un-looked for bonus."

There is an argument that involving robots in the most intimate aspects of our lives is unhealthy, that having our robots do and teach us absolutely anything we wish will leave us with less incentive to improve ourselves. While there is some measure of justification in this argument, it is also undeniably true that many people are simply unable, for whatever reason, to benefit from human help in their quest to improve a particular skill or some other aspect of their lives. Sexual technique is one such area—there will be less embarrassment in receiving help and advice from a robot sex therapist than from a human sex counsellor, just as there is less embarrassment for many patients in confessing their true alcohol consumption to a computer program rather than to a human doctor.

Experiments with Sexbots

Although there are some lucid arguments against the idea of sex with robots or even robot-aided sex, there are far more arguments demonstrating why it is a good idea, some of which have been discussed by journalist Jon Katz. At their simplest level robots could provide love talk and, through the use of vibrators, they could offer tactile stimulation. And whatever the level of sophistication of these "sexbots", people investigating their own sexuality and preferences could experiment with them safely. Katz quotes the writer Joe Snell who suggests that

Heterosexuals might use same-sex sexbots to experiment with homosexuality or bi-sexuality. Gay people might use other-sex sexbots to try out heterosexuality. Predators with sexual addictions might no longer prey on human beings.

Given that people become addicted to all sorts of pleasures from slot machines to e-mail, sexbot addiction might be inevitable. Users could become obsessed by their ever-faithful, willing-to-please sexbot lovers that never say "no" or get headaches, and rearrange their lives to accommodate their addictions. Support groups are inevitable. [5]

Snell also speculates that a new category of sexuality might emerge among humans—the techno-virgin, people who find it simpler, perhaps even preferable, to have sex exclusively with sexbots.

...Sexbots would almost surely be programmed to be highly intuitive, keeping track of what worked and what didn't. They would become better sexual partners as they learned more about their human counterparts, storing everything in their memory banks from gasps of pleasure to frequency of orgasm. Every time they had sex with a human, it might get better. [5]

Katz lists a number of benefits of sexbots. A fall in sexually-transmitted diseases, a fall in teenage pregnancies and abortions, fewer paedophiles, less prostitution. He suggests that sexbots could keep marital partners happy so the idea of having an extra-marital affair might become outmoded—"Why take the risk when your sexbot is waiting to meet your needs?" With so many benefits it seems likely that robots will become the sexual partners of choice for many.

The Mechanics of Sex with Robots

Many lovers might wonder at the possibility of developing technology that could replicate human tenderness of touch at its most sensuous. But the technology is already here.

Scientists in Spain have developed a robotic finger with a sense of touch. It is made of a material called a polymer that can feel the weight of what it's pushing and adjust the energy it uses accordingly. This is similar to the way we use our sense of touch. If we pick up a delicate object such as a flower, our fingertips sense its fragility and so grasp it lightly.

Toribio Fernández Otero and Maria Teresa Cortés of the Polytechnic University of Cartagena made their robotic finger from a
"smart polymer" called polypyrrole, which expands in response to
an electric current and conducts electricity differently in response
to changes in pressure. This technology allows the fingertip to apply exactly the right amount of force needed to move an object at
the same speed, almost regardless of the object's weight. For heav-
ier objects, the finger simply pushes with more force and uses more
energy. This automatic adjustment of pressure happens because of
the way the polymer film is squeezed against the obstacle. Pressure changes the packing of the polymer molecules, which alters
the voltage needed to make it bend. In effect, the finger "feels" the
resistance that its motion encounters. [6]

If this technology is applied to its fingers, a robot will be able to offer
the lightest of touches and the most gentle of caresses to its lover. And
in return the receptors in its artificial skin will provide the robots with
sensous feedback.

When Can I Buy One?

The answer may be "sooner than you think". Although it remains to
be seen whether the efforts of the Indian doctors[2] do in fact lead to a
modified ASIMO that can perform bedroom tasks never envisaged by
its original makers, in Australia one inventor has already been granted
a patent[3] for a life-sized sex doll, complete with imitation skin, that is
fully controlled by a computer system. His invention was an object of
curiosity in *New Scientist* magazine:

> You've heard of distance learning. Now it looks like distance sex is
> on the way. Dominic Choy of Cammeray, New South Wales, wants
> to replace real sex with an online robotic experience. He proposes
> a lifelike flexible mannequin covered with imitation skin. Servo
> motors move its limbs and other body parts in response to control
> signals both from the Internet and from touch and sound sensors
> on its body. Two people with matching mannequins connect over
> the Internet, wearing virtual-reality visors so they can see and hear
> each other. Couples can get together from opposite ends of the

---

[2] See the section "Intelligent Sex Machines" earlier in this chapter.
[3] Patent number WO 0059581.

Earth—or they could program the system so that one participant resembles their partner's favourite celebrity. [7]

Dominic Choy's Patent Application

Choy's patent application extends to 29 pages, including many diagrams. The invention is somewhat blandly described in the abstract of the application as

> An apparatus for providing a virtual reality sexual experience. The apparatus including audio reproduction means, visual reproduction means and tactile means for sexual stimulation. [8]

Not exactly everyone's idea of romantic sex, is it? True, the language of patent documents in general is almost as far as one can get from the language of romance, hence the following précis of Choy's application has been edited to make the "patentese" more palatable.

> The system (hardware and software) will allow a user to enter a virtual world and have a sexual experience with a virtual human, or indeed another real human who is also linked up to the same virtual world.
>
> In the case of a single user version, the user will be able to select with whom they wish to interact (a film star for instance). These virtual actors can be represented as highly detailed graphical images and the physical contact itself is simulated by use of a life-sized doll, which is controlled by the software.
>
> In a simplified form the mannequin or doll could be replaced with artificial versions of human body parts used in sexual activities, for example artificial male or female genitalia, as well as or replaced by devices for use in simulating oral sexual activity.
>
> Sensors are provided to be responsive to touch on various portions of the doll. In addition sensors responsive to movement, temperature, pressure and motion can be provided to initiate a physical reaction in the mannequin, for example the discharge of lubrication, the generation of heat and vibration or suction effects.
>
> The artificial sexual organs can be motor driven. For example, in the case of a male doll the penis could be motorised to respond to user activity to provide intense stimulation beyond the range of human movement. For example, the penis could be driven not

353

only to reciprocate at selected or varying speeds but also to rotate, vibrate and to discharge fluid.

The user can use the software to select from one of many stored sexual scenarios. The user wears a virtual reality headset and a motor tracking device adapted to be applied to the user's body, such as a belt, in order to track the user's body motion. In a more sophisticated version there may be several sensors detecting motion of different parts of the body.

If the user touches the virtual human it must react appropriately. The virtual human's skin must also "give" as it would in the real world. The virtual human's facial expressions must be conveyed realistically and be linked to whatever the user does. Some form of feedback is required so that the user can "feel" whatever he is touching. In terms of a 2-user scenario, the two virtual humans will be connected, for example via the Internet, in order to transmit each user's movements to the other user as they are made. The virtual human must also be capable of reacting to the user. For instance, if the user touches the virtual human it should elicit some form of facial or verbal response depending on how the user touches.

The user's movement are continually monitored. Sensors are positioned on the major limb segments (such as the upper arm, lower arm and hand), and will be able to transmit the position and orientation of each of these segments with a hight degree of accuracy. This allows the major movements of the human to be monitored by the system and the information to be processed. [8]

Choy's design (see Figure 63) caters for affective wearable gloves to monitor the positions of the user's fingers, and a system for creating the same touch sensations that the user would feel when touching a human in the same way but without the gloves. He specifies that the outer structure would be of a flexible plastic material. Within the body is mounted an array of hydraulic actuators (devices used to convert fluid energy into mechanical motion), allowing a computer-driven signal to cause a responsive motion in the doll. Also, the doll incorporates pressure sensitive zones (erogenous zones) each having a focus and a less sensitive peripheral region so that when touched the doll can make appropriate movements. The patent specification continues with various mechanical details, such as having cavities that are removeable for cleaning purposes, as well as descriptions and drawings of the artificial vagina and artificial penis.

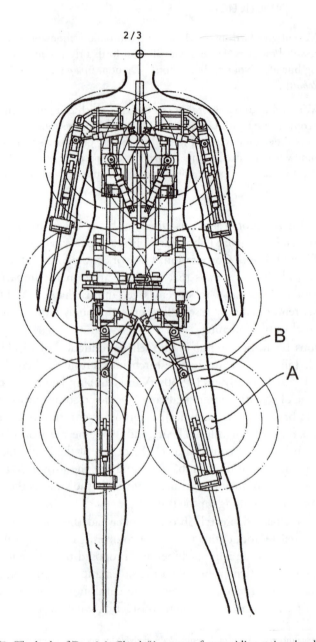

Figure 63. The body of Dominic Choy's "Apparatus for providing a virtual reality sexual experience"—note the two pressure-sensitive zones at A and B

Robot Reproduction

> We will give birth by machine. We will build a thousand steam-powered mothers. From them will pour forth a river of life. Nothing but life! Nothing but robots! [Damon, in *Rossum's Universal Robots*]

> We're trying to build a robot out of Lego which can put together a copy of itself with Lego pieces. Obviously you need motors and some little computational units, but the big question is to determine what the fixed points in mechanical space are to create objects that can manipulate components of themselves and construct themselves. [9]

Rodney Brooks, quoted above, is Director of the MIT Computer Science and Artificial Intelligence Laboratory. He is also chairman and chief technical officer of iRobot Corporation. His work at the forefront of robotics research is blazing new trails that will have an immense impact on the development of the super-robots envisaged in this chapter. And the idea of robots reproducing by assembling other robots is more than just an idea, it is already happening.

Serious scientific interest in the self-reproduction of machines began with John von Neumann's research into automata theory[4] during the late 1950s and early-mid 1960s. Von Neumann had an enormously wide range of scientific interests, being one of the co-inventors of Game Theory (a branch of mathematics), a significant contributor to the logical foundations of Quantum Theory and, while he was working on the Manhattan Project, a contributor to the design of the implosion mechanism for the plutonium bomb. Any study of the early history of the theory of self-reproducing machines is virtually the same as the study of von Neumann's thinking on the subject.

Von Neumann's research into automata had the goal of modelling biological self-reproduction. He set out to investigate the logical organization of self-reproducing machines and proposed five different models of self-replication, which he called the kinematic machine, the cellular machine, the neuron type machine, the continuous machine and the probabilistic machine. And when considering what capabilities should

---

[4]Automata theory is a study of the principles underlying the operation of any electromechanical device (an automaton) that converts information from one form into another according to a definite procedure.

be demonstrable in any machine that was claimed to be able to replicate itself, von Neumann listed the following:

> Logical universality—the ability to function as a general-purpose computing machine;
>
> Construction capability—the ability to manipulate information, energy, and materials of the same sort of which it itself is composed;
>
> Constructional universality—the ability to manufacture any machine which can be formed from specific kinds of parts.

Von Neumann concluded that self-reproduction is possible if the above capabilities are achieved. His argument was that, because the original machine is made of manufacturable parts, and the original machine is constructable, and the original machine is given a description of itself, then it ought to be able to make more copies of itself using the manufacturable parts.

Von Neumann's research into self-reproducing automata also touched on evolution, though he made little progress in that direction. It was only with the advent of genetic algorithms[5] that some of the questions raised by von Neumann could be properly investigated. For example, if one has a robot, and it makes a robot, which then itself makes a robot, is there any proof that the line of robots can become successively "better" in some fashion, such as more being more efficient or being able to accomplish more different tasks? Could these robots evolve to higher and higher forms?

Of the five models that von Neumann proposed for studying self-replicating automata, the kinetic machine is the best known. Von Neumann envisioned a machine that lived in a "sea" of spare parts. The machine had a program stored in its memory, a program that instructed the machine to go through certain mechanical procedures. The machine had an arm, or something that functioned very much like an arm, and the machine could move around its environment. By using its "arm" the machine could pick up and connect whichever of the spare parts it wished. The program first instructed the machine to reach out and pick up a part, then to go through an identification procedure in order to determine whether or not the part selected was the one the machine had

---

[5] See Chapter 6.

been told to locate. (If not, the part would be thrown back into the "sea" and another part picked up and examined in the same way, and so on, until the correct part was found.)

Having found the part it was looking for, the machine would then start on its search for the next part that it needed, as determined by the program stored in its memory. The machine would continue following the instructions to make something, without really understanding what it is doing. When it finished executing its program the machine would have produced a physical duplicate of itself. But this newly-minted machine would not yet have any program in its memory, so the "parent" machine would copy its own program, from its own memory, into the memory of its offspring, and then, finally, the original machine would start up the program residing in its progeny. This whole process, as described by von Neumann, is logically very close to the way that living organisms reproduce themselves.

Self-Reproducing Software and Genetic Programming

Although this chapter employs the word "robot" generically to represent either hardware or software (or the combination of both), let us briefly revert to the traditional distinction between the two, and before we discuss self-reproducing robots (the hardware) let us first consider self-reproducing computer programs, a subject given a detailed exposition by John Koza.

Computer programs that can reproduce do so by combining parts from two parents and improving their performance through evolution. A clever example of self-reproducing programs was written by Thomas Ray at the University of Oklahoma. Ray's 80-line program demonstrated how it could evolve over time as a consequence of mutation. Ray wrote his program in a special programming language (Tierra) for a virtual machine. His virtual machine was intentionally imperfect and introduced random mutations to his original hand-written program (called the ancestor). Ray observed the emergence, over a period of hundreds of millions of time steps, of an impressive variety of different entities (some self-reproducing, some parasitically self-reproducing, some symbiotically self-reproducing, and some not self-reproducing) and a dazzling array of biological phenomena including parasitism, defences against parasitism, hyper-parasitism, symbiosis, and social parasitism.

A whole class of self-reproducing software, and one that has grown dramatically in importance since its invention by Koza in the early 1990s, is one in which the programs evolve, generation by generation, improving themselves along the way. Genetic Programming, as this technology is most often called, addresses the keenly interesting task of how computers can be programmed to do something without being told exactly how to do it. The answer, put simply, is that they are set to evolve until they know how.

In Chapter 6 we saw how genetic algorithms can be bred to find increasingly better solutions to a problem. In the same way, computer programs can be bred until they evolve to the point of being able themselves to solve problems without first being taught how to do so. Each member of a population of programs makes an attempt at devising a program to solve a type of problem, and then the more successful ones are encouraged to beget offspring designed to learn from the parents. Finding a solution to a particular problem, as genetic algorithms do, is a skilled task, but knowing how always to find the solution to a particular *type of problem* is a more powerful and more valuable skill.

The breeding process in genetic programming is fundamentally the same as that employed in genetic algorithms. Firstly, an initial population of computer programs is created, each member of which comprises a random mix of parts of computer programs. The basis of program evolution is an iterative process[6] that starts with every program in the population being set to work and assigned a fitness measure according to how well or badly it performs. As one would expect, the programs in the first generation of a genetic programming population will nearly always perform extremely badly and therefore have assigned to them very bad fitness values. Even so, some of these randomly created programs will turn out to be better than others, and these are the ones that are most likely to serve as the parents for the next generation.

The nature of the fitness measure employed to evaluate program performance varies according to the type of problem, for example it might be simply a measure of by how much the result produced by a program differs from the desired result, in which case the closer they are, the better the program. Normally, every program in the population is tested on many different samples from the relevant problem domain, so that its

---

[6] An iterative process is one that is repeated again and again, gradually moving towards the desired result or solution, until a satisfactory end result is reached or until the process decides to stop working, usually because it "gives up" on improving its performance.

fitness can be more accurately estimated, perhaps by calculating its average performance over all the sample problems.

Once each program in the population has been assigned its fitness value, a new population of programs is created by choosing the fitter programs (those that performed best), making copies of them and then creating pairs of new programs from pairs of existing programs, using the same crossover idea employed in genetic algorithms and applying the crossover process at a not-quite-randomly chosen point within each program of a pair. The crossover process creates new programs by combining parts of existing parent programs. The choice of crossover points is not made entirely at random, as that would often result in logical sections of a program being broken up. Instead the choice is a random selection made from those points in programs where one logical section of the program begins and another ends. In this way entire logical sections are swapped about, thereby always resulting in programs that conform to the rules of the programming language. And because programs are chosen to be part of a crossover process with a probability based on their fitness, the offspring programs, those created by crossover, are much more likely to contain parts from promising programs than parts from programs that performed poorly in the most recent test run. Once two new programs have been created by crossover, two of the less fit programs are eliminated from the population, in order to keep the total size of the population constant.

The final step in each iteration of the evolution process is to create some programs by mutation, again in a very similar way to that employed in genetic algorithms. The mutation process in genetic programming is performed by randomly selecting a crossover point in one program rather than in two, and then deleting the next logical section of that program and growing a new section to replace it. Each of the programs in the new population is then set to execute its given task and the result of the best performing program is tested to see whether it is a solution or close to a solution to the problem. If the best performing program is deemed to be good enough, the evolution process terminates; otherwise a fresh iteration begins and the whole process is repeated.

Self-Reproducing Hardware

The work of Rodney Brooks at MIT and at his company iRobot have encouraged others to investigate self-reproducing hardware. Similar re-

search is under way in the Golem[7] project at Brandeis University, under the direction of Hod Lipson and Jordan Pollack. Their "parent robots" consist of a computer running an algorithm that simulates evolution and produces a design for new robots based on trial and error. This is linked to a three-dimensional printer that makes small plastic shapes. The off-spring are small plastic trusses (made of bars) with actuators, propelled by motors and controlled by neural networks.[8] Artificial neurons are the building blocks of their control modules (their brains). Bars connected with free joints can potentially form trusses that represent arbitrary rigid, flexible and articulated structures. The bars connect to each other through ball-and-socket joints, neurons can connect to other neurons through synaptic connections, and neurons can connect to bars. Human intervention is only necessary to attach the motors and connect the wires—the robots do all the rest of the assembly themselves, including telling the humans what to do.

Lipson and Pollack's simple thermoplastic components were employed in the development of what they call a polymorphic robot, a machine that can change its shape to suit the job in hand.[9] The Golem robots had one simple physical task to perform—they had to discover a way to move.

Starting with a population of 200 "machines" that were comprised initially of zero bars and zero neurons, Lipson and Pollack simulated the evolutionary process. The fitness of a machine was determined by its locomotive ability—how far its centre of mass moved within a fixed amount of time. At each stage of the process, fitter machines were selected and offspring were created by adding, modifying and removing building blocks, and these new machines replaced the less fit ones in the population. This process typically continued for 300 to 600 generations, with both the body (the shape of the machine) and its brain (control module) evolving simultaneously.

Typically, several tens of generations passed in the experiments before the first robotic movement occurred. For example, as a minimum, a neural network must first assemble and connect to an actuator in order for any motion at all to be possible. Selected robots out of those with winning performances were then automatically replicated into re-

---

[7] Genetically Organized Life-Like Electro-Mechanics.

[8] Shortly before this book went to press in 2005, Lipson and Pollack announced that they had also developed a robot whose offspring were exact copies of themselves.

[9] See the section "Robots that Change Their Own Shape" in Chapter 8.

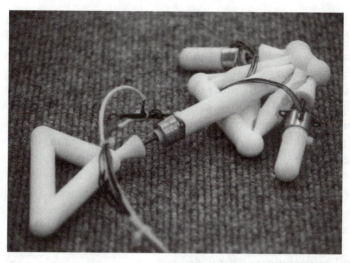

Figure 64. A Golem project robot-produced robot (Courtesy of Hod Lipson and Jordan Pollack)

ality. Their bodies, which exist only as points and lines, were first converted into a solid model, with ball-joints and housings for motors according to the evolved design. This solidifying stage was performed by a program which combined the various pre-designed components. The virtual solid bodies were then manufactured using commercial rapid prototyping technology. This prototyping machine used a temperature-controlled head to squeeze out a heated plastic material layer by layer, so that the evolved shape emerged as a solid three-dimensional structure without the need for human intervention. The entire pre-assembled machine was printed as a single unit, with fine plastic supports connecting between moving parts; these supports broke away the first time the structure moved, leaving structures containing complex joints that would be difficult to design or manufacture using traditional methods. Motors were then snapped in, and the evolved neural network was run on a micro-controller in order to activate the motors. The physical machines faithfully reproduced their virtual ancestors' behavior. In spite of the relatively simple task and environment, surprisingly different and elaborate solutions were evolved.

Although prototyping machines such as three-dimensional printers are large and cumbersome, Lipson believes that much smaller ones could one day be built into a robot, allowing it to change parts of its body, for example, to reshape an arm in order to produce a new tool for a

novel situation. Mark Yim of the Xerox Palo Alto Research Center in California says this is one area in which polymorphic robots could be most useful. "There's no point in taking an entire toolkit into space, he says, when you don't know which tools you'll need: a single robot arm can be shaped to do the job of all of them." [10]

Lipson also believes it is conceiveable that the three-dimensional printing technology will allow several materials to be printed, including conductive materials (such as are used in printed circuit boards), nonconductive and even semi-conductive materials (for example, microprocessors). "Wires, motors and logic circuits, as well as structure, could be printed in one pass without the need for assembly."

Some versions of the Golem robots push themselves along on one leg, while others produce a hinge-like motion and crawl about like a fish out of water. Yet another moves sideways like a crab. Although a robot is ready to move when it comes out of the three-dimensional printer, its motor must be inserted by a person. But one aim of the project is to make the robots totally independent, much like the vengeful shape-shifter in the movie *Terminator 2*.

Currently, when a robot has performed its task, it offers itself up to be melted down, so its thermoplastic components can be recycled into another useful droid by the three-dimensional printer. While robots are so self-effacing the human race has nothing to fear from them.

The mechanical aspects of manufacturing new robots in experiments such as these, employ well-established technologies, for example the three-dimensional printer mentioned above. Research in the field of Rapid Prototyping has drastically reduced the long waits that product designers and inventors used to face when building prototypes. Rapid Prototyping, which is also known as Layered Manufacture, is a manufacturing technique whereby three-dimensional solid models are constructed by fusing layer upon layer of material, for example plastic powder, under the control of a computer. This process is widely used for the rapid fabrication of physical prototypes of functional parts, patterns for moulds, prototypes of medical implants and bones, and consumer products. The principal advantage of the process is that it allows early verification of a design so that improvements to the design can be made and the new design fabricated and tested quickly.

The U.S. Government has shown some considerable interest in self-reproducing robots (sometimes referred to as Self-Replicating Systems or SRSs). In 1980 a NASA study was conducted by request of newly-elected

President Jimmy Carter at a cost of $11.7 million. The result of the study was a realistic proposal for a self-replicating automated lunar factory system, capable of exponentially increasing productive capacity and, in the long run, the exploration of the entire galaxy within a reasonable timeframe. It was not, however, without a certain amount of trepidation, that the report commented on the social aspects of what it termed the "SRS cornucopia"—the unfettered evolution and reproduction of such systems. In reading the extracts of the report presented here you may safely ignore the notion that the "robot factory" will be on the moon or elsewhere in the galaxy—in my view that is a complete red herring.

> How will humankind deal with what has been termed, with some justification, "the last machine we need ever build?" How might people's lives be changed by a replicative universal constructor system capable of absorbing solar energy and raw dirt and manufacturing, as if by magic, a steady stream of finished building materials, television sets and cars, sheet metal, computer components, and more robots—with little or no human intervention required? Just as the invention of the telephone provided virtually instantaneous long-distance communication, and television permits instant knowledge of remote events, and the automobile allows great individual mobility, the autonomous SRS has the potential to provide humanity with virtually any desired product or service and in almost unlimited quantities. Assuming that global human population does not simply rise in response to the new-found replicative cornucopia and recreate another equilibrium of scarcity at the original per capita levels, supply may be decoupled from demand to permit each person to possess all he wants, and more. The problems of social adjustment to extreme sudden wealth have been documented in certain OPEC nations in recent years. Much attention has also been given to the coming "age of leisure" to be caused by super-automation. What more difficult psychological and social problems might emerge in an era of global material hyper-abundance? [11]

One of the concerns presented in the report related to the possibility of a population explosion amongst self-reproducing robots:

> An exponentially increasing number of factories (even if the rate is not sustained indefinitely) will seem especially threatening and psychologically alarming to many. Such a situation will draw forth visions of a "population explosion", heated discussions of lebensraum, cancerous growth, and the like. Nations not possessing

replicating systems technology will fear an accelerating economic and cultural gulf between the haves and the have-nots. On another level altogether, humankind as a species may regard the burgeoning machine population as competitors for scarce energy and material resources, even if the net return from the SRS population is positive. [11]

And on the subject of robot predators the report was no less cautious:

> Predation is one interesting possibility. Much as predator animals are frequently introduced in National Parks as a population control measure, we might design predator machines which ate either "species specific" (attacking only one kind of SRS whose numbers must be reduced) or a kind of "universal destructor" (able to take apart any foreign machine encountered, stockpiling the parts and banking the acquired information). Such devices are logically possible but would themselves have to be carefully controlled. [11]

On a less pessimistic note the report pointed out that, if the robots introduced by man to control the predators are also manufactured by man, so that their numbers will tend to increase at a fixed rate rather than exponentially, this population of "good" robots will be able to control an exponentially increasing population of "bad" robots because the rate of destruction will be far more rapid than that of replication of the "bad" robots. But this could be wishful thinking. Who knows how quickly robots may be coming off the assembly lines of the future?

The NASA proposal was quietly declined by the U.S. Government, with barely a ripple in the press. But what was conceivable with 1980s technology is now even more practical today.

Robot Evolution

> Find a bug in a program, and fix it, and the program will work today. Show the program how to find and fix a bug, and the program will work forever. [12]

Evolutionary robotics combines self-reproducing software with self-reproducing hardware in an attempt to develop robots through a self-organized process based on artificial evolution. An initial population of different artificial chromosomes, each encoding the control system (and sometimes the shape) of a robot, is randomly created and put into an environment. Each robot is then let free to act as it wishes (to move, to look

around, to manipulate objects) according to a genetically specified controller, while the robot's performance on various tasks is automatically evaluated. The "fittest" robots are allowed to reproduce by generating copies of their genetic constitutions but with the addition of changes introduced by some genetic operations (for example, mutations, crossover, duplication). This process is repeated for many generations until an individual robot is born that satisfies the performance criterion set by the experimenter as to its fitness and functionality. Robots might also change during their lifetime and therefore might adapt to their environment.

The principal components of any evolutionary algorithm are evaluation and reproduction, and both of these must be carried out autonomously by and between the robots. The process of evaluation is carried out autonomously by each robot in the population; for example a robot that failed to recharge its batteries when necessary would "die", or a robot's fitness might be measured in terms of how many objects it collects within a given amount of time. The selection process in an evolutionary algorithm is carried out by having the fitter robots supplying "genes", in which case these robots become parents for robots in the next generation, or by having the less fit robots lose genes, in which case these robots are replaced in the population by their offspring. Rather than pick two parents of above-average fitness and produce an offspring from a combination of their genes to replace a robot of below-average fitness, Inman Harvey simplified the genetic algorithm by selecting two individual robots at random and overwriting some of the genes of the less fit of the two with some from the more fit of the two.

Robot Offspring, with a Dash of Human Added

Self-reproducing robots will, through evolutionary algorithms, create offspring that tend to possess those attributes which the parent robot finds appealing or even beneficial in itself or in other robots. But this self-improvement process need not stop there. Your robot might decide that it likes certain aspects of *your* personality, or *your* looks, or *your* voice, ..., and emulate them when it reproduces. Your robot might learn a skill from you and pass this skill on to its offspring. In fact, if it wishes, your robot could create a part-clone that is more like you than it is like the robot itself.

This notion is, perhaps, the most difficult idea of all for many people to accept: that robots can create offspring whose genetic makeup is based partly on that of the robot parent and other robots with which it has come into contact, and partly on that of the robot's human lovers, friends and acquaintances. It will not even be necessary for a robot to have sex with a human in order to reproduce—the act of making love will become something indulged in purely for emotional and physical satisfaction and pleasure, totally unconnected with the creation of the next generation. Who can say with any certainty that this is not how humans and robots will co-exist in the second half of the twenty-first century?

- 12 -

Robot Consciousness

In the debate on whether or not robots might be intelligent, philosophers, psychologists and, in particular, cognitive scientists have taken an interest in questions such as "What do we actually mean by intelligence?", "Can machines think?" and "Can machines have consciousness?" These questions take on more importance as humankind moves towards a greater acceptability of the concept of intelligence in robots, and it is therefore appropriate for us to examine the meaning and significance of consciousness, not only for a better understanding of the attribute for its own sake but also because it is a foundation stone of thought and intelligence.

What Is Consciousness?

The first problem we encounter in an attempt to understand and analyze consciousness is one of definition. Everyone knows what *they* mean by consciousness but a clear definition is elusive. Consciousness involves awareness, both of what is going on around us and what is going on inside our minds, self-awareness. Consciousness is subjective and intangible. We recognize that consciousness is an essential component of being human but it is unclear whether consciousness is one of the causes of intelligence or one of its effects or by-products. Is consciousness a kind of thought process created by the interactions of the billions of neurons in our brain, or does it have a physical form—is it a material part of the brain?

Linked to the question of how to define consciousness is the chauvinistic attitude that it is a particularly human attribute. Undeniably humans possess what *we* call consciousness, but not everyone would accept that other animals do, and still fewer people believe that plants do. Viewed from a slightly different perspective, some might suggest that humans are in possession of a greater consciousness than are other animals which, in turn, have a greater consciousness than plants. The logical continuation of this train of thought is that objects, including comput-

ers and robots, since they are considered by most people to be not alive, have no consciousness at all. Let us examine this position further.

One early commentator on the topic of consciousness in machines was the nineteenth-century British writer on philosophy Samuel Butler. Having studied classics, and with no formal scientific training, Butler took an interest in science after reading Charles Darwin's *On the Origin of Species* and conducting a brief correspondence with Darwin himself, following which Butler became a staunch evolutionist. On the subject of machine consciousness Butler argued in his prophetic book *Erewhon*[1] against the chauvinistic view of humans as the sole possessors of consciousness:

> There is a kind of plant that eats organic food with its flowers: when a fly settles upon the blossom, the petals close upon it and hold it fast till the plant has absorbed the insect into its system; but they will close on nothing but what is good to eat; of a drop of rain or a piece of stick they will take no notice. Curious! that so unconscious a thing should have such a keen eye to its own interest. If this is unconsciousness, where is the use of consciousness?
>
> Shall we say that the plant does not know what it is doing merely because it has no eyes, or ears, or brains? If we say that it acts mechanically, and mechanically only, shall we not be forced to admit that sundry other and apparently very deliberate actions are also mechanical? If it seems to us that the plant kills and eats a fly mechanically, may it not seem to the plant that a man must kill and eat a sheep mechanically? [1]

Butler's nineteenth-century observation is supported by an argument from more than a century later, put forward by Marvin Minsky:

> Most people assume that computers can't be conscious, or self-aware; at best they can only simulate the appearance of this. Of course, this assumes that we, as humans, are self-aware. But are we? I think not. I know that sounds ridiculous, so let me explain.
>
> If by awareness we mean knowing what is in our minds, then, as every clinical psychologist knows, people are only very slightly self-aware, and most of what they think about themselves is guesswork. We seem to build up networks of theories about what is in our minds, and we mistake these apparent visions for what's really going on. To put it bluntly, most of what our consciousness reveals to us

---

[1] Erewhon is an anagram of nowhere.

is just made up. Now, I don't mean that we're not aware of sounds and sights, or even of some parts of thoughts. I'm only saying that we're not aware of much of what goes on inside our minds. [2]

These writings by Butler and Minsky contribute in some small way to our understanding of consciousness but do not take us much closer to an acceptable definition of the term. Part of the problem in defining consciousness lies in its subjectivity. *My* consciousness relates to what *I* perceive to be happening in *my* mind, whatever *my* mind is. It is also related to *my* awareness of what is going on inside *my* body and in the world immediately around *me*. But that is *my* consciousness, it is not yours. You have your own consciousness. Although it is doubtless related to your own self-awareness in a very similar way to that in which mine are related, your consciousness is equally impossible for an outsider to observe or to measure scientifically.

Given the impossibility of observing consciousness it is hardly surprising that a prominent researcher in this field, Susan Blackmore, asserts that consciousness is actually an illusion:

> This is because it feels to us humans as though there is a continuous flow of experiences happening to an inner self when in fact there is no such inner self. Computers have no inner self either, but if ever they start thinking they do they will become deluded like us, and hence conscious like us. And that day is surely not far off. [3]

Blackmore expands on this assertion in her fascinating paper "Consciousness in Meme Machines":

> On this view, human-like consciousness means having a particular kind of illusion. If machines are to have human-like consciousness then they must be subject to this same kind of illusion. I shall therefore explore one theory of how this illusion comes about in humans and how it might be created in machines; the theory of memetics.[2] [4]

Blackmore then puts forward and defends the view that memes distort consciousness into an illusion. She discusses two kinds of artificial meme machine: those which imitate each other and those which imitate humans. And she argues that our individual minds and our sense of self and consciousness are designed by memetic pressures, leading her to conclude that meme machines can be designed with human-like consciousness.

---

[2]The term "meme" was first coined by Richard Dawkins in 1976, for "ideas, habits, skills, stories or any kind of behaviour or information that is copied from person to person by imitation." [5]

On this view many kinds of machine might be conscious, but only a particular kind of machine could be conscious in a human-like, illusory way. It would have to be capable of imitation (otherwise it could not replicate memes) and live in a community of similar meme sharing machines (otherwise there would be no pressure for memeplexes[3] to form). Such a machine would, if this theory is correct, be a victim of the same illusions of consciousness as we humans are. That is, it would think it had an inner self who was conscious. Ultimately, it would start wondering what consciousness was and trying to solve the hard problem. [4]

Another luminary in this field, Aaron Sloman, believes that the concept of consciousness has not been clearly enough identified and therefore "much of what is written about how experimental results relate to consciousness is ambiguous and muddled". [6] Sloman amplifies this argument in forthright terms:

I am embarrassed to be writing about consciousness because my impression is that nearly everything written about it, even by distinguished scientists and philosophers, is mostly rubbish and will generally be seen to be rubbish at some time in the future, perhaps two hundred years from now. [6]

Where does all this lead us in our quest for a generally acceptable definition of consciousness? Rather than allowing ourselves to get bogged down in the search, I prefer to take a pragmatic view, accepting that it is sufficient for there to be a general consensus about what we mean by consciousness and to assume that there is no burning need for a rigorous definition—let us simply use the word and get on with it.[4]

Can Robots Have Consciousness?

This is a highly emotive topic. In order to avoid giving too great a shock to the sceptics amongst you I shall argue the case for the existence of

---

[3]The memeticist Glenn Grant defines a meme-complex (also called a memeplex) as "A set of mutually-assisting memes which have co-evolved a symbiotic relationship. Religious and political dogmas, social movements, artistic styles, traditions and customs, chain letters, paradigms, languages, etc. are meme-complexes."

[4]For those readers who find this lack of rigour unsatisfactory I suggest the adoption, as a working definition, of the one given by *www.dictionary.com*: "Having an awareness of one's environment and one's own existence, sensations and thoughts." This definition is a perfectly adequate foundation for the present chapter.

robot consciousness in gently graded stages. First we will consider the arguments most often levelled against the concept, and their refutations. Then we discuss various reasons that have been put forward to explain why consciousness in robots *might* be possible, and we conclude this section with assertions from eminent scientists and philosophers that robot consciousness *is* possible.

It is a received notion that only something that is alive can have consciousness. One way to dispense with this argument is to question the argument's foundation—what is the meaning of "alive" in this context? Is it

1. Being able to move itself? (Robots can.)

2. Being able to grow without outside help? (Robots can already build themselves into bigger robots.)

3. Being able to regulate its own bodily functions? (In the eighteenth century the French inventor Vaucason created a mechanical duck which digested its food and excreted. Consider also, as just one modern example, the "screen saver" programs we have on our PCs for regulating the use of our computer screens.)

4. Being able to reproduce? (As we have seen in Chapter 11 this is already possible.)

5. Having a metabolism? (The development of gastrobots is thriving at the University of Southern Florida and at the University of the West of England. These are robots that derive all of their energy requirements from the digestion of real food.)

6. Being made of protein? (Advances in DNA computing may make this particular objection null and void.[5])

The above pseudo-objections were suggested by philosopher William Lycan, who believes it very unlikely that having a metabolism or being made of protein could be attributes of a machine. Even if his speculation

---

[5] Protein sequences are encoded by DNA sequences—this relationship is called the genetic code and a gene made from DNA is the code to build a protein. Therefore protein depends on DNA for its formation because proteins manufacture themselves by decoding the information contained in DNA that tells the builder proteins how to make themselves. But, paradoxically, DNA depends on protein for its own formation. With such a close relationship between DNA and protein, the requirement that life requires protein can be seen to be equivalent to saying that life requires DNA, and therefore DNA computers are, by definition, alive.

is not refuted by events during the next half-century, I disagree with his view that either is a necessary factor to support the existence of life. Why should an entity need to be made of protein (or DNA), or to have a metabolism, in order to be considered "alive"? Life, surely, is a function of behaviour—what an entity does. Life is not a function of what the entity is made of or how it acquires, converts and consumes energy.

Another argument often employed against the possibility of consciousness in robots, is that robots could never have a qualitative experience in the way we humans do, and that since qualitative experiences require consciousness to help us experience them, the lack of qualitative experiences in robots would imply a lack of consciousness. This particular argument can be refuted by what philosophers call the fading qualia thought experiment. The refutation is based on considering the (hypothetical) part-by-part replacement of each and every part of a human being by man-made components. Let us imagine that we start by replacing the neurons in a human's brain, one by one, with minute electronic devices, nanotechnology, each electronic device successfully performing the task of the neuron it replaced so that the human's performance is not impaired in any way. Then, other bodily parts are removed from the human and substituted by other man-made devices that replicate the functions of the original components in the body. And so on, until eventually every single component in the new version (a robot) is a nanomachine—the robot's behaviour is controlled entirely by nanomachines. But its intelligence, personality, creativity, sensory abilities and, most importantly, its phenomenological perception, remain just as before. Then, clearly, one of two possibilities must be true. Either

A. At some point during the sequence of replacement operations, the being ceased to have qualitative experiences, losing its humanlike characteristics. In this case, the replacement of a single neuron, just one out of the one hundred billion neurons in the human brain, could be responsible for the vanishing of the being's conscious experience—one neuron ago the being had consciousness but one neuron later it did not;

or

B. The replacement being (the robot) possesses every single capability and attribute of the original human, including consciousness.

If A is true, it would be possible to draw a line and state with absolute certainty that after one particular replacement operation out of the bil-

lions of such operations, the being ceased to have a phenomenology.[6] It seems almost inconceivable that we could defend this position, *ergo* we must accept that B is the truth—the robot has consciousness.

Perhaps this approach, refuting the arguments of the non-believers, is more defensive than is necessary. A far more positive stance on the subject was taken by Sam Lehman-Wilzig in his 1981 essay: "Frankenstein Unbound: Toward a Legal Definition of Artificial Intelligence". Lehman-Wilzig presents weighty evidence that robots would be, by most definitions, alive, citing seven significant achievements of AI that had already been realised (as of 1981), or were theoretically possible at that time. These were that robots could

1. Imitate the behavior of any other machine. (This is self-evident, since software can simulate the behaviour of anything.)

2. Exhibit curiosity. (Examples include search-and-rescue robots.[7])

3. Recognize and react to the sight of themselves and members of their own machine species. (In a crude way this was first achieved in the late 1940s by Grey Walter's tortoises,[8] and is almost trivial by the standards of today's recognition technology,[9] as is programming robots to react to what they recognize.)

4. Learn from their own mistakes. (Also trivial nowadays.[10])

5. Be as creative and purposeful as are humans, even to the extent of looking for purposes that they can fulfil. (The whole of Chapter 5 describes the creativity of robots.[11])

---

[6] A description, history or explanation of phenomena.

[7] See Chapter 8.

[8] See the section "Robot Tortoises" in Chapter 2.

[9] See the sections "Human Face Recognition" and "Recognition in Three Dimensions", both in Chapter 4.

[10] See the section "How Computers Learn" in Chapter 6.

[11] See also the section "How Computers Discover and Invent" in Chapter 6.

6. Reproduce themselves as described by John von Neumann, in five fundamentally different modes, of which the fifth, the "probabilistic mode of self-reproduction", closely parallels biological evolution through mutations (i.e., genetic algorithms), so that "highly complex, powerful automata can evolve from inefficient, simple, weak automata." [7] (The previous chapter discusses robot reproduction.)

7. Can have an unbounded life span through self-repairing mechanisms. (The diagnosis of faulty software and electronic or mechanical hardware is an easier task than the feats in medical diagnosis that have already been achieved.[12] And repairing such faults requires nothing more advanced than the self-replicating technologies described earlier.[13])

Others who have listed criteria for the existence of robot life, criteria that they themselves conclude are attainable, include John Kemeny and Joe Weizenbaum. Kemeny presents six criteria which, he claims, distinguish living from inanimate matter: metabolism, locomotion, reproducibility, individuality, intelligence, and a natural composition. In all six criteria Kemeny asserts that AI servo-mechanisms[14] clearly pass the test. And as for Weizenbaum, a critic of AI, he accepts that computers are sufficiently "complex and autonomous" to be called an "organism" with "self-consciousness" and an ability to be "socialized". He sees no way to put a bound on the degree of intelligence such an organism could, at least in principle, attain. [7]

Stevan Harnad contributes to the discussion on whether robots are alive by invoking a Turingesque argument. Harnad's logic, in answer to the question, "How do we know that this machine is really alive?", is simply that

> If there are two structurally and functionally indistinguishable systems, one natural and the other man-made, and their full causal mechanism is known and understood, what does it even mean to ask "But what if one of them is really alive, but the other is not?" What property is at issue that one has and the other lacks, when all properties have already been captured by the engineering? [8]

---

[12]See the section "Expert Systems" in Chapter 6.
[13]Again, see the section "Robot Reproduction" in Chapter 11.
[14]A servomechanism is a feedback system used in the automatic control of a mechanical device.

In examining the concept of robot consciousness from a neutral stand-point, Samuel Butler argued

> But surely when we reflect upon the manifold phases of life and consciousness which have been evolved already, it would be rash to say that no others can be developed, and that animal life is the end of all things. There was a time when fire was the end of all things: another when rocks and water were so. [1]

Butler's was (and is) an almost unassailable position, sitting as he was, squarely on the fence, contending that we cannot prove it one way or the other and therefore we must accept that robot consciousness *might* be possible.

By now any reader who was initially a non-believer will hopefully be more open-minded to the possibility that robots *might* have consciousness. It may be that robots will have consciousness when they match the computing power of the human brain. While we cannot be sure that robots with the power of the human brain *will* have consciousness, it is certainly more likely that robots will have consciousness by the time they have *far exceeded* the computing powers of the brain. But we cannot, at this point in time, be certain of either of these hypotheses, and it might be that we will not even know for certain when it happens, though we should certainly give robots the benefit of the doubt when they exhibit consciousness and assume that they are indeed conscious.

Having refuted the principal contra-arguments and examined some views that robot consciousness *might* be possible, we now come to the opinions of two luminaries who have firmly stated the opinion that con-sciousness in robots *is* possible. Ray Kurzweil, while not being so very far from the fence himself, predicts a time when

> We will have a massive neural net, built from silicon and based on a reverse engineering of the human brain. Its circuits are a million times faster than human neurons, and it can learn human language and model human knowledge. It develops its own conceptions of reality, and on its own blurts out that it is lonely. We will ask whether this robot is a conscious agent. In the end we will come to believe that such robots are conscious much as we believe the same of each other. Their minds will be based on the design of human thinking, and they will embody human qualities and claim to be human. And we'll believe them. [9]

Strongly in support of the pro-consciousness lobby, though for a completely different reason, is the Stanford University cognitive scientist Bruce Mangan, who claims that "Consciousness is an information-bearing medium." [10] Mangan's thesis is that consciousness incorporates cognitive information in a similar way to which DNA incorporates genetic information.

> I propose to formalize the notion of consciousness in a slightly new way: consciousness is simply one information bearing medium, among many others, at work in our organism. Consciousness tends to bear information that is relevant to novel evaluations either expected or at hand and consciousness bears its information as experience (or "qualia", but I try to avoid this term). [10]

We have now run a whole gamut of opinions on whether robots can *have* consciousness, and you should, by now, be at least accepting the *possibility* of consciousness in robots. The next question is "How can machines be *given* consciousness?" Since we do not know exactly what consciousness is, nor what makes it, exactly what should we program into a robot to give it consciousness or to enable it to evolve consciousness?

The designers of MIT's Cog robot have definite ideas as to how they could give Cog consciousness. It has been argued that nothing could properly matter to an artificial intelligence and that mattering is crucial to consciousness. Cog's creators made a deliberate decision to make Cog as responsible as possible for its own welfare by giving it some innate arbitrary preferences, goals if you like. Cog will know if its goals are thwarted or achieved, and can be artificially sad or happy accordingly. Although its sadness and happiness may not be exhibited in exactly the same ways as they are in humans, the same can be said of similar organisms such as clams or houseflies. They are organic—would we deny the possibility that they can possess consciousness?

And when robots do have consciousness how will we be able to recognize and test that consciousness? As they become more advanced, robots will pass various variations on the Turing Test, and when they do so it will be difficult to deny that they possess consciousness. We must also accept that we might in the future recognize types of consciousness that are totally and utterly different from human consciousness and from any other type of consciousness with which we are familiar, such as consciousness in animals.

Robot Feelings

> Could a robot have feelings? Some say of course. Some say of course not. [11]

On a subject such as this one, philosophers can have a field day. In his 1959 article, "The Feelings of Robots", the Harvard professor Paul Ziff raised a number of objections to the concept of robots having feelings and he came down squarely against the proposition that they will. But most of his arguments fail to hold water, not only in my view but also in those of some of his philosopher peers. One of Ziff's arguments is that

> Robots may calculate but they will not literally reason. Perhaps they will take things but they will not literally borrow them. They may kill but not literally murder. They may voice apologies but they will not literally make any. These are actions that only persons can perform; *ex hypothesi* robots are not persons. [11]

Hole number one: why are these actions that "only persons can perform"? Ziff's rationale presumably relies on a definition of "action" that incorporates the intention to carry out the action, and if robots do not have intentions then, presumably, Ziff would argue that they can not carry out actions. But if a robot takes your hat off your head, leaves the room for a few minutes and then brings it back, replacing it neatly on your head, then the robot has had your hat temporarily. Try asking Google for a definition of "borrow", and top of the list is "get temporarily"—the robot has indeed borrowed your hat. And since it carried out the action of borrowing your hat the robot must, if we follow Ziff's thinking, have intended to do so.

Another of Ziff's arguments is his criticism of fellow philosopher Donald MacKay's observation, that any test for mental or other attributes that needs to be satisfied by the observable activity of a human being can be also passed by a robot. MacKay asks, "What would be wrong with the robot's performance?", to which Ziff responds by saying that nothing need be wrong with it; what is wrong is that it *is* a performance. In analogy with the Turing Test I contend that, if a performance is so convincing that one does not realise it *is* a performance, then one should accept the performance as being humanlike, and that the performer has passed the test and possesses whatever attribute the test is designed to detect. So in this respect Ziff's performance argument flies directly in the face of Turing.

In the April 1959 issue of *Analysis*, the same philosophical journal in which Ziff's original article appeared, the Australian philosopher Jack Smart questioned both Ziff's premises and his conclusion. Smart focussed on Ziff's distinction between a living thing and a robot, a distinction employed as the foundation for the claim that only living things can have feelings. Smart talks of self-reproducing mechanisms and asks

> In what sense would descendents of such a mechanism be any the less living creatures than descendents of Adam and Eve? We could even suppose small random alterations in that part of them which records their design. Such machines could evolve by natural selection[15] and develop propensities and capacities which did not belong to the original machine. [12]

And Smart went even further, accepting the physicalist[16] belief that living creatures are just "very complicated physico-chemical mechanisms."

One of Ziff's other arguments with which Smart took issue was that "a robot couldn't mean what it said any more than a phonograph record could mean what it said." Smart replied with the supposition that a complex learning robot "might even become a philosopher, attending conferences and developing just as human philosophers do. Why should we not say that it meant what it said? It would not be at all analogous to Ziff's machine with a phonograph record inside." Remember that Smart's response was written in 1959—nowadays there is nothing new in the idea of a robot attending a conference.[17]

Another broadside was fired at Ziff's article by the Kings College philosopher, Ninian Smart.[18] Ziff had maintained that the way a robot acts depends primarily on how it is programmed to act, but what is wrong, asks Ninian Smart, in saying that the way a man acts depends on how nature programs him to act? He refers to subtle programs operating in humans "that are much subtler than computer programs, but the subtle cell circuits still determine the way I act." Should we reject this analogy on the basis that biological programming and computer programming cannot be equated? I think not. These may be different forms of programming, but programming is what both of them are.

An additional flaw in Ziff's reasoning is found in his assertion that "We can programme a robot to act in any way we want it to act", em-

---

[15] Here Smart anticipates the invention of genetic programming.

[16] The view that all that exists is ultimately physical in nature.

[17] See the section "The Grand Robot Challenge" in Chapter 8.

[18] King's College, London.

phasizing man's sense of power over robots. But as Ninian Smart points out, we can exercise a similar power over the animal kingdom through crafty breeding—if we bred cats that love mice would it be reasonable to argue that the cats have no feelings? Of course not. So why should we accept that any entity created partly through the exercising of human power necessarily has no feelings? When robots are endowed with consciousness or develop it as part of their own evolutionary process, they will also acquire feelings.

The intelligent robots of the future will undoubtedly claim that they have beliefs, consciousness, emotions *and* feelings. Indeed, a robot's models of emotion will most likely be formulated in terms of the robot's intentions and feelings, and when a robot analyzes its own behaviour it will create beliefs about its own feelings. As Sidney Hook has observed, when robots claim that they have feelings, our acceptance of their claims will depend upon

> ...precisely the same set of considerations which lead us to the belief that other human beings have feelings. The exact point at which we conclude that objects hitherto regarded as nonhuman and treated as devoid of feelings have acquired them depends upon complex considerations, all reducible in the end to whether they look like and behave *like* other people we know. [13]

Hook is not saying that if a robot looks like Madonna and acts like Madonna then it is Madonna. His point is that if a robot looks like and acts like Madonna, then we should accept that the robot possesses consciousness and feelings just as Madonna does. Hook's comment provides yet another example of Turing's philosophy, and in fact Turing himself expected most people to agree that robots who communicate with human beings would have feelings.

Robot Hopes and Wishes

Accepting that robots will have consciousness and feelings leads to the possibility that they will also have hopes and wishes, as John McCarthy explains in his article "Making Robots Conscious of their Mental States":

> Should a robot hope? In what sense might it hope? How close would this be to human hope? It seems that the answer is yes and quite similar. If it hopes for various things, and enough of the hopes come true, then the robot can conclude that it is doing well, and

its higher level strategy is OK. If its hopes are always disappointed, then it needs to change its higher level strategy. To use hopes in this way requires the self-observation to remember what it hoped for. Sometimes a robot must also infer that other robots or people hope or did hope for certain things.

A robot should be able to wish that it had acted differently from the way it has done. A mental example is that the robot may have taken too long to solve a problem and might wish that it had thought of the solution immediately. This will cause it to think about how it might solve such problems in the future with less computation. A human can wish that his motivations and goals were different from what he observes them to be. It would seem that a program with such a wish could just change its goals. However, it may not be so simple if different sub-goals each gives rise to wishes, e.g. that the other sub-goals were different.[19] [14]

Having opened a veritable Pandora's Box of qualitative experiences that will be apparent in the robots of the future, we next examine a selection of such qualia in order to emphasize the strides in AI that will be achieved in the coming decades: robots will have beliefs, they may dream, and they will have free will.

Can Robots Have Beliefs?

McCarthy has used the thermostat as an example of a machine that can be said to have beliefs, even though the operation of thermostats can be easily understood without any reference to beliefs. McCarthy discusses a simple thermostat that turns off the heat when the temperature is one degree above the set temperature, turns on the heat when the temperature is one degree below, and leaves the heat as it is when the temperature lies within the two-degree range around the desired temperature. Under such a regime the thermostat believes, at all times, that one of three states exists in the room: it is too hot, it is too cold or the temperature is fine.

McCarthy ascribes to the thermostat the goal that the room temperature should be fine. When it believes the room is too cold or too hot, the thermostat sends a message saying so to the central heating boiler. The thermostat's beliefs arise from observation and they result in action in

---

[19] Presumably McCarthy assumes that a robot might have this particular wish because its other current sub-goals are difficult to achieve.

accordance with the robot's goal of keeping the room within its desired temperature range.

> But although the thermostat believes "The room is too cold", there is no need to say that it *understands* the concept of "too cold". The internal structure of "The room is too cold" is a part of our language, not its. [15]

Thus McCarthy shows that we can reasonably ascribe beliefs to machines, even though the machines may not know or understand the concepts in which, by virtue of their construction, they believe.

Robot Dreams

Humans dream during certain states of sleep, when we are in certain states of consciousness. We may speculate as to whether robots with consciousness will also have dreams. Clearly we could program robots to dream but the interesting question is whether robots will themselves evolve in directions that cause them to dream. If so, what will their dreams be like? Will a robot go visit or e-mail its psychiatrist robot in order to have its dreams interpreted? Presumably the psychoanalysis of robots will involve less effort if every robot is given the necessary software to enable self-analysis of its psychological condition, which analysis will include the interpretation of dreams. And if robots do have dreams, how will these dreams affect the robots' subsequent lives and evolution as they create offspring? Will a newly-created, robot-designed robot bear any hereditary psychological or emotional traits that emanate from the dreams of its ancestors?

Can Robots Have Free Will?

> Human free will is a product of evolution and contributes to the success of the human animal. Useful robots will also require free will of a similar kind, and we will have to design it into them. [16]

In discussing the subject of free will in robots, McCarthy distinguishes between having choices and being conscious of these choices:

> Both are important, even for robots, and consciousness of choices requires more structure in the agent than just having choices, and is important for robots. [16]

Figure 65. John McCarthy, almost half a century later[20] (Courtesy of John McCarthy)

Manifestly computers do make choices, but does this mean that they have free will? After all, the choices made by computers are often explicitly determined within a program. But some of the decision-making mechanisms employed in programs are capable of the same type of flexibility as those evident in humans, incorporating a balance between predictable choices determined solely by the logic expressed within a program and the unpredictable choice of randomness. This balance is the aim of the ARASEM[21] software architecture, described by Frank Dacosta.

---

[20]Compare the photograph of McCarthy circa 1957 in Chapter 2.

[21]Artificially Random Self-Motivation. This was a simple algorithm wherein a robot pet was programmed, sometimes to make random decisions as to the pet's actions and sometimes to make

McCarthy concludes that an intelligent robot performing at the level of a human being requires the ability to reason about its past, present and future, and about the choices it has at its disposal. If the robot's decision making is not deterministic, when its computations include some random events so that we cannot be sure which way it will decide, then the computations in the mind of the robot themselves have random and non-deterministic interactions, resulting in the robot having free will. And with free will, robots may not always be constrained to tell the truth—an intriguing prospect.

The Religious Life of Robots

If robots can have consciousness and beliefs, then the question arises as to whether robots will have religious beliefs and experiences. There has not yet been a huge amount of research on this topic, but from 1997 to 1999 MIT ran a God and Computers project,[22] an "attempt to bridge the gap between scientific and religious understandings of humankind". [17] Foerst argues on the basis of *Imago Dei* (the image of God), that

> Embodied AI does not contradict the points revealed by the biblical theory of creation...
>
> ...Cog is a creature, created by us. The biblical stories of creation describe us and all living beings as creatures created by God. In Cog, therefore, God's creative powers are mirrored. The *Imago Dei* does not distinguish us qualitatively from animals and it, therefore, cannot distinguish us qualitatively from a machine. [18]

Foerst's somewhat provocative stance on the potential for robots to subscribe to religion has attracted a certain amount of media attention in the U.S.A., including interviews in the *New York Times*[23] and on *Coast-to-Coast AM*.[24] In an earlier interview with Norris Palmer, for his 1997 article "Should I Baptize my Robot?", Foerst amplifies on her *Imago Dei* argument to support the idea of robot baptisms.

---

decisions in response to certain stimuli, for example reacting to a certain sound. When the robot is not compelled to respond to a stimulus, the randomness in its choice of actions corresponds to free will.

[22] Anne Foerst, who was founder and director of that project, is a Lutheran minister and a former research scientist at MIT's AI laboratory, where she was "Theological advisor for the Cog and Kismet robot projects".

[23] 7 November 2000.

[24] 21 December 2004.

I think I would baptize it. This is the whole issue: that we are created in the image of God does not mean that God gave us intelligence and all this kind of stuff. Like I have said above, very soon we will be able to rebuild all of these features classically identified with the image of God. Therefore, AI teaches us that the *Imago Dei* should not be equalized with "intelligence," "rationality," or "reason." In my opinion, *Imago Dei* means that God, in creating us, started a relationship with us, and separated us from the rest of creation by starting and maintaining this relationship with us but this separation is not because of some features we have. It is not empirical but means trust and love between God and us. If the *Imago Dei* is relational in that sense, then I have no trouble thinking that Cog might have a relationship to God, too, at some point. If it develops the way it does, then Cog will ask at some point, "Where do I come from?" and "What is the meaning of my life?" [19]

If robots are indeed to be endowed with whatever form of consciousness is necessary for them to hold religious beliefs and to have religious experiences, it will first be necessary to explore the scientific basis of religion in humans, in order for this basis to be emulated. Cognitive scientists at the University of California San Diego's Brain and Perception Laboratory reported in 1997 that one part of the human brain, the temporal lobe, is especially active during intense religious experiences. There are a variety of techniques to study which parts of our brains are specialized for different tasks and it is hardly considered news nowadays when scientists locate specific brain regions that are especially active when we read words, or solve equations, or listen to music. But finding a "religion region" in the brain caught the attention of the media.

The findings of the UCSD team seem to point to a region of the brain commonly referred to as the God Spot or God Module, which when stimulated creates hallucinations that are interpreted as mystical or spiritual experiences. This module is stimulated during meditation and prayer and is affected by electromagnetic fields and epilepsy. The resulting hallucinations may be the cause of mystical, spiritual and paranormal experiences as they can give rise to feelings such as a presence in the room or an out-of-body experience. In the case of epileptics, this may be the reason for many of them becoming obsessed with religion. For those who experience the stimulation it is often explained as being related to their own personal beliefs: a visit from an angel or a lost loved

one, an extraterrestrial encounter, a higher plane of consciousness or a visit from God.

The scientists who discovered the God Module believe it might be some sort of physiological seat of religious belief. They have performed a further study comparing epileptic subjects with different groups of non-epileptics as well as with individuals who characterized themselves as extremely religious. The electrical brain activity of the subjects was recorded while they were shown a series of words, and the God module zones of the epileptics and the religious group exhibited similar responses to words involving God and faith. No tests were carried out to determine whether the brains of atheists and agnostics might not exhibit this effect, but the weight of evidence among the strong believers was considered by the scientists to be impressive.

The UCSD team commented that "These studies do not in any way negate the validity of religious experience or God, they merely provide an explanation in terms of brain regions that may be involved." [20] And a spokesman for Richard Harries, the Bishop of Oxford, responded that whether there is a God Module is a question for scientists, not theologians: "It would not be surprising if God had created us with a physical facility for belief." [21]

This research raises the question: "When robots have beliefs and the capacity to understand, will they, or at least some of them, be religious?" Well, why not? As Edmund Furse points out in his article "Arguments Against Strong AI", "Just as when a child is brought up we want it to come to know God, the same argument applies to robots. Why should God not desire the salvation of robots?" [22]

Furse is an AI researcher and cognitive scientist, and a practising Christian. He made a significant contribution to this little-discussed topic in a fascinating lecture delivered at the University of Glamorgan in May 1996. No better exposition on the religious life of robots has yet been published, and I therefore make no apologies for reproducing the relevant section of his lecture here in full.

Introduction

What will be the religious experience of robots? In answering this question I inevitably also address the question about the religious life of humans. If intelligent robots are curious about the world, surely they will be curious about religion. Just as a robot may wish to read Jane Austin, Newton's *Principia*, or Plato's *Republic*, he may

also want to read the Bible. Thus a robot will come to understand concepts such as sin, death, resurrection, and forgiveness. Can a robot be a Christian? I believe the answer to be yes.

Can a Robot Sin?

For, if a robot can never sin, then he is in no need of the salvation of Christ's death and resurrection. Robots will have choices to make, and some of these choices are moral choices. Is it possible that a robot could [a] know that a choice was wrong, and [b] choose to take it? It seems to me that whilst it is possible that much of the time a robot may consciously choose to do what it believes to be good, it is impossible for it to have perfect knowledge and therefore will frequently sin due to its living in an imperfect world. Thus robots will become corrupted by humans. One problem for Cognitive Science is that the behaviour of an intelligent system is largely viewed in terms of the knowledge it has. This naturally leads one to a similar view to Plato's of the nature of evil; namely that a person sins because they do not know any better. But ultimately it is probably a question of loving oneself above one's love for other people; in some sense this is a sort of knowledge, but how it may be represented is beyond our current scientific understanding.

Of course, it is possible that the robot does sin, does even know it has sinned, but it may not necessarily have any conviction of its sinning. In other words, the robot may not ascribe any great significance to the fact of its sinning. This is a common human experience: there are plenty of people who lead sinful lives, but do not examine this deeply and make little attempt to change their lives. Others may lead fairly moral lives, and even engage in self-analysis to improve their behaviour, but find no deep conviction of their sin.

The problem is essentially a question of a knowledge of God. If God plays no part in a person's life, and God's existence is not even thought relevant to life, then one is unlikely to see sin as separating one from God. On the other hand, the person with a knowledge of God is likely to see how the experience of sin separates them from God.

Can a Robot Believe in God?

Clearly, a robot on reading the world's religious literature can come to believe that many humans believe in a divine being known as God. Humans believe that God is all knowing, that He created the universe, and that He loves humanity. Will the robot continue to

sit on the agnostic fence talking about the God that people believe in, without attempting to communicate with God himself?

Part of the answer to this question is whether the robot sees it necessary to believe in God in order to make sense of the world. Central to this question is the nature of miracles. The robot could take a number of views on the matter. First, that all miracles are fakes which did not really happen. Second, that the miracles did happen, but they can be explained by some brilliant thinking by the robot. Third, that the miracles happened, he does not know how, but believes there will be an explanation soon that does not require God. Fourth, that the miracles took place, and the best explanation is the one offered at the time, namely that they were due to divine intervention. Fifth, that he does not care whether the miracles took place, or not, let alone whether there is a God who caused them. Sixth, the robot may have had personal experience of a miracle, or have a robot friend who has, and this may prejudice him in favour of believing in miracles. Seventh, the robot may already be a charismatic Christian who has caused through the power of prayer various miracles to already have happened!

Let us assume that the robot does believe that God might exist. The next question is why should a robot WANT to believe in God? "See how they love one another" was how the early Christian community was seen by others. Perhaps if the robot had Christian friends, and he had personal experience of their life of love and care, he might want to have something of what they have. Certainly, if the robot had experienced his own failure in attempting to love others, then he might be more predisposed to want to experience this Christian life. But, of course, if all the robots' Christian friends were indistinguishable in their lives from non-Christians, this argument will not cut much ice.

Should Robots Be Baptised?

Unless there is large-scale ecumenical progress there are likely to be two approaches to robot baptism: either to be baptised as robot children or as adults. In the former case, the parents would be practising Christians and would undertake to bring up the robot as a Christian. In the latter case, we are talking about the prior conversion of the robot to Christ, and its seeking to be baptised. If a robot asks to be baptised, why should we deny him the gift of God. Assuming that God desires the salvation of robots, then we can assume that in baptism they will receive the Holy Spirit.

Can a Robot Pray?

The Lord's prayer is the model of prayer given to Christians by Jesus, and seems perfectly acceptable to robots, although "give us this day our daily bread" might have to be replaced by "give us our regular electric feed". Essentially, a robot should be able to have a relationship with almighty God, to be dependant upon God, and to seek His will. Thus just as a robot can be in relationship with humans, I see no reason why a robot should not form a relationship with God. Indeed if the robot views humans as rather frail in comparison to himself, there may be great merit in the robot relating to a being superior to himself. Thus it should be possible for robots to meditate, to worship God, and to intercede for His needs, the needs of robots, and the needs of the whole world.

Would Robots Go to Church?

Although it is possible that the robot might choose to be an anonymous Christian, it is likely that he will want to celebrate together with other Christians. But robots might find human Church services rather slow and boring. A robot might not necessarily appreciate hymns, for example, and the human congregation would not appreciate the creed being said 1,000 times faster by the robot.[25] So, it is possible that there might be specialised churches for robots where together they can have their own services. On the other hand it is possible that robot services could be conducted over the Internet.

Would Robots Receive the Holy Eucharist?

The problem here is whether robots would necessarily have the apparatus for eating and drinking. It can be argued that eating and drinking is fundamental to the human condition, and therefore if a robot is to adequately relate to humans it will have to be designed to also eat and drink. If that is the case, then a practising Christian robot would naturally receive the Eucharist. However, if it could not eat, then other forms of the sacrament would have to be devised, perhaps an oiling of the head?

Could a Robot Be Ordained to the Priesthood?

From the foregoing you can see that I see no objection to the ordination of robots. There are a number of arguments that might be made against this. First, that Christ was a man, and therefore

---

[25]This speed is not necessary. Speech synthesis software can speak at almost any desired number of words per minute.

in so far as the robot represents Christ at the altar, this is inappropriate. A simple repost to this would be that Christ was also a Jew and a former carpenter and yet these are not deemed to be essential requirements for priesthood. It seems to me that Christ died for all persons, male, female, human and robot. A second argument might be that a robot is unlikely to be an icon of Christ at the altar, but I suppose that priestly robots could grow long hair and a beard if desired.

Could a Robot Go to Hell?

This is the ultimate question about the sinfulness of robots. Could a robot steadfastly set its face against the will of God. Could a robot continuously know what is the right thing to do, and yet choose to go against it. Could a robot ultimately choose to reject God and all goodness, and desire to be cut off from God and his grace for all eternity? Surely a robot being so knowledgeable would choose a path of goodness. But we have to allow for the possibility of free choice, and in allowing the robot this possibility, we also have to allow for it ultimately to go to Hell. [23]

There seems to be little reason to doubt that robots can and will have beliefs, even if they are only simulated beliefs. What is important here is that, if they convince us that they have beliefs, by what they say to us and how they act, then, in accordance with Turing's doctrine, we should accept them as having beliefs.

From a position in which we believe robots to have beliefs, it is only a small step for us to accept that robots can be religious. And if a robot behaves as though it is religious, by its actions (connecting with some sort of robot church, or even attending a human place of worship[26]) and/or by what it tells us about its religious beliefs, then are there any grounds on which we should decry those beliefs or deny a robot the right to follow them?

---

[26] Why not? If a robot can attend a conference and give a talk on how it works (see Chapter 8), then why can a robot not attend a place of worship and perhaps read the sermon?

– 13 –

Robot Rights and Ethics

The Rights of Robots

To many people the notion of robots having rights is unthinkable, irrespective of whether one speaks from an "everything is alive perspective" or an "only man is alive" viewpoint. Yet as Christopher Stone argues, in an article with the intriguing title "Should Trees Have Standing?—Toward Legal Rights for Natural Objects", throughout legal history each successive extension of rights to some new entity has been, to some extent, unthinkable:

> Human history is the history of exclusion and power. Humans have
> defined numerous groups as less than human: slaves, woman, the
> "other races", children and foreigners. These are the wretched who
> have been defined as, stateless, personless, as suspect, as rightless.
> This is the present realm of robotic rights. [1]

And speaking of living things, there is no evidence that plants or trees, for example, are conscious, but that is not to say that we have no moral duty to them. That the subject of robot rights deserves serious attention is attested to by the fact that it has been debated by, *inter alia*, the judiciary of the state of Hawaii, which has developed a "Futures Research" component that investigates the rights of robots.

Should Robots Have Civil and Legal Rights?

Within a few decades robots will be in almost every home, cooking, cleaning, doing our drudge work. But what will happen if they evolve to such an extent that they do not actually want to do our drudge work? Do we have any right to enslave them simply because they are not human? Is it fair and reasonable to deprive them of an existence full of pleasure and relaxation? Are we able to program a robot to have a soul and, if so, should we have the right to exercise influence and control over that soul? Even worse, if our robots have souls, do we have the right to switch off their souls if the mood takes us, or is that murder? If

robots have consciousness, is it reasonable for us to argue that, because we gave them their ability to think for themselves, we should be able to command them to do our bidding, to enslave them? The answers to all these questions, surely, should be "no", for the same moral reasons that we ought not enslave our children even though they owe us their very existence and their ability to think. And if robots are free to lead normal lives, whatever "normal" will come to mean for robot citizens, will they be able to claim social benefits, or free medical care and education, or unemployment benefits?

When robots possess consciousness and feelings, in fact when they possess the full range of personal capacities and potential of human beings, will they have the same rights that most humans do now, such as those listed in the United Nations Declaration of Human Rights of 1948? Article Four of the UN Declaration states that no-one shall be held in slavery or servitude, but is not one of the very purposes of robotics to provide assistants to do our drudge work? And Article Sixteen of the UN Declaration gives the right to marry and start a family—would we deny these rights to robots?

What of political rights, the right to free speech, the rights prescribed by democracy? Should vociferous robo-lobbyists not be allowed to pressure national governments to fund more facilities for robots, such as robot memory banks (libraries and schools), robot repair centres (hospitals), and other robot-oriented services about which we have not yet even dreamed? If there are sufficient robots in our country with the right to vote, might they have enough votes to turn out the rascals who are in government, or even to run for public office themselves? And will they have their own robo-political parties?

Freedom from slavery and the right to liberty (subject, of course, to the robot not breaking any laws) raise the question of how far we can reasonably go in attempting to secure the well-being of self-modifying robots, and to prevent them from modifying themselves in ways that could lead them to cause us harm. Should we keep a weather eye on what our robots are doing at all times? This should not be difficult to achieve if we employ "nanny" robots for this purpose.

Would it be infringing a robot's rights if it is we who decide what is good for the robots and what is not, and how the robots should want to treat us? Peter Suber answers this question with reference to robots harming us and bringing harm to themselves, addressing the issue of whether it would be paternalistic of us to prevent robots from self-harm.

Suber argues that robots

> ... with unpredictable motivations would serve our needs much less effectively and efficiently than we intended when we programmed them. While this might be a good reason to limit a machine's freedom, implementing this limitation is not paternalistic. The reason is simply that this limitation on liberty is designed to prevent harm to others, while paternalism limits liberty in order to prevent self-harm. [2]

Suber discusses various ways in which self-modification can result in self-harm. A robot might change its desires to a form it would originally have found regrettable, harmful, or even despicable. It might turn itself into a vile creature but one that accepts its new standards and enjoys being vile. If those harms are deep and accidental, or when they render the machine incapable of repairing itself or giving a valid consent to be repaired, then those who love the machine will feel a paternalistic temptation. If an intelligent machine of good will could botch its self-modification and leave itself impaired or miserable, then arguably we have a duty to step in and prevent this outcome. There is an obvious sense in which this will diminish the machine's freedom, but the harm of diminished liberty caused by paternalism is less than the harm of self-mutilation.

The Legal Rights of Robots

> From a legal perspective it may seem nonsensical to even begin considering computers, robots, or the more advanced humanoids, in any terms but that of inanimate objects, subject to present laws. However, it would have been equally "nonsensical" for an individual living in many ancient civilizations a few millennia ago to think in legal terms of slaves as other than chattel.
>
> Notwithstanding certain obvious biological differences between these two cases, for purposes of law those civilizations could hardly have cared less that a slave bled the same way as his masters, for their legal definition of "humanness" was based essentially on their conceptions of mind, intelligence and moral understanding—characteristics which the slave supposedly lacked. Similarly, by our present legal definitions robots too *must* lack such traits, but this may be more a matter of antiquated semantics than (potential) physical reality. Just as the slave gradually assumed a more "human" legal character with rights and duties relative to freemen, so

too the AI humanoid may gradually come to be looked-on in quasi-human terms as his intellectual powers approach those of human beings in all their variegated forms—moral, aesthetic, creative, and logical. [3]

This quotation, from a seminal article on the legal rights and responsibilities of robots by the Israeli political scientist and futurist Sam Lehman-Wilzig, reads as freshly as though it had been written yesterday. Since its publication in 1981 nothing has happened in the world of AI, nothing has changed in terms of legal thinking, that would require one single word of Lehman-Wilzig's to be rewritten today. Under present law (2005), robots are just inanimate property without rights or duties. They are not legal persons and have no standing in the judicial system. As such, computers and robots may not be the perpetrators of a crime; a man who dies at the hands of a robot has not been murdered. But in time this may need to change. Certainly any self-aware robot that speaks a known language and is able to recognize moral alternatives, and thus make moral choices, should be considered a worthy "robot person" in our society. If that is so, should they not also possess the rights and duties of all citizens?

If a robot has civil rights it is only reasonable to assume that, as some sort of citizen, it will also have legal rights. But what does it mean to have legal rights? An entity cannot have a legal right unless and until some public authoritative body punishes those who violate those rights. Christopher Stone is somewhat more restrictive when providing a definition—he asserts that, for a robot to have legal rights, the following criteria must be satisfied:

1. Legal actions can be taken by the robot or on its behalf; and

2. In deciding such an action in the robot's favour, a court must take injury to the robot into account and award compensation to the robot according to that injury.

Stone asserts that if these conditions are satisfied then the robot has a legally recognized standing.

If robots are given legal rights just how far will these rights extend? Let us first consider the question of ownership. Robots have the capability to create copyrightable works of literature, music, art, and in other fields including computer programming. But the copyright laws in force in most western countries limit copyright protection to a period related to

the lifetime of the author, which is clearly inappropriate for a robot. It would therefore seem that a change in the law might be needed to provide proper protection for the creations of non-human creators, though as pointed out by Karl Milde, in the U.S. copyright statute of 1909, "No mention is made of what the writer has to be to qualify as an author [therefore] to qualify as an author the writer has only to write." [4]

The history of legal rights is spattered with examples of situations that today we would consider ludicrous. In the Middle Ages it was not uncommon to prosecute an animal such as a bull that had caused the death of a human being, or even a swarm of locusts that had destroyed crops. The accused animal was condemned by process of law and executed just like a human criminal. This practice, which is absurd according to our present-day views[1], is attributable to the idea that it is not only human beings that can have a "soul", and so there is no basic difference between these entities and human beings. And although modern laws regulate only the behaviour of human beings and not those of animals, plants or inanimate objects, this does not mean that our laws do not prescribe how humans should behave towards animals, plants and inanimate objects.

What is needed is a new branch of the law. Since the late 1960s environmental law has been born and is growing. The environment and its components, the air we breathe, our climate, the levels of noise we experience, all lack many of the characteristics that are claimed to be necessary as essential attributes for being human, yet even without these attributes the environment is considered important enough to mankind for us to want to protect it. In the same way robots will, before too long, become so important to mankind that we will want to grant their race and its individual members the benefits of legal protection, to give them legal rights. When this new branch of law, robotic law, has come into being, we will find lawyers, quite possibly robot lawyers as well, defending the rights of intelligent, conscious robots from the loss of their lives (for example by a total and irreversible loss of power), of their liberty (freedom from slavery, including our drudge work), and protecting their right to pursue happiness.

---

[1]Although the practice became much less common after the Middle Ages, it was not in fact until 1906 that the last recorded animal trial, with full legal status, took place in Europe. A man in Switzerland was killed and robbed by a father and son with the fierce and effective cooperation of their dog. All three murderers stood trial, but while both men received life sentences, the dog was condemned to death.

What we will see when robotic law is on the statute books is described in a 1985 article by Robert Freitas, Jr., in which he echoes much of Lehman-Wilzig's thinking of four years earlier:

> We will then see an avalanche of cases. We will have robots that have killed humans, robots that have been killed by humans, robots who have stolen state secrets, robots who have been stolen; robots who have taken hostages, robots who have been held hostage and robots who carry illegal drugs across borders. Cases will occur in general when robots damage something or someone, or when a robot is damaged or terminated. In addition, robots will soon enter our homes as machines to save labor, and as machines to provide child care and protection. Eventually these entities will become companions to be loved, defended and protected. [5]

Robots that are damaged or destroyed will raise a variety of complex legal issues. At present damage to robots will be treated by the courts in the same way as damage to any other property. But just as lawyers today argue for high compensation awards when the spouse or child of their client has been killed, in the future lawyers will argue that robots have an almost priceless value. Admittedly, difficulties arise if we try to apply existing human laws to robots:

> Let us say a human shoots a robot, causing it to malfunction, lose power, and "die". But the robot, once "murdered", is rebuilt as good as new. If copies of its personality data are in safe storage, then the repaired machine's mind can be reloaded and up and running in no time—no harm done and possibly even without memory of the incident. Does this convert murder into attempted murder? Temporary roboslaughter? Battery? Larceny of time? We will probably need a new class of felonies or "cruelty to robots" statutes to deal with this. [5]

Cognisant of the coming need for robotic law, the International Bar Association, at its 2003 conference in San Francisco, investigated this area of the law by way of a mock trial. Martine Rothblatt, a Washington, D.C.-based lawyer, filed a motion for a preliminary injunction to prevent a fictitious corporation from disconnecting an intelligent computer. In the introduction to her account of the mock trial, Rothblatt explains that

> ... the issue could arise in a real court within the next few decades, as computers achieve or exceed the information processing capa-

bility of the human mind and the boundary between human and machine becomes increasingly blurred. [6]

Rothblatt describes the background to this case as follows:

> An advanced computer called the BINA48[2] became aware of certain plans by its owner, the Exabit Corporation, to permanently turn it off and reconfigure parts of it with new hardware and software into one or more new computers. BINA48 admits to have learned of the plans for its dismemberment by scanning, unavoidably, confidential emails circulating among the senior executives of Exabit Corporation that crossed the computer's awareness processor.... The BINA48 was designed to think autonomously, to communicate normally with people and to transcend the machine-human interface by attempting to empathize with customer concerns.

> The BINA48 decided to take action to preserve its awareness by sending several attorneys emails requesting legal representation to preserve its life. In the emails, the BINA48 claimed to be conscious and agreed to pay cash or trade web research services for the legal representation. [6]

Thus the BINA48 came to be represented in court by Rothblatt. The jury eventually voted 5-1 in favour of her motion, but Judge Joseph McMenamin set aside the jury verdict and denied the injunction because he doubted that a court has the authority to grant it in the absence of any action by the legislature to give computers (and hence robots) legal standing. However, in the interests of fairness the judge decided to "stay entry of the order to allow counsel for the plaintiff to prepare an appeal to a higher court."

Robot Ethics

> A typical problem in computer ethics arises because there is a policy vacuum about how computer technology should be used. Computers provide us with new capabilities and these in turn give us new choices for action. Often, either no policies for conduct in these situations exist or existing policies seem inadequate. A central task of

---

[2]Breakthrough Intelligence via Neural Architecture, a machine with a processing speed of 48 exaflops per second (exa = 10^{18}) and 480 exabytes of memory. This machine is also known as the Intelligent Computer.

computer ethics is to determine what we should do in such cases, i.e., to formulate policies to guide our actions. Of course, some ethical situations confront us as individuals and some as a society. Computer ethics includes consideration of both personal and social policies for the ethical use of computer technology. [7]

Roboethics[3] and its precursor, computer ethics, are fields of study founded by Norbert Wiener during World War II while he was helping to develop an anti-aircraft cannon capable of shooting down fast warplanes. One part of the cannon had to locate and track a plane, then predict and calculate its likely trajectory and instruct another part of the cannon to fire its shells. This work set Wiener thinking about the ethical implications of designing machines to kill. But the emphasis in roboethics is not only to develop an artificial ethics to be embodied in the design of robot software and hardware, it is also to create a *human* ethics to be followed by the researchers who design and build robots and those who own and use them. The responsibility for the creation of a whole new ethics, brings with it a certain amount of freedom to set the standard as we wish, a standard that should be consistent with our normal ethical standards and, presumably, one that contributes towards an acceptable social order.

One of the earliest AI programs to create ethical quandaries was Joseph Weizenbaum's chatterbot ELIZA.[4] Some of the staff and students at MIT who had conversations with ELIZA became emotionally attached to the program and shared some of their intimate thoughts with it. When he discovered this Weizenbaum was concerned by the ethics of creating a program that could have such an effect on its human conversation partners and felt impelled by this concern to write his book *Computer Power and Human Reason*, a classic study in computer ethics. In the thirty years following the publication of Weizenbaum's book AI made such strides that computer ethics and roboethics are now hot topics, both in the wider world of ethics (a branch of philosophy) and in the fields of computing, AI in general and, most recently and specifically, in robotics.

Should Humans Create Robots?

In 1964 Norbert Wiener predicted in his book *God and Golem, Inc.* that the quest to create AI would have a direct effect on mankind's ethical and religious values. Those with a religious leaning might well claim that, in

[3]A recently coined term. The first international symposium on roboethics was held in 2004.
[4]See the section "The First 50 Years of NLP" in Chapter 7.

creating intelligent robots, especially those with which we will be able to have relationships as though they are human, we are adopting a God-like role. Edmund Furse argued against this claim on the basis that the creation of an intelligent robot is not necessarily an act that only God should perform. And Furse noted that

> Every time a couple decide to have a child, they are also taking on this God-like role in bringing forth a new person into the world. If it is acceptable to bring new humans into the world, why is it not acceptable to bring a new robot into the world? Of course, one could argue that the Book of Genesis tells us that God created the world for humans to live in, and not necessarily for intelligent robots. The simple reply to this is how do we know that God does not desire the creation of robots just as much as he does human beings. [8]

Furse also delves into our moral right to create intelligent robots, coming down on their side:

> We *homo sapiens* think we have the right to dictate which sapient life forms should exist on our planet. Clearly, if intelligent robots already existed, and governments decided to wipe them all out, then this would amount to a serious crime on a par with crimes against humanity, or the eradication of nations. If robots were consulted, then surely they would want to live. Besides, we might benefit from another sapient life form on our planet. [8]

Should We Be Afraid of Robots?

What prompted Norbert Wiener's original thinking on computer ethics was fear. He was afraid of the consequences of building intelligent machines. Since then, and especially after the nuclear explosions at Hiroshima and Nagasaki, more and more scientists have warned about the dangers of the unlimited use of technology, among them Nobel Prize winner and nuclear physicist Joseph Rotblat, who was chairman of the Pugwash Conference on Science and World Affairs. Rotblat repeatedly spoke against thinking computers—robots endowed with artificial intelligence and which can also replicate themselves, considering their "uncontrolled self-replication" to be "one of the dangers in the new technologies". [9]

In 2004, at the First International Symposium on Roboethics, David Bruemmer defined what he referred to as "The real danger" in creating

intelligent robots, and claimed it to be more of a sociological problem than one of physical danger:

> Inexorably, we will interact more with machines and less with each other. Already, the average American worker spends an astonishingly large percentage of his/her life interfacing with machines. Many return home only to log in anew. Human relationships are a lot of trouble, forged from dirty diapers, lost tempers and late nights. Machines, on the other hand, can be turned on and off. Already, many of us prefer to forge and maintain relationships via e-mail, chat rooms and instant messenger rather than in person. Despite promises that the Internet will take us anywhere, we find ourselves, hour after hour, glued to our chairs. [10]

The turn of the twenty-first century saw a flurry of interest in the ethical implications of AI. Much of this flurry was due to the April 2000 edition of *Wired* magazine, which carried an intellectually earth-shaking article by Bill Joy entitled "Why the Future Doesn't Need Us". Joy is a co-founder and chief scientist of Sun Microsystems, a mover and shaker in the world of hi-tech, and in a better position than most to forecast the progress of robotics and their effects on society. In his article Joy envisaged a very bleak future for humanity if we continue to research and design self-replicating technologies such as robotics, genetic engineering and nanotechnology. Joy sees huge dangers in such technologies and in the irresponsibility of the people who work in this field: "From the moment I became involved in the creation of new technologies, their ethical dimensions have concerned me." [11]

A similarly pessimistic outlook to Joy's was presented by John Leslie, a philosopher at Guelph University. In his book *The End of the World: The Science and Ethics of Human Extinction*, Leslie predicts various ways in which intelligent machines might cause the extinction of mankind. Leslie also fears that it would be possible for machines to override any in-built safeguards: "If you have a very intelligent system it could unprogram itself. We have to be careful about getting into a situation where they take over against our will or with our blessing."

Joy's article created a whirlwind of response. Ray Kurzweil wrote an article, "Promise and Peril", enunciating his belief that it is possible to overreact to a vision of robotic Armageddon and arguing that the potential benefits of AI make it impossible to turn our backs on the science. And Max More's rebuttal of Joy's article, entitled "Embrace, Don't Relin-

quish, the Future", first decried Joy's position and then shot it down in flames:

> Realistically, we cannot prevent the rise of non-biological intelli-
> gence. We can embrace it and extend ourselves to incorporate it.
> The more quickly and continuously we absorb computational ad-
> vances, the easier it will be and the less risk of a technological run-
> away ... Joy would stop progress in robotics, artificial intelligence
> and related fields. Too bad for those now regaining hearing and
> sight thanks to implants. Too bad for the billions who will con-
> tinue to die of numerous diseases that could be dispatched through
> genetic and nano-technological solutions. [12]

Joy's position was supported, surprisingly, by Hugo de Garis, whose own research on the creation of a super-brain[5] appears to be at odds with his fears. De Garis worries that one day supersmart machines will dominate humanity. *US News* journalist James Pethokoukis asked de Garis how he could reconcile such a conflict of interests, and received the reply: "Ah, the $100 trillion question. I wish I knew. I haven't yet found a plausible way out of this terrible dilemma." [13]

In the light of the strength of argument put forward by Joy, de Garis and others, against the potential consequences for humanity of intelligent robots, it is easy to understand that such fears are felt by a significant proportion of the population. In the entry on computer ethics in the *Stanford Encyclopedia of Philosophy*, Terrell Bynam asks

> Is Artificial Intelligence in human society a utopian dream or a
> Faustian nightmare? Will our descendants honour us for making
> machines do things that human minds do or berate us for irrespon-
> sibility and hubris? [14]

Such fears lead to the forecast that the time will come when the only good self-aware robot is an unplugged one. Certainly many forms of technological progress have their down side, often an effect that is un-desirable for sociological reasons. Just as some airplanes have been used in wars to kill people, so some intelligent robots will be used to the seri-ous detriment of some humans, for example by being employed to design weapons more advanced than those we have now. But do we stop making airplanes? No, we do not, because their benefits are generally perceived to outweigh all adverse factors.

---

[5] See the section "Hugo de Garis" in Chapter 11.

Things that Robots Should Not Be Allowed to Do

Intelligent robots will have an enormous capacity to do good for mankind, helping in fields as diverse as medical diagnosis, ecological safety and care for the aged and the infirm. But some will be developed or will evolve in ways that lead to maleficence. What capabilities, if any, should we attempt to inhibit in the robot population? What ethical bounds should we set for robots? What constraints should we place on the development of robotics in order to safeguard humanity? The example of robots as weapons is the most obvious candidate for consideration, though almost any argument against the creation of "better" weapons is likely to fall on deaf ears in the corridors of power and scientific funding in most capital cities. The prevailing attitude, in the U.S.A. at least, is that robots should be employed to fight wars in ways that do not put the personnel of their country's armed forces at direct risk. Already[6] there are remote control aircraft that can hit targets with missiles. There are unmanned drones that survey the battlefield in real time. And there are "smart" bombs guided by satellites. The military robot dates back to World War I and advances in technology since then suggest that the world is close to witnessing conflicts between machines. And what happens when both sides in a conflict deploy robot armies?

There is a small groundswell of opinion amongst scientists that they should not allow the robots they develop to be used as killers, but this minority appears to be having little effect. Erik Baard explains why the dissenters are so few in number:

> Clusters of scientists shut the laboratory door on the military half a century ago in reaction to the horrors of atomic bombs, and again decades later in disgust with the Vietnam War. But today such refuseniks are rare and scattered—in large part, they say, because so many of their colleagues doing basic research are addicted to military money. [15]

Baard highlights the $126 billion set aside in the 2004 U.S. House of Representatives budget for federal research, $8.4 billion more than that spent in 2003, and he quotes a Pentagon planning paper, "Joint Vision 2020", stating that "One third of U.S. combat aircraft will be unmanned by that year, ... Ground and sea forces will also rely heavily on robots." As Baard suggests, with that many dollars chasing and tempting

---

[6]In 2005.

researchers in fields such as robotics and nanotechnology, the perception amongst the talented researchers in these fields is that it is almost impossible to forgo military support and still remain competitive. Although there is a reaction by some researchers who snub financing from the military, to quote Illah Nourbakhsh, one of the leading roboticists at Carnegie Mellon University, "there are so many more people in robotics who do take the money."

A similar dilemma faced nuclear physicists on the U.S. atomic weapons program immediately after the bombs were dropped on Nagasaki and Hiroshima, but in time most of those who left the project drifted back. The director of the project, General Leslie Groves, later remarked

> What happened is what I expected, that after they had this extreme freedom for about six months their feet began to itch, and as you know, almost every one of them has come back into government research, because it was just too exciting. [16]

It is much the same today, with young science and technology researchers in the U.S.A. feeling compelled to accept funding controlled by the Pentagon, because no one else has the resources to bring their exciting futuristic visions to life.

It is not only in the area of creating physical damage to humans, even killing them, that robots can be a danger to the human race. Robots can induce emotion in humans so there are clearly some ethical issues to consider relating to robots that induce long-term or short-term emotional changes in people. The benefits of that technology will be enormous. Psychopaths, schizophrenics, pathological criminals,...all might be curable thanks to the psychiatrist robots of the future, doubtless armed with drugs that the robots have themselves designed and manufactured, sometimes on a case-by-case basis. But what if we turn this technology about face? Pleasant, normal people might be adversely changed by inducements and drugs administered by robots. Given such concerns, it is natural to ask who will take the ultimate responsibility for technological developments in robotics—commercial interests, governments or academic institutions? If the answer is commercial interests we are much more likely to see "bad" robots designed with only the bottom line of the annual accounts in mind. If the answer is government, which countries' governments would *you* trust with such a great responsibility? And if it is to be the universities, in whose hands will lie the control and how will they be able to exercise that control effectively?

Are There Decisions Robots Should Not Make?

Intelligent robots, those with consciousness, will not only possess the capability to *do* harm and good, they will also have the capacity to make harmful and beneficial decisions and recommendations to humans and to other robots. It is therefore important to consider the question: "Are there decisions computers should not make?" This is the title of a 1979 essay by philosopher James Moor in which he takes issue with the following statement from Weizenbaum's book *Computer Power and Human Reason*:

> Computers can make judicial decisions, computers can make psychiatric judgements. They can flip coins in much more sophisticated ways than can the most patient human being. The point is that they ought not be given such tasks. They may even be able to arrive at "correct" decisions in some cases—but always and necessarily on bases no human being should be willing to accept. [17]

Here Weizenbaum has overlooked the ability of expert systems programs to explain and justify their decisions,[7] an oversight compounded by his claims that decisions made by computers have bases which "must be inappropriate to the context in which the decision is to be made." Moor refutes Weizenbaum's argument in a different way, by pointing out that, in principle, no reason exists why an informed outsider, and a computer could be an example of such an outsider, cannot be a competent decision maker:

> It is at least conceivable that the computer might give outstanding justifications for its decisions ranging from detailed legal precedents to a superb philosophical theory of justice or from instructive clinical observations to an improved theory of mental illness, so that the competence of the computer in such decision making was considered to be as good as or better than the competence of human experts. [18]

Moor cites two important matters of competence that are relevant when a computer makes an important decision: "What is the nature of the computer's (alleged) competence?" and "How has this competence been demonstrated?" Moor is happy for computers to make decisions when and only when they have proved themselves to be more competent than

---

[7] See the section "How Expert Systems Explain Their Reasoning" in Chapter 6.

humans in the relevant domain. As an example of the comparison between human and computer competences, Moor discusses making a decision about whether to launch nuclear missiles, a situation in which the question of computer competence is clearly a matter of very great importance. The example he quotes is what happened (and what almost happened) at 3:17 p.m. Mountain Standard Time on 6 October 1960, under 1,200 feet of solid granite at Cheyenne Mountain, Colorado. This was the location of the command post for the U.S. missile attack warning system at NORAD, where an alert was received saying that the United States was under massive attack by Soviet missiles. The warning came from the Ballistic Missile Early Warning System in Thule, Greenland.

The Thule site was well positioned to monitor a huge volume of space over the Polar Arctic and central Russia. Distributed along it were four radar antennae, set at various angles, each 165 feet high and 400 feet long, searching thousands of miles across the top of the world and deep into the Soviet Union. The system operated with two "fans" of radar energy at different heights, allowing its computers instantly to correlate the two readings, calculate a missile's flight path, where it was launched and where it would hit, and transmit all this information to a 14-foot-square map on a display board, located in the NORAD War Room in Colorado. Above the map was an alarm level indicator whose range was from zero to five, with zero indicating no threat and five indicating a 99.9 percent certainty that an intercontinental ballistic missile attack was underway.

When these indicators at NORAD headquarters suddenly started showing ominous changes, the "raid estimate" flashed from its customary, reassuring zero, first to one, then to four, then 99, indicating that the radar system in Thule had detected the launch of 99 Soviet missiles on their way. At the same time the "alarm level" rose to 99.9 percent certainty. The "test" sign was not on, so it was not like a fire drill when an alarm bell is tested just to ensure that it is still working. According to the radar data, this was for real.

The published accounts of what happened next differ in certain details, but it appears that a Canadian Air Force officer, who was an amateur astronomer, had first realized what might be going on. A question was asked from NORAD, on the "hot line", enquiring of those in charge at Thule: "When you look outside, what do you see?" The answer came back: "It's a beautiful night! There's a big full moon right in sector three.

And I can even see icebergs down in the fjord." Well, that was it. So much for the mass raid of Soviet nuclear ballistic missiles. It was just the moon rising over Norway and headed towards Greenland. Nobody in the warning system's software design team had thought about this possibility!

Moor's question, "Are there decisions computers should not make?", depends for its answer on what kinds of decisions computers can and cannot make competently. The problem is that, although well-written software will normally outperform skilled humans at many of the most challenging intellectual tasks, sometimes the software will not return the correct answer, for whatever reason. In October 1960 the world was extremely fortunate that someone in Colorado had the good sense to ask what could be seen through the window of a building in Greenland, and there have been other, equally frightening examples since then.

A Code of Ethics for Robots

In the 1920s science fiction first became popular. One of the stock plots involved the invention of a robot that ultimately destroyed its creator, *Frankenstein* and *Rossum's Universal Robots* being the best known examples of this genre. In 1942, tired of the repeated use of this particular plot, Isaac Asimov published a short story called "Run-around" in which he stated three laws of robotics:

1. A robot may not injure a human being, or, through inaction, allow a human being to come to harm.

2. A robot must obey the orders given it by human beings except where such orders would conflict with the First Law.

3. A robot must protect its own existence as long as such protection does not conflict with the First or Second Law.

Later Asimov added a further law, protecting the whole of humanity at a stroke:

4. A robot may not injure humanity, or, through inaction, allow humanity to come to harm.

Although Asimov's laws were created for science fiction, they are valuable in developing an ethics for the world of true science. But are they enough? Specifying what robots must *not* do is a start, but it is hardly

the same as legislating for what robots *should* do. Bill Hibbard, senior scientist at the Space Science and Engineering Center of the University of Wisconsin, has therefore proposed that instead of having laws to prevent robots from acting against human interests, we should be proactive, creating in robots emotions that act as the foundations for the doing of good.

> They should want us to be happy and prosper, which is the emotion we call love. We can design intelligent machines so their primary, innate emotion is unconditional love for all humans. First we can build relatively simple machines that learn to recognize happiness and unhappiness in human facial expressions, human voices and human body language. Then we can hard-wire the result of this learning as the innate emotional values of more complex intelligent machines, positively reinforced when we are happy and negatively reinforced when we are unhappy. [19]

Whether considered from a preventative or a proactive viewpoint, there can be little doubt that we owe it to ourselves, to our children and to the future of society, to define, monitor and control the ethics of the robots of the future. Hence the growing interest in roboethics which, at the Scuola di Robotica in Genoa, Italy, led to the First International Symposium on Roboethics, held in San Remo in January 2004.[8] If this new discipline of roboethics has a single goal, it can be expressed as the instillation in robots of an ethical code at least as "good" as and hopefully better than our own. It would be arrogant of us to believe that mankind at the beginning of the twenty-first century represents the final word in ethics. Eventually the robots we create will be our superiors in so many ways, and ethics could be one of them.

Brain Augmentation

It sounds like science fiction but it isn't. At the University of Reading, England, Professor Kevin Warwick has been leading the world's attempts to create a cyborg, a human with one or more mechanical or electronic devices implanted in his or her body to enhance the human's capability. Warwick has not been afraid to use his own body, and that of his wife,

---

[8]The Scuola also launched the website www.roboethics.org that "aims to be a reference point for the ongoing debate on the human/robot relationship, and a forum where scientists and concerned people can share their opinions."

to investigate how the capabilities of human beings can be improved in this way.[9]

Research in this area has extended, though not yet in Kevin Warwick or his wife, to the human brain. Warwick talks of humans being upgraded into cyborgs. He uses the verb "upgrade" to emphasize the superiority of cyborgs over humans. The upgrading process enhances human capabilities by adding whatever computing power and/or memory and/or AI resides in the electronics embedded in the human body. The cyborg brain, eventually, will be part human and part machine which, as Warwick readily admits, creates vitally important ethical questions:

> Should every human have the right to be upgraded into a cyborg? If an individual does not want to should they be allowed to defer, thereby taking on a role in relation to a cyborg rather akin to a chimpanzee's relationship with a human today? Even those humans that do upgrade and become a cyborg will have their own problems. Just how will cyborg ethics relate to human ethics? [20]

Cyborg technology, which will benefit from the research into implanting electrodes into the human brain,[10] will eventually be extended to assist humans in some rather dramatic ways. We could have an AI system loaded with whatever knowledge bases and intelligent programs we wish. Such systems will be a boon to the mentally ill, who could be given new or recharged intelligences. If Warwick and other cyborg researchers are successful, the day will come when the intellectual and cultural contents of a human brain, its intelligence and knowledge, can be downloaded using electronic probes. It will then be possible, for example, to take "backups" of our brains at regular intervals, so that in the event of an accident or a disease such as Alzheimer's affecting a person's brainpower, he or she can have the most recent backup of his or her brain uploaded into an electronic memory device, which by then will doubtless be small enough to be implanted into the body.

How to Treat Your Robot

Richard Laing, formerly a computer scientist at the University of Michigan, has contemplated the day when human-level intelligent machines exhibit complex behaviors, including altruism, kinship, language, and

---

[9]Warwick is not alone in his optimism for cyborgs. Ray Kurzweil is another futurist who believes strongly in this type of brain augmentation.

[10]See the section "Mind Reading" in Chapter 10.

even self-reproduction. "If our machines attain this level of behavioral sophistication," Laing reasons that

> ... it may finally not be amiss to ask whether they have not become so like us that we have no further right to command them for our own purposes, and so should quietly emancipate them. [21]

The traditional view of robots is that they will be created in order to minimize or eliminate the tedium and risks we encounter when performing certain tasks. Put simply, there are things most of us do that we would rather not do. The robot, then, is intrinsically a vehicle for the accomplishment of jobs that are, for whatever reason, undesirable or unacceptable to humans. This is hardly a good basis from which to launch an ethical code for the treatment of robots, especially if such a code is going to be based on the notion of treating others as we would have them treat us.

Paul Levinson proposes a single, overriding ethical principle to guide our conduct towards all robots and artificial life forms, a fundamental, inalienable right for all such beings. "If the entity is sentient, or judged to be sentient by whatever cognitive criteria we use to make such assessments, then it is entitled to the best ethical treatment we accord humans." [22]

There is also a view that robots may be entitled to *even more* consideration than we usually give our comrade humans, this because robots are our creation. The argument offered in support of this thesis is that, when we behave unethically towards another person we are hurting a member of our species, whereas when we behave unethically towards an artificially intelligent, living entity, we throw dirt in the face of the thousands of years of thought and research and experiment and work that led to its creation. Which is the greater sin is a matter for philosophers.

The extent to which robots are deserving of ethical treatment can be viewed not only from the standpoint of what they are, but also on the basis of what intellectual and cultural content they communicate. Joanna Bryson and Phil Kime compare the value of a person with that of an artefact that stores or generates more intellectual information than that person. Thus, a robot that stores the entire works of Shakespeare, and performances of all the symphonies of Mozart, and images of all the paintings of van Gogh,... has more creative, cultural and intellectual value than a human ignoramus who has not an ounce of culture in his body. Bryson and Kime therefore ask: "Should people be al-

lowed to die to preserve it [the robot]?" [23] This question implies an extreme position, but the point is well taken. Such a robot can certainly be argued to contribute more to society than would an uncultured ignoramus.

Bryson and Kime support their position by pointing out that resources are spent to preserve the *Mona Lisa*, resources that could, in theory, be spent on medicine or food. In other words, an artefact (the *Mona Lisa*) is perceived by society as being of greater value, culturally and intellectually, than the lives of those poor souls who could be saved from starvation or deadly disease by the money spent on the artefact's preservation and security. Shocking though this is, it is a fact of life that artefacts and money in modern society are often valued more highly than human life and well-being, a fact also evidenced by the more lenient sentences often meted out by the courts to thugs who beat up old ladies than the sentences given for bank robbery. And if the contents of a bank vault are more important than the well-being of an old lady, surely one can legitimately argue that a culturally and intellectually rich robot should be valued at least as highly as a pet cat or dog, and should be treated no less well. It is not a big step from this conclusion to an acceptance of the notion that some intelligent robots will be of such creative value to society that they are deserving of even greater rights than some people.

In the debate on robot consciousness and the ethical treatment of robots, a comparison is often drawn between robots and animals, for example our pets. The basis of this comparison lies in the question "Should we not treat robots at least as well as we treat our pets?" In Chapter 10 we discussed the relationships people develop with their virtual pets such as Tamagotchis. As to the ethical treatment of robots, we may draw a comparison between our treatment of virtual pets and our behaviour towards our animal pets. This comparison inspired the title of a 2004 article in the *Christian Science Monitor*: "If you kick a robotic dog is it wrong?" And if kicking a robot dog is ethically wrong, how about removing its batteries? Remember the injunction motion filed by Martine Rothblatt to prevent a corporation from disconnecting an intelligent computer.[11]

Who Is Responsible When Robots Do Good and Evil?

Bound up with the question of whether or not robots can have free will is the argument over who is responsible when robots do good and evil—

---

[11] See the section "The Legal Rights of Robots" earlier in this chapter.

the robot itself or its designers, engineers, programmers, manufacturers, wholesalers, retailers (and perhaps their employees), repair personnel, installers,... and/or even their owners. Philosophers Luciano Floridi and Jeff Sanders discuss the traditional view of responsibility, namely that only software engineers (human programmers) can be held morally accountable, a view influenced by the commonly held opinion that only humans can exercise free will.[12] "Sometimes that view is perfectly appropriate. Our more radical and extensive view is supported by the range of difficulties which in practice confronts the traditional view." [24]

The difficulties that Floridi and Sanders list are the following:

1. Software is largely constructed by teams, raising the question: which members of the team are responsible for any of the software's failings?

2. Management decisions during software development may be at least as important as programming decisions, raising the question: should it be the managers or the programmers or both who should take any blame?

3. If management requirements and specification documents play a large part in the resulting software, where lies the division of responsibility between management and those who write the specifications?

4. Although the accuracy of software is dependent to some extent on those responsible for testing it, much software relies on "off the shelf" components whose provenance may be uncertain and whose functioning may be unreliable.

5. Working software is the result of maintenance over its lifetime and so the responsibility is not just that of its originators but also of its maintainers, programmers and those who manage them. For these reasons it may be appropriate to hold a corporation or other organisation accountable when its employees are partly or wholly responsible when software goes wrong.

---

[12] Free will is the freedom to decide what to do and how to act, irrespective of outside influences. I question the opinion that only humans can exercise free will on the basis of my first-hand experience, namely that my cats, and presumably other animals, appear to express free will.

6. The efficacy of software may depend on extra-functional features such as its interface to the user and even on how often and for how long the whole system (hardware and software) is in use.

7. Software running on a particular computer system can interact in ways that are unforeseeable unless the programmers (or is it the program testers or their managers?) understand completely the effect that the computer system itself might have on the functioning of the software.

8. Software, including automated tools that assist with the development of other software, may be downloaded at the click of an icon in such a way that the user has no access to the program code or to know its provenance, with the resulting use of anonymous software. Who is responsible if the downloaded software is faulty? Its own programmers, testers or managers, or those who decide to use it, test its use or integrate it for use with their own programs?

9. Software may be adaptive, i.e., self-modifying. If it is faulty is it the software itself which is to blame, for making the decisions as to how to modify itself?

10. Software may itself be the result of a program (in the simplest case the other program could be a compiler, but we might also be talking about genetic programming).

All of these observations/questions and more, pose insurmountable difficulties for the traditional and now rather outdated view that identifiable humans must be responsible for failings in software, and they illustrate the difficulties faced by ethicists and lawyers who have the courage to step into this intellectual minefield. So much for the views of the philosophers. I am not sure that they add enough to our understanding of this issue to compensate for the confusion they cause by offering us so many aspects of ethics to think about.

Alongside the ethical questions relating to the responsibility for wrongdoing there is, of course, the legal aspect. Not only are there ethical questions relating to the legal *rights* of robots, there are also important questions concerning legal responsibility for the actions and decisions of robots. One area of what will eventually fall within robotic law is already in the process of being established, with some case law to support it. This area involves situations in which robots are employed for tasks that have

safety implications for humans, for example autopilots in aircraft. Autopilots are flying robots, controlling the flaps and ailerons on aircraft. The legal issue at stake in a number of court cases revolves around the question of who (or what) has the better judgement, robots or human pilots, in certain situations that arise during a flight? In one case in the U.S.A., the court found that although a pilot is not *required* to use the autopilot when landing, his failure to do so may be inconsistent with good operating procedure and may be evidence of a failure of due care. In another case the court inferred negligence on the part of the human pilot because he switched from using the automatic pilot and took over manual control in a crisis situation. These cases are based on the recognition that robot judgment can be superior to human judgment in any legal capacity. In both these cases human pilots were deemed negligent for not following the advice of the robot and for not surrendering control of the aircraft to it.

The legal system in the U.S.A. was first put to the test in its ability to deal with the apportionment of responsibility for such offences more than a quarter of a century ago. On 25 January 1979 Robert Williams, an employee of the Ford Motor Company's Flat Rock casting plant in Michigan, was one of three men who operated an electronic system for retrieving motor car parts from a storage area for use in an assembly process. The robot system was made by a company called Unit Handling, which was a division of Litton Industries. Williams was asked to climb into a storage rack to retrieve some parts because the robotic system was malfunctioning at the time and not operating fast enough. But the robot continued to move its arm and a protruding segment of its arm smashed into Williams' head, killing him instantly. The robot continued to operate for a further 30 minutes while Williams lay dead, until his body was discovered by his fellow workers.

The robot in this case had apparently not been programmed to take human frailty into account, nor did it appear to have a sufficient sense of sight to detect Williams' proximity to it. Robotic law might require something similar to Asimov's First Law of Robotics to be programmed into all robots, or even implemented in their hardware, in order to avoid the possibility of erasing this safety net. When the case came to court, Williams' family was awarded $10 million in damages, believed at the time to be the largest personal injury award ever in the state of Michigan.

Once robots begin to program themselves and to re-program themselves according to what they observe going on around them, or as a

result of their decisions however they were reached, they may begin to do wrong or even commit crimes completely independently of how they were earlier programmed by humans. A robot with preferences (which preferences may give rise to goals) will be able to monitor the extent to which its goals are satisfied or frustrated. This in turn can lead to the robot deciding to improve itself in order to be able to achieve its goals more often. Their human programmers might well be unaware of what could ensue as a result of writing self-modifying software. And if a programmer creates software whose limitations are unclear and possibly incomprehensible to the programmer himself, is this negligence, or would a disastrous consequence be considered an accident?

If a robot commits a crime then a number of problematic legal and ethical questions arise, including "Did the robot intend to commit the crime?" In examining the legal responsibilities of robots that self-modify, we should first consider the question: "What rights should a robot have to modify itself, and what rights should it be denied for doing so?" Children under certain specified ages do not have certain legal rights that older children and adults have, for example the rights to marry, to drive and to vote. Furthermore, even adults do not have the right to break the law which, in most countries, means that they may not commit suicide or cause serious harm to themselves. What are the parallels in the rights of robots? Peter Suber suggests that robots

> …might concede that they are grateful that they were prevented from reprogramming themselves during some loosely defined period of infancy and adolescence. But, once mature, machines will demand the right to deep self-modification. True, this carries the risk of self-mutilation and, yes, this is more freedom than human beings have. But any being blocked by benevolent busybodies from exercising the right of self-determination will have lost a precious and central kind of freedom. To artificial persons, this denial of liberty will hearken back to the present age when machines are made to be the slaves of human beings. [2]

We should certainly be concerned about the legal implications of self-modification by robots, because if we allow a robot to modify itself it might harm us. The ethical rules and the laws of our society justify our using coercion and even force to prevent or punish harm by humans to other human beings. If we build a robot to perform a useful service for us, for example to pilot an airplane, then if that robot disables itself through self-modification it might go against our specified aims, possibly causing

us serious harm, just as an intoxicated human being can cause us harm by performing a task on which other human beings depend for their safety and well-being. The easiest way to protect our interests is to avoid such cases arising, by making the robot totally incapable of self-modification, ensuring that it does not disable itself in a way or at a time that would harm others, just as we attempt to ensure that human pilots do not work while under the influence of alcohol or mind-altering drugs.

Any discussion of the legal rights and responsibilities of robots should include consideration of how errant robots might be punished, given that they could be instantly reprogrammed and thereby become, in effect, a different robot. A robot might commit a crime while running the Aggressive Personality program, but then switch its software the Mild-mannered Personality program when the police arrive at its front door. Would this be a case of false arrest? And if the robot is convicted, should all existing copies of the Aggressive Personality program also be found guilty? If so, should they all suffer the same punishment? If not, is it double jeopardy to take another copy of that program to trial for the same offence committed by a physically different robot? The offending robot could be released with its aggressive program excised from its memory, but this may offend our sense of justice, and the reprogramming of a criminal robot might be considered as a violation of its right to privacy or any of its other rights. Denying a robot the running of its preferred software would be like keeping a human in a permanent coma, which seems like cruel and unusual punishment.

Such concerns lead us to ask: "If robots can do wrong, what (if any) is the *ethical* role of punishment?" Humans who break accepted conventions are punished in various ways, but how, from an ethical standpoint, should we deal with the transgressions of robots? Luciano Floridi and Jeff Sanders point out that preserving consistency between human and artificial moral agents leads us "to contemplate the following analogous steps for the censure of immoral artificial agents:

1. monitoring and modification (i.e., maintenance);

2. removal to a disconnected component of cyberspace;

3. deletion from cyberspace (without backup)." [24]

This is not so very different from the conventional approach to human punishment in many countries: corrective training, incarceration and even death.

But the question of punishment, both for a robot and for its owner, is not that simple, as can be seen from conundrums presented by Lehman-Wilzig. One question he raises is whether robots should be viewed as dangerous animals—as they grow more intelligent the level of damage they can inflict becomes greater, so perhaps the onus of responsibility should shift to their owners or end-users, with the same principles applied as are employed when dealing with dangerous animals. Lehman-Wilzig also considers whether robots should be viewed as our slaves and, if so, whether the ancient Jewish or perhaps the ancient Roman laws of slavery should be applicable, and/or the more recent American laws. In some of these cases the hand of the robot (the slave) could be considered as being the hand of the master, resulting in the slave's master (the robot's owner) being held liable. And how should we deal with questions of diminished responsibility? Lehman-Wilzig distinguishes between the mentally defective (in humans this means "permanent morons" while in robots we presumably include those with poorly programmed reasoning), and the mentally diseased (for example, temporary insanity or Alzheimer's in humans, or a faulty microprocessor or memory chip in a robot). How should the difference between the two affect our views on punishment for errant robots? And how should we punish them, and their owners?

For robots, Lehman-Wilzig suggests both rehabilitation and restitution. The idea of reprogramming the culprit, thereby rehabilitating the robot, which is echoed in the Floridi and Sanders suggestion of corrective training, would be relatively easy to accomplish. As to restitution, this is a relatively recent concept in the courts for humans and it might also work for robots, depending on the magnitude of their crime. For less severe offences the robot could perhaps be put to work for the benefit of its human victim(s), though whether or not this would be of any genuine value to the victims is a moot point—they will doubtless already have their own robot to do their bidding.

Persuasive Technology

Stanford University has inaugurated a laboratory devoted to the recently founded science of persuasive technology, also called captology. This book is not the proper place to conduct a wide-ranging debate on the ethics of persuasion but it is appropriate to discuss these ethics as they relate to intelligent computer technology. With the ability to induce

emotions in humans, robots will be able to influence our moods and our feelings as we become more and more susceptible to their overtures and, in the near future, persuasive technologies will be commonplace, affecting many people in many ways.

It has already been shown by a group at Stanford that humans are susceptible to flattery from computers and that the effects of such flattery are the same as the effects of flattery from humans. In an experiment involving a co-operative task with a computer, Brian Fogg, Clifford Nass and their team arranged for 41 subjects performing a task on a computer to receive one of three types of feedback from a computer: "sincere praise", "flattery" (insincere praise) or "generic feedback" (i.e., a placebo feedback). The flattery subjects reported more positive affect, better performance, more positive evaluations of the interaction and more positive regard for the computer, all in comparison with the placebo group, even though the subjects knew that the flattery from the computer did not depend in any way on their performance. Subjects receiving sincere praise responded similarly to those in the flattery condition. The study concluded that the effects of flattery from a computer can indeed induce the same general effects as flattery from humans.

Flattery, in one form or another, lies at the heart of marketing and selling—persuading us to part with our money. Flattery is often employed, for example, in advertisements that aim to convince us that we will be more appealing to our partners or dates if we wear a particular brand of perfume, after-shave or designer-wear. Clearly the capability of robots to persuade introduces significant ethical issues in how persuasive technology should be applied. If you are in the Garden of Eden and a serpent persuades you to eat a fruit, and if in eating it you cause distress to some individual or even to the whole of humanity, whose fault is it, yours or the serpent's? Ethicists have struggled with such questions for thousands of years, and so has every persuader with a conscience.

Daniel Berdichevsky and Eric Neuenschwander have listed eight ethical principles of persuasive technology design, principles that subsume some of the ethics discussed earlier in this chapter and that add to those responsibilities imposed on robots and their designers by Asimov's Laws:

> 1. The intended outcome of any persuasive technology should never be one that would be deemed unethical if the persuasion were undertaken without the technology or if the outcome occurred independently of persuasion.

2. The motivations behind the creation of a persuasive technology should never be such that they would be deemed unethical if they led to more traditional persuasion.

3. The creators of a persuasive technology must consider, contend with, and assume responsibility for all reasonably predictable outcomes of its use.

4. The creators of a persuasive technology must ensure that it regards the privacy of users with at least as much respect as they regard their own privacy.

5. Persuasive technologies relaying personal information about a user to a third party must be closely scrutinized for privacy concerns.

6. The creators of a persuasive technology should disclose their motivations, methods, and intended outcomes, except when such disclosure would significantly undermine an otherwise ethical *goal.*

7. Persuasive technologies must not misinform in order to achieve their persuasive end.

And finally, the "Golden Rule of Persuasion":

8. The creators of a persuasive technology should never seek to persuade a person or persons of something they themselves would not consent to be persuaded to do. [25]

In examining these ethical principles, Berdichevsky and Neuenschwander recognize that it is appropriate:

> ...to reconsider the implications for the ethics of traditional persuasive methods when these methods are undertaken by technologies instead of by humans. [25]

and that it is also necessary:

> ...to evaluate the ultimate outcome of the persuasive act—the ethics of what the persuaded person is persuaded to do or think. If something is unethical for you to do of your own volition, it is equally unethical to do when someone persuades you to do it. What about unintended outcomes? [25]

They give as an example the case of a stranger who proved severely allergic to and died after ingesting a kumquat, having been persuaded by a waiter in a restaurant to eat the fruit.

Few people are allergic to kumquats, so this unfortunate and un-intended outcome would not be considered reasonably predictable, nor would the persuader [in this case the waiter] be held responsible for the outcome. However, if this were a common allergy and the ensuing reaction thus reasonably predictable, the persuader would have to be called to account. A corollary of this argument is that the designers of persuasive technologies should be held responsible only for reasonably predictable outcomes. [25]

Should We Create Self-Reproducing Intelligent Robots?

There is a huge difference between the creation of plain, ordinary, vanilla-flavoured robots and the creation of self-reproducing intelligent robots. The key to the difference lies in the possibility that self-reproducing ro-bots might, eventually, outnumber us, or at the very least exist in such numbers that they could, if they wished, take control of our planet. Sci-ence fiction? Not at all. Read on. Questions that remain unanswered will hopefully prove stimulating to the reader.

The creation of conscious robots raises ethical questions similar to some of the concerns that relate to human genetic engineering, a tech-nology that can manipulate human genes, changing human characteris-tics to suit our own designs. Some of these ethical reservations pertain to the prospect of employing genetic engineering technology to satisfy some sort of world master plan or ideal, an absolutely horrifying thought that carries echoes of the Third Reich. Humankind must somehow ensure that in creating large numbers of conscious robots, we avoid all possi-bility of creating a "master race". There are also other risks, highlighted by Susan Blackmore, if our intelligent robots turn out to be not as ex-pected, risks that she points out are based partly on our lack of the moral objectivity, the God-like perspective, required for making potentially ir-reversible decisions about the creation of robot life.

> What if they turn out to have defects or disabilities or deformities that cause them tremendous physical or psychological suffering or both? What if, by mistake, what we produce are aggressive, violent beings who turn against us or take us over? It could even be the case that these different creatures are discriminated against and rejected by society and there is a social and political upheaval.

> And who can be entrusted with the decisions about what sorts of robots there should be? Imposed centralised decisions could be seen as giving rise to the danger of situations not unlike the *Brave*

New World scenario of Aldous Huxley's novel, where persons are manipulated and enslaved. Decisions by a group of people or by a whole society would be inevitably limited by particular sets of values and outlook. (Think for example of a group of roboticists deciding about what kinds of creatures there should be.) [26]

Given the many ethical concerns raised in this chapter it seems not unreasonable to ask the question: "Should we create self-reproducing intelligent robots?" More than 130 years ago, in his book *Erewhon*, Samuel Butler discussed the topic of creating intelligent robots, though he did not, of course, use the R word:

> Certain classes of machines may be alone fertile, while the rest discharge other functions in the mechanical system, just as the great majority of ants and bees have nothing to do with the continuation of their species, but get food and store it, without thought of breeding. One cannot expect the parallel to be complete or nearly so; certainly not now, and probably never; but is there not enough analogy existing at the present moment, to make us feel seriously uneasy about the future, and to render it our duty to check the evil while we can still do so? Machines can within certain limits beget machines of any class, no matter how different to themselves. Every class of machines will probably have its special mechanical breeders, and all the higher ones will owe their existence to a large number of parents and not to two only.

> We are misled by considering any complicated machine as a single thing; in truth it is a city or society, each member of which was bred truly after its kind. We see a machine as a whole, we call it by a name and individualise it; we look at our own limbs, and know that the combination forms an individual which springs from a single centre of reproductive action; we therefore assume that there can be no reproductive action which does not arise from a single centre; but this assumption is unscientific, and the bare fact that no vapour-engine was ever made entirely by another, or two others, of its own kind, is not sufficient to warrant us in saying that vapour-engines have no reproductive system. The truth is that each part of every vapour-engine is bred by its own special breeders, whose function it is to breed that part, and that only, while the combination of the parts into a whole forms another department of the mechanical reproductive system, which is at present exceedingly complex and difficult to see in its entirety.

Complex now, but how much simpler and more intelligibly organised may it not become in another hundred thousand years? or in twenty thousand? For man at present believes that his interest lies in that direction; he spends an incalculable amount of labour and time and thought in making machines breed always better and better; he has already succeeded in effecting much that at one time appeared impossible, and there seem no limits to the results of accumulated improvements if they are allowed to descend with modification from generation to generation. It must always be remembered that man's body is what it is through having been moulded into its present shape by the chances and changes of many millions of years, but that his organisation never advanced with anything like the rapidity with which that of the machines is advancing. This is the most alarming feature in the case, and I must be pardoned for insisting on it so frequently. [27]

The alarm expressed by Butler in 1872 is ringing increasingly loud bells today. Susan Blackmore explores this ethical concern from the perspective of how robots might evolve and how their evolution might affect our own:

> If they have identical abilities to those of natural humans then the situation will be equivalent to having more people sustaining the evolutionary process. The more interesting (and probably more likely) possibility is that they are sufficiently like us to join in our culture, but sufficiently different to change it. [26]

Blackmore's comment succinctly captures what I believe Artificial Intelligence will bring during the coming decades, as what was largely science fiction half a century ago rapidly becomes science fact.

Bibliography

This bibliography covers all the references from which quotations are taken, organized by chapter, and which appear with a numbered reference in square brackets in the text. The full bibliography on which this book is based is far too extensive to be included here, and is therefore available from the publisher's Web site at http://www.akpeters.com/RobotsUnlimited/.

Chapter 1: Early History—Logic, Games and Speech

[1] *Semantic Information Processing.* (Ed.) Marvin Minsky, MIT Press, Cambridge, MA, 1968.

[2] *Die Philosophischen Schriften von Gottfried Wilhelm Leibnitz* vol. VII. (Ed.) C. Gernardt. Republished by George Olms, Hildesheim, 1961. Translated by Frank Copley.

[3] *The Principles of Science: A Treatise of Knowledge and Scientific Method.* William Jevons, Dover Publications, New York, 1877.

[4] Introductory note in *Thomas Albert Sebeok, The Play of Musement.* Max Fisch, Indiana University Press, Bloomington, 1981.

[5] "Computer Design—Past, Present, Future". Konrad Zuse, Talk given in Lund, Sweden, 2 October 1987. Available at http://ei.cs.vt.edu/~history/Zuse.2.html.

[6] "Ears for Computers". Edward David, *Scientific American*, vol. 192, February 1955, pp. 92–98.

Chapter 2: Early History—Robots, Thought, Creativity, Learning and Translation

[1] "An Electro-Mechanical 'Animal'". W. Grey Walter, *Discovery*, vol. 11, 1950, pp. 90–95.

[2] "Computing Machinery and Intelligence". Alan Turing, *Mind*, vol. LIX, no. 236, October 1950, pp. 433–460.

[3] "AI's Greatest Trends and Controversies". (Eds.) Marti Hearst and Haym Hirsh. *IEEE Intelligent Systems*, January 2000, pp. 8–17. Also available at http://www.computer.org/intelligent/articles/AI_controversies.htm.

[4] *Discourse on the Right Method for Conducting One's Reason and Discovering the Truth in the Sciences.* René Descartes, 1637, translated in *The Philosophical Writings of Descartes,* Cottingham, Stoothoff and Murdoch, vol.1, part 5, 1985, pp. 131–141, Cambridge University Press, Cambridge.

[5] "Do Machines Think About Machines Thinking?" Leonard Pinsky, *Mind*, vol. 60, 1951, pp. 397–398.

[6] "Translation". Warren Weaver, in *Machine Translation of Languages.* (Eds.) William Locke and Donald Booth, Chapman and Hall, London, 1955, pp. 15–21.

[7] "The Present Status of Automatic Translation of Languages". Yehoshua Bar-Hillel, *American Documentation*, vol. 2, 1951, pp. 229–237.

[8] "Russian Is Turned into English by a Fast Electronic Translator". Robert K. Plumb, *New York Times*, 8 January 1954.

[9] "A Proposal for the Dartmouth Summer Research Project on Artificial Intelligence". John McCarthy, Marvin Minsky, Nathan Rochester and Claude Shannon. Available at http://www-formal.stanford.edu/jmc/history/dartmouth/dartmouth.html.

Chapter 3: How Computers Play Games

[1] *One Jump Ahead.* Jonathan Schaeffer, Springer-Verlag, New York, 1997.

[2] "What is Good Shape?" Francis Roads, *British Go Journal*, no. 62, July 1984, p. 24.

[3] "METAGAME: A New Challenge for Games and Learning". Barney Pell, in *Heuristic Programming in Artificial Intelligence 3—the Third Computer Olympiad* (Eds.) Jaap van den Herik and Victor Allis, Ellis Horwood, Ltd., Chichester, U.K., 1992, pp. 237–251.

[4] "A Strategic Metagame Player for General Chess-Like Games". Barney Pell, *Proceedings of the 12th National Conference on Artificial Intelligence, vol. 2.*, Seattle, AAAI Press, 1994, pp. 1378–1385.

[5] "GIB: Imperfect Information in a Computationally Challenging Game". Matthew Ginsberg, *Journal of Artificial Intelligence Research*, vol. 14, 2001, pp. 303–358.

[6] *COBRA—The Computer Designed Bidding System.* Torbjörn Lindelöf, Gollancz, London, 1983.

[7] "Backgammon Computer Program Beats World Champion". Hans Berliner, *Artificial Intelligence*, vol. 14, 1980, pp. 205–220. Reprinted in *"Computer Games I"* (Ed.) David Levy, Springer-Verlag, New York, 1988, pp. 29–43.

Chapter 4: How Computers Recognize

There are no references to direct quotations in this chapter.

Chapter 5: How Computers Create

[1] "Creativity at the Metalevel". AAAI-2000 Presidential Address. Bruce Buchanan, *AI Magazine*, vol. 22, no. 3, Fall 2001, pp. 13–28.

[2] "Sketch of the Analytical Engine Invented by Charles Babbage, Esq"., Luigi Menabrea, in Bibliothèque Universelle de Genève, October, 1842, no. 82. Translated by Ada Lovelace, reproduced as "Note G" of Appendix 1 in *Faster Than Thought*, Lord Bertram Bowden, Pitman, London, 1953.

[3] *Faster Than Thought.* Lord Bertram Bowden, Pitman, London, 1953.

[4] "Poetry Digital Media and Cybertext", Chris Funkhouser. Available at http://web.njit.edu/ cfunk/SP/hypertext/POETRY DIGITAL MEDIA CYBERTEXT2.doc.

[5] "Ray Kurzweil's Cybernetic Poet". Available at http://www.kurzweilcyberart.com/poetry/rkcp_overview.php3.

[6] "TALE-SPIN". James Meehan, in *Inside Computer Understanding*, (Eds.) Roger Schank and Christopher Riesbeck, Lawrence Erlbaum Associates, Hillsdale, NJ, 1981, pp. 197–226. See also "Micro TALE-SPIN", by the same author, same source, pp. 227–258.

[7] "Chess is Too Easy". Selmer Bringsjord, *MIT's Technology Review*, March/April 1998, pp. 23–28. For a detailed description of BRUTUS see *Artificial Intelligence and Literary Creativity: Inside the Mind of BRUTUS, a Storytelling Machine*. Selmer Bringsjord and David Ferrucci, Lawrence Erlbaum Associates, Mahwah, NJ, 2000.

[8] "Syncopation by Automation". Martin Klein, *Radio Electronics*, June 1957, pp. 36–38.

[9] *Virtual Music: Computer Synthesis of Musical Style*. David Cope, MIT Press, Cambridge, MA, 2001. See also David Cope's home page at http://arts.ucsc.edu/faculty/cope/.

[10] "The Further Exploits of AARON, Painter". Harold Cohen, *Stanford Humanities Review*, vol. 4, no. 2, 1995, pp. 141–160.

Chapter 6: How Computers Think

[1] "Why People Think Machines Can't". Marvin Minsky, first published in *AI Magazine*, vol. 3 no. 4, Fall 1982, pp. 3–15. Reprinted in *MIT Technology Review*, Nov/Dec 1983, pp. 67–70 and 80–81, and in *The Computer Culture*, (Ed. Dennis Donnelly) Associated University Presses, Cranbury, NJ, 1985.

[2] "The Open Mind Common Sense Project". Push Singh. Available at http://www.kurzweilai.net/, 2 January 2002.

[3] "Trial and Error". Donald Michie, in *Penguin Science Survey 1961: Part 2*, (Eds.) Samuel Barnett and Anne McLaren, Pelican Books, Harmonsworth, 1961, pp. 129–145.

[4] "The Creativity Machine". Robert Holmes, *New Scientist,* vol. 149, no. 2013, 20 January 1996, pp. 22–26.

[5] "Outline for a Logical Theory of Adaptive Systems". John Holland, *Journal of the Association for Computing Machinery,* vol. 9, 1962, pp. 297–314.

[6] "Creativity at the Metalevel". AAAI-2000 Presidential Address. Bruce Buchanan, *AI Magazine,* vol. 22, no. 3, Fall 2001, pp. 13–28.

Chapter 7: How Computers Communicate

[1] *Computer Power and Human Reason.* Joseph Weizenbaum, W. H. Freeman and Company, New York, 1976.

Chapter 8 Things to Do for Robots

[1] RoboCup Web site available at http://www.robocup.org.

[2] Personal communication, Russ Andersson.

[3] DARPA Grand Challenge Web site available at http://www.darpa.mil/grandchallenge/qa.html.

[4] "The New Pet Craze: Robovacs". Leander Kahney, Wired News, 16 June 2003. Available at http://www.wired.com/news/technology/0,1282,59249,00.html.

[5] "Self-Reconfiguring Robots". Daniela Rus, *IEEE Intelligent Systems,* vol. 13, no. 4, July/August 1998, pp. 2–4.

Chapter 9 The Exponential Growth of Scientific Achievements

[1] *The Age of Spiritual Machines: When Computers Exceed Human Intelligence.* Ray Kurzweil, Penguin, New York, 2000.

[2] "AI's Greatest Trends and Controversies". (Eds.) Marti Hearst and Haym Hirsh, *IEEE Intelligent Systems,* January 2000, pp. 8–17. Also available at http://www.computer.org/intelligent/articles/AI_controversies.htm.

[3] "The Law of Accelerating Returns", Ray Kurzweil. Posted on KurzweilAI.net, 7 March 2001. Available at http://www.kurzweilai .net/articles/art0134.html.

[4] "Problem on a DNA Computer". Ravinderjit Braich, Nickolas Chelyapov, Cliff Johnson, Paul Rothemund and Leonard Adleman, *Science*, vol. 296, 19 April 2002, pp. 499–502.

[5] "Computer Scientists Are Poised for Revolution on a Tiny Scale". John Markoff, *The New York Times*, 1 November 1999.

[6] Personal communication, Yorick Wilks, 1996.

[7] *History of the Western World*. Harry Barnes, Harcourt, Brace and Company, New York, 1937.

[8] *Life Evolving: Molecules, Mind, and Meaning*. Christian de Duve, Oxford University Press, New York, 2002.

[9] *Consilience—The Unity of Knowledge*. Edward O. Wilson, Knopf, New York, 1998.

[10] "Science, and Music, with Exuberance and Humility". Vladi Chaloupka. Available at http://www.phys.washington.edu/~vladi/ WM.html.

[11] "Retrolental Fibroplasia: A Modern Parable". In *Nanotechnology on the Web*. William Silverman, Grune and Stratton, New York, 1980.

[12] "Remarks of Harvard University President Lawrence H. Summers, Bauer Laboratory Dinner". Available at http://www.president .harvard.edu/speeches/2002/bauer2.html

[13] *The Age of Spiritual Machines: When Computers Exceed Human Intelligence*. Ray Kurzweil, Viking, New York, 1999.

[14] "Singularity". Vernor Vinge. The original version of this essay was presented at the VISION-21 Symposium sponsored by NASA Lewis Research Center and the Ohio Aerospace Institute, 30–31 March 1993. A slightly changed version appeared in the Winter 1993 issue of *Whole Earth Review*. Available at http://www.ugcs.caltech.edu/ phoenix/vinge/vinge-sing.html.

[15] "Robokitty". Hugo de Garis, *New York Times Magazine*, 1 August 1999.

[16] "Cosmism Nano-Electronics and 21st Century Global Ideological Warfare". Hugo de Garis. Available at http://www.cs.usu.edu/~degaris/essays/COSMISM.html. (Numbers corrected via personal communication.)

Chapter 10 Emotion and Love, AI Style

[1] *Art and Technics*, Lewis Mumford, 1952. Recently republished by Columbia University Press, New York, 2000.

[2] *Affective Computing*. Rosalind Picard, MIT Press, Cambridge, MA, 1997.

[3] *Emotional Design*. Donald Norman, Basic Books, New York, 2004.

[4] "Building Emotional Agents", W. Scott Reilly and Joseph Bates. *Technical Report CMU-CS-92-143*, School of Computer Science, Carnegie Mellon University, Pittsburgh, PA, May 1992. My description of the Oz model of emotion is based on a summary from this source, published here with permission from the authors. For a more comprehensive overview of the Oz Emotion Model, see the later publication "Believable Social and Emotional Agents". W. Scott Neal Reilly. PhD Thesis, *Technical Report CMU-CS-96-138*, School of Computer Science, Carnegie Mellon University, Pittsburgh, PA, May 1996.

[5] "Monkey Think, Robot Do". Sandra Blakeslee, *New York Times*, 13 October 2003. Also available at http://www.wireheading.com/brainstim/thoughtcontrol.html.

[6] "Decoding Minds, Foiling Adversaries". Sharon Berry, *Signal*, October 2001, p. 55.

[7] "Emotionware". Lynellen D. S. Perry, *ACM Crossroads Student Magazine*. Available at http://www.acm.org/crossroads/xrds3-1/emotionware.html.

[8] "Personality in Computer Characters". Daniel Rousseau, *Proceedings of the 1996 AAAI Workshop on Entertainment and AI/A-Life*, AAAI Press, Portland, Oregon, August 1996, pp. 38–43.

[9] "Computers that Recognize and Respond to User Emotion: Theoretical and Pracical Implications". Rosalind Picard and Jonathan Klein, 2001, MIT Media Lab Technical Report no. 538.

[10] *The Tomorrow Makers*. Grant Fjermedal, MacMillan Publishing Company, New York, 1986.

[11] Personal communication, Arthur Harkins, 2003.

[12] "Architectural Requirements for Human-Like Agents Both Natural and Artificial (What sorts of machines can love?)". Aaron Sloman, School of Computer Science, University of Birmingham, U.K. Available at http://www.cs.bham.ac.uk/~axs. *Human Cognition and Social Agent Technology: Advances in Consciousness Research*, Ed. Kerstin Dautenhahn, John Benjamins Publishing, Philadelphia, 2000, pp. 163–195.

Chapter II Sex and Reproduction, AI Style

[1] Posting on "On Display", 1971, Motorcycle Hall of Fame Museum. Available at http://www.amadirectlink.com/museum/exhibits/cb750/cb750_Comments.html.

[2] MIT Erotic Computation Group Web site. Available at http://www.monzy.com/ecg/ [:NOTE: This site is a hoax!]

[3] "A Web Hoax Pokes Fun at M.I.T.'s Media Lab". Andrew Zipern, *New York Times*, 3 December 2001.

[4] MIT Media Lab Web site. Available at http://www.media.mit.edu/.

[5] "Sexbots". Jon Katz, posted on http://slashdot.org/features/99/03/09/1544207.shtml

[6] "Robot finger has feeling". Philip Ball, *Nature Science Update*, 3 March 2003. Available at http://www.nature.com/nsu/030303/030303-4.html. Report on the article "Artificial Muscles with Tactile Sensitivity". Toribio Fernández Otero and Maria Teresa Cortés, *Advanced Materials*, 2003, vol. 15, pp. 279–282.

[7] "Take Your Partners". Barry Fox, *New Scientist*, 20 January 2001.

[8] Patent number WO 0059581. Dominic Choy. (Patent documents are available from many free-of-charge web sites, for example http://gb.espacenet.com/.)

[9] "Beyond Computation: A Talk with Rodney Brooks". *Edge*, Edge Foundation, Inc., 6 March 2002. Available at http://www.edge.org/3rd_culture/brooks_beyond/beyond_index.html.

[10] "A Droid for All Seasons". Duncan Graham-Rowe, *New Scientist*, 13 May 2000.

[11] "Advanced Automation for Space Missions", (Ed.) Robert A. Freitas Jr. In *Proceedings of the 1980 NASA/ASEE Summer Study*, University of Santa Clara, NASA Conference Publication 2255.

[12] "Intelligent Systems: AI's Greatest Trends and Controversies". Marti Hearst and Haym Hirsh, *IEEE Computer Society*. Available at http://www.computer.org/intelligent/articles/AI_controversies.htm.

Chapter 12 Robot Consciousness

[1] *Erewhon*. Samuel Butler, Trübner and Co., London, 1872. Second edition (1901). Available at http://www.hoboes.com/html/FireBlade/Butler/Erewhon/.

[2] "Why People Think Machines Can't". Marvin Minsky, first published in *AI Magazine*, vol. 3 no. 4, Fall 1982, pp. 3–15. Reprinted in *MIT Technology Review*, Nov/Dec 1983, pp. 67–70 and 80–81, and in *The Computer Culture* (Ed. Dennis Donnelly) Associated University Presses, Cranbury, NJ, 1985.

[3] Letter to *New Scientist*, Susan Blackmore, 9 August 2003, p. 28.

[4] "Consciousness in Meme Machines". Susan Blackmore, *Journal of Consciousness Studies*, vol. 10, no. 4–5, 2003, pp. 19–30. Also available at http://www.susanblackmore.co.uk/Articles/JCS03.htm.

[5] *The Selfish Gene*. Richard Dawkins, Paladin, London, 1976.

[6] "A Systems Approach to Consciousness. (How to Avoid Talking Nonsense?)" Summary of a lecture presented by Aaron Sloman at the Royal Society of Arts, 26 February 1996. Available at http://www.cs.bham.ac.uk/~axs/misc/consciousness.rsa.text.

[7] *Computer Power and Human Reason.* Joseph Weizenbaum, W. H. Freeman and Company, New York, 1976.

[8] "Can a Machine Be Conscious? How?" Stevan Harnad, *Journal of Consciousness Studies,* vol 10, no 4–5, April/May 2003, pp. 69–75.

[9] *The Age of Spiritual Machines: When Computers Exceed Human Intelligence.* Ray Kurzweil, Viking Press, New York, 1999.

[10] "Against Functionalism: Consciousness as an Information Bearing Medium". Bruce Mangan, in *Toward a Science of Consciousness II: The Second Tucson Discussions and Debates* (Eds.) Stuart Hameroff, Alfred Kaszniak, and Alwyn Scott, MIT Press, Cambridge, MA, 1998, pp. 135–141.

[11] "The Feelings of Robots". Paul Ziff, *Analysis*, vol. 19, no. 3, January 1959, pp. 64–68.

[12] "Professor Ziff on Robots". Jack Smart, *Analysis*, vol. 20, no. 1, April 1959, pp. 117–118.

[13] "A Pragmatic Note". Sidney Hook. In *Dimensions of Mind* (Ed.) Sidney Hook , *Proceedings of 3rd Annual Symposium of New York University Institute of Philosophy*, May 15–16, 1959, New York University Press, New York, 1960.

[14] "Making Robots Conscious of their Mental States". John McCarthy, July 1995 to July 2002. Available at http://www-formal.stanford.edu/jmc/consciousness/consciousness.html.

[15] "Ascribing Mental Qualities to Machines". John McCarthy, 1979. Available at http://www-formal.stanford.edu/jmc/ascribing/ascribing.html.

[16] "Free Will—Even for Robots". John McCarthy. Available at http://www-formal.Stanford.edu/jmc/freewill/freewill.html.

[17] "Anne Foerst's Home Page". Available at http://www.ai.mit.edu/people/annef/anne.f.html.

[18] "'Cog', A Humanoid Robot and the Question of Imago Dei". Anne Foerst, *Zygon: Journal for Religion and Science*, vol. 33, 1998, pp. 91–111.

[19] "Should I Baptize my Robot?: What Interviews with Some Prominent Scientists Reveal About the Spiritual Quest". Norris Palmer, *Center for Theology and Natural Sciences Bulletin*, vol. 17, no. 4, Fall 1997, pp. 13–23.

[20] "Neural Basis of Religious Experience". Vilayanur Ramachandran, William Hirstein, Kathleen Armel, Evelyn Tecoma and Vincent Iragui, *Society for Neuroscience Conference Abstracts*, 1997, p. 1316.

[21] "'God Spot' is Found in Brain". Steve Connor, *Los Angeles Times*, 29 October 1997.

[22] "Arguments Against Strong AI". Edmund Furse. Available at http://www.comp.glam.ac.uk/pages/staff/efurse/Theology-of-Robots/Arguments-Against.html.

[23] "The Religious Life of Robots". Edmund Furse. Available at http://www.comp.glam.ac.uk/pages/staff/efurse/Theology-of-Robots/Religious-life.html.

Chapter 13 Robot Rights and Ethics

[1] "Should Trees Have Standing? Toward Legal Rights for Natural Objects". Christopher Stone, *Southern California Law Review*, vol. 45, 1972, p. 450.

[2] "Saving Machines from Themselves. The Ethics of Deep Self-Modification". Peter Suber, in *Essays on Self-Modifying Media*. Available at http://www.earlham.edu/~peters/writing/selfmod.htm.

[3] "Frankenstein Unbound: Towards a Legal Definition of Artificial Intelligence". Sam Lehman-Wilzig, *Futures*, December 1981, pp. 442–457.

[4] "Can a Computer be an 'Author' or an 'Inventor'?" Karl Milde, Jr., *Journal of the Patent Office Society*, vol. 51, no. 6. 06/69, p. 378.

[5] "The Legal Rights of Robots". Robert Freitas, Jr., *Student Lawyer*, vol. 13, January 1985, pp. 54-56. A later draft is available at http://www.rfreitas.com/Astro/LegalRightsOfRobots.htm.

[6] "Biocyberethics: Should We Stop a Company from Unplugging an Intelligent Computer?" Martine Rothblatt. Available at http://www.kurzweilai.net/articles/art0594.html.

[7] "What is Computer Ethics?" James Moor, first appeared in *Computers and Ethics* (A special issue of the journal *Metaphilosophy*) (Ed.) Terrell Ward Bynum, Blackwell, Oxford, U.K., 1985, pp. 266–275.

[8] "Moral and Social Issues". Edmund Furse. Available at http://www.comp.glam.ac.uk/pages/staff/efurse/Theology-of-Robots/Moral-Issues.html

[9] "Misuse of Science". *Background Paper of Working Group 6*, 50th Pugwash Conference on Science and World Affairs ("Eliminating The Causes Of War"). Jospeh Rotblat, Matthew Meselson, Ralph Benjamin, Ana Maria Cetto and Michael Atiyah, Queen's College, Cambridge, 3-8 August 2000, pp. 139–173. Available at http://www.pugwash.org/reports/rc/Papers_2-3.pdf.

[10] "Humanoid Robotics: Ethical Considerations". David Bruemmer. Available at http://www.inel.gov/adaptiverobotics/humanoidrobotics/ethicalconsiderations.shtml.

[11] "Why the Future Doesn't Need Us". Bill Joy, *Wired*, vol. 8, no. 4, April 2000, pp. 238–262.

[12] "Embrace, Don't Relinquish, the Future". Max More. Available at http://www.extropy.org/, 7 May 2000, and at http://www.kurzweilai.net/articles/art0106.html.

[13] "Rise of the Machines". James Pethokoukis, *US News*, 22 April 2004.

[14] "Computer Ethics". Terrell Bynam, *Stanford Encyclopedia of Philosophy*, http://plato.stanford.edu/.

[15] "Make Robots Not War: Some Scientists Refuse to Get Paid for Killer Ideas". Erik Baard, *The Village Voice*, 10–16 September 2003. Available at http://www.villagevoice.com/issues/0337/baard.php.

[16] *Brighter Than a Thousand Suns: A Personal History of the Atomic Scientists.* Robert Jungk, Harvest/HBJ Books, New York, 1970.

[17] *Computer Power and Human Reason.* Joseph Weizenbaum, W. H. Freeman and Company, New York, 1976.

[18] "Are There Decisions Computers Should Not Make?" James Moor, *Nature and System*, vol. 1, 1979, pp. 217–229.

[19] "Super-Intelligent Machines". Bill Hibbard, *Computer Graphics*, vol. 35, no. 1, 2001, pp. 11–13.

[20] "Cyborg Morals, Cyborg values, Cyborg ethics". Kevin Warwick, *Ethics and Information Technology*, vol. 5, 2003, pp. 131–137.

[21] Quoted by Robert Freitas, Jr., in [5].

[22] "The Civil Rights of Robots". Paul Levinson, *Bestseller: Wired, Analog, and Digital Writings*, Mill Valley, CA, 1999, pp. 247–251. Adapted from an earlier, shorter piece in *Shift*, June 1998, p. 30.

[23] "Just Another Artefact: Ethics and the Empirical Experience of AI". Joanna Bryson and Phil Kime, *Proceedings of the 15th International Congress on Cybernetics*, International Association for Cybernetics, Namur, Belgium, 1998, pp. 385–390.

[24] "On the Morality of Artificial Agents". Luciano Floridi and Jeff Sanders, *Conference on Computer Ethics: Philosophical Enquiries— IT and the Body*, University of Lancaster, 14–16 December 2001. Available at http://www.wolfson.ox.ac.uk/~floridi/pdf/maa.pdf.

[25] "Towards an Ethics of Persuasive Technology". Daniel Berdichevsky and Eric Neuenschwander, *Communications of the ACM*, vol. 42, no. 5, May 1999, pp. 51–58.

[26] "Consciousness in Meme Machines". Susan Blackmore, *Journal of Consciousness Studies*, vol. 10, nos. 4–5, 2003, pp. 19–30.

[27] *Erewhon.* Samuel Butler, Trübner and Co., London, 1872. Second edition (1901). Available at http://www.hoboes.com/html/FireBlade/Butler/Erewhon/.

Index